European Railway Signalling

compiled by a Project Group
under the auspices of
the Institution of Railway Signal Engineers

general editor: Colin Bailey

A & C Black · London

First published 1995
A & C Black (Publishers) Limited
35 Bedford Row, London WC1R 4JH
ISBN 0-7136-4167-3

© 1995 Institution of Railway Signal Engineers

A CIP catalogue record for this book
is available from the British Library.

All rights reserved.
No part of this publication may be reproduced,
stored in a retrieval system or transmitted
in any form or by any means, electronic,
mechanical, photocopying, recording or otherwise,
without the prior permission in writing
of A & C Black (Publishers) Limited.
Designed by Simon Borrough
Typeset in Melior 10pt
by Create Publishing Services Limited, Bath
Printed in Great Britain by Ebenezer Baylis & Son
Limited, Worcester

Contents

Preface by Michel Walrave — iv
Preface by A. C. Howker — v
Acknowledgements — vi
Abbreviations — vii

1 **Railway Signalling Principles** — 1
  D. Hotchkiss

2 **Train Detector Systems** — 121
  F. de Vilder

3 **Switch Operating and Proving Systems** — 143
  M. Vallez

4 **Signals** — 160
  W. Böhm

5 **Interlocking Cabin** — 182
  H. Lindenberg

6 **Block Systems** — 221
  F. de Vilder

7 **Radio Systems in Signalling** — 235
  A. Fisher

8 **Internal and External Safety Conditions** — 240
  J. Pizarro

9 **Automatic Train Protection and Control** — 265
  J. Catrain

10 **High Speed Line Signalling System** — 282
   J. Poré

11 **Rail Traffic Management** — 311
   A. Exer

12 **Level Crossing Protection** — 343
   P. Middelraad

13 **Other Safety Systems** — 356
   J. Hammargren

14 **The Future** — 362
   J. Catrain

Sources — 369
Index — 371

Full contents details are given on the first page of each chapter.

Preface
by Michel Walrave
Chief Executive of the International
Union of Railways (UIC)

Three years ago, Jacques Catrain, then Chairman of the IRSE, took the particularly appropriate initiative of suggesting this book. The result is a complete record of all the different railway signalling systems, in the broadest sense of the term, that can be found across Europe, with the (deliberate) exception of urban transit systems. This excellent compilation has been produced jointly by railway engineers and industry in 12 European countries.

With the wealth of information it contains, this publication is bound to constitute a reference, the sole of its kind to date, for scientists and academics, but this is not its basic purpose. It is intended primarily for the engineers who have to build tomorrow's railways in Europe. By showing the various solutions adopted in the different countries to the problems of railway signalling, and highlighting both the original features and common characteristics of the different techniques, *European Railway Signalling* will be of enormous assistance in guiding development of suitable solutions to ensure interoperability over national networks, in particular in the case of the European high speed rail system.

Efforts to achieve technical harmonization in the signalling, and in the telecommunications, sectors are notoriously difficult, since they are in direct contrast to decades of national tradition. They are, however, no longer a Utopian dream. Spurred on by the European Union, the railways are forging ahead with developing high speeds and are increasingly aware of the potential offered by the international market. Procurement contracts are now being extended to bidders from other countries in the EU area. Industry, although still a firm believer in the advantages of competition, is reorganizing on a multinational basis and is equally anxious to move towards common standards and specifications in order to reduce production costs and thus enhance competitiveness
and export potential, on a world scale in particular.

As emerges clearly from the last chapter on future prospects, the ETCS project in hand at UIC and largely based on work undertaken in partnership
between France and Germany as part of the DEUFRAKO M project, is entirely in phase with one of European railways' main lines of research, centred as it is on the design of new mutually compatible operating systems offering high standards of performance. The aim is to repond to the problem of technical harmonization whilst illustrating the enormous potential for innovation and progress inherent in railway operations. This time there is more at stake than simply defining interfaces to enable one design of rolling stock to be worked over several networks, as an easy and swift way of overcoming immediate problems. The focus is

Preface
by A. C. Howker
President, Institution of Railway Signal Engineers, 1994/1995

instead future-orientated and the goal to develop an overall traffic control system using IT and data transmission techniques to the hilt in order to optimize railway operations internationally in performance, cost and safety terms.

Given their current difficulties and the drop in traffic not solely due to the recession of the early 1990s, Europe's railways desperately need to press on with applying the most recent technology on the widest possible scale so that a combination of this and improvements in their organisation may enable them to strengthen their competitive position in lasting fashion and claim a place in the overall transport system that is more in keeping with the general interest and EU trends.

Creating Europe is not just a task for the politicians, civil servants or lawyers. Engineers should also play a major role, particularly the railway sector where technology is a leading factor and where international co-operation, an age-old tradition despite national disparities, is of increasingly vital importance.

As Chief Executive of the UIC, I am particularly delighted with this publication which will serve to add enlightenment on the subject of railway signal engineering and will assist all those with a fervent interest in this demanding subject in obtaining a clearer understanding of others' viewpoints.

The two previous books, *Railway Signalling* and *Railway Control Systems*, concentrated on signalling systems which followed British practice, but the majority of the Institution's membership based in countries other than Great Britain, there always was the intention to publish books which describe the signalling principles and practices of other administrations. This is the first of such books and with the opening of the Channel Tunnel it is appropriate that it deals with European signalling. The varied traffic requirements of European railways and hence the associated signalling practices described in this book will be of benefit to both professional and interested students of railway signalling. I am sure that it will be referred to for many years to come.

Acknowledgements

On behalf of the Institution of Railway Signal Engineers this book has been elaborated at the European level under the management of a Project Group.

Each participating country was represented by a National Editor leading a National Team who supplied each of the Chapter Editors with information and data regarding the country's approach to railway signalling.

The Chapter Editors collected all the information and collated it into a general situation underlining the main differences between the various signalling systems.

A Railway Checker of the concerned country and belonging to the Railway Administration was in charge of ascertaining the correctness of all the information in respect of a particular railway.

The IRSE wishes to express its warm thanks to railway administrations, signalling manufacturers and individual members for their support and valuable co-operation in this approach to the production of *European Railway Signalling*.

The Institution, as a body, is not responsible for the views and opinions expressed by individual authors.

Project Group

Chairman	J. Catrain – GEC Alsthom
IRSE	C. Bailey – General Editor
ÖBB	H. Steindl / J. Berger / H. Pfleger
FS	G. Cerullo
SNCF	J. P. Guilloux
SBB	H. Hayoz
BR	D. Hotchkiss, Prism Engineering
DB AG	E. Schuster

Railway Administrations
ÖBB – NMBS/SNCB – SBB – DB AG – RENFE – SNCF – BR – FS – CFL – NSB – NS – BV

Signalling Manufacturers
ABB Signal A.B. – ABB Señal – Abengoa – ACEC Transport – Alcatel Austria – Ansaldo – Caerail – CSEE Transport – Dimetronic – GEC Alsthom – Integra Siemens – SASIB – SEL Señalizacion – Siemens – TIFSA – Westinghouse Signals Ltd

National Editors

Austria	**W. Böhm** – W. Weithofer – H. Kramer – J. Berger – P. Wegenstein
Belgium	**M. Vallez** – F. de Vilder – L. Brabant – R. Vanhauwaert – E. Léchevin – J. Verschaeve
Switzerland	**A. Exer** – H. Hayoz – R. Kilchenmann – H. R. Zehnder – H. Althaus – B. Antweiler – E. Otz – T. Czermely – C. Zufferey – W. Markwalder – C. Kupper – H. Schwinnen – B. Stamm – F. Hausel
Germany	**H. Lindenberg**– R. Heibsmann – H. Dewald – P. Wegener – W. Eichner – E. Schuster – A. Knight – R. Natterer – K. Lennartz – K. Dolch – U. Kern – J. Stutzbach – R. Berger – J. Polz – G. Stoll – Prof. Dr. H. Walther
Spain	**A. Lozano** – J. Pizarro – A. V. Sancho – M. Hernandez – J. A. Hutado de Mendoza – J. Juardez – A. Garrido – L. T. Rojas – C. Manzano – L. Escriba – A. Puyol
France	**J. Poré** – J. Rigaud – Y. Gruère – F. van Deth
Great Britain	**J. Francis**
Italy	**G. Astengo** – G. Cerullo – M. Musiani – A. Altobelli – G. Bonfigli – P. Pezzi – G. Gritti – M. Francassi – E. Verardo
Luxembourg	**H. Werdel** –F. Groos – A. Mathieu
Norway	**R. Hortman**
Netherlands	**P. Middelraad** – R. M. Ebbs – J. G. M. van Keulen – G. H. Koning – M. H. van der Werf
Sweden	**C. Fällman** – I. Karlsson – J. Hammargren

Chapter Editors
D. Hotchkiss – F. de Vilder – M. Vallez – W. Böhm – H. Lindenberg – A. Fisher – J. Pizarro – J. Poré – A. Exer – P. Middelraad – J. Hammargren – J. Catrain

Railway Checkers

ÖBB	H. Steindl	BR	D. Hotchkiss, Prism Engineering
NMBS/SNCB	F. de Vilder	FS	G. Cerullo
SBB	H. Hayoz	CFL	H. Werdel
DB AG	E. Schuster	NSB	R. Hortman
RENFE	J. Pizarro	NS	P. Middelraad
SNCF	J. P. Guilloux	BV	C. Fällman

Abbreviations

Railway Adminstrations
ÖBB   – Austria
NMBS/SNCB   – Belgium
SBB   – Switzerland
DB AG   – Germany
RENFE   – Spain
SNCF   – France
BR   – Great Britain
FS   – Italy
CFL   – Luxembourg
NSB   – Norway
NS   – Netherlands
BV   – Sweden

A
AC   : alternating current
AC–TC   : alternating current track circuit
AF–TC   : audio frequency jointless track circuit
ARC   : automatic route calling
ARS   : automatic route setting
ATC   : automatic train control
ATO   : automatic train operation
ATP   : automatic train protection
ATR   : automatic train routing
AVE   : *Alta Velocidad*
Az   : axle counter

B
BR   : British Railways (Great Britain)
BV   : Ban Verket (Sweden)

C
CCTV   : closed circuit television
CFL   : Chemins de Fer Luxembourgeois
CTC   : centralized traffic control

D
DB AG   : Deutsche Bahne AG*
DC   : direct current
DC–TC   : direct current track circuit
DLM   : data link module

E
EBP   : *Elektronische Bedien Post* (electronic control post)
ECS   : external communication system
EPROM   : erased programmed read out memory
ESTW   : *Elektronisches Stell–Werk* (electronic interlocking)

F
FDM   : frequency division multiplex
FFSK   : fast frequency shift keying
FM   : frequency modulation
FS   : Ente Ferrovie dello Stato (Italy)
FTGS   : remote fed and coded jointless audio frequency track circuit

G
GWS   : gateway system

H
HVI–TC : high voltage impulse track circuit
Hz   : hertz

I
ICE   : high speed train
IECC   : integrated electronic control centre
IRC   : interlocking remote control
ISM   : system monitor

*References in the text to German railways are to Deutsche Bundesbahn (DB) which was a self-regulatory body at the time that the text was drafted. Since then, Deutsche Bundesbahn (DB) has been turned into a public limited company, the Deutsche Bahn AG (DB AG). The official regulator for German railways is the new Federal Railway Office (Eisenbahnbundesamt, EBA).

K
kV : kilovolt(s)

L
LAN : local area network
LX : level crossing
LZB : *LinienZug Beeinflussung*

M
MPM : multi processor module
MHz : megahertz

N
Ni Cd : nickel cadmium
NMBS/SNCB : Nationale Maatschappij der Belgische Spoorwegen/Société Nationale des Chemins de Fer Belges (Belgium)
NS : Nederlandse Spoorwegen (Netherlands)
NSB : Norges StatsBahner (Norway)
NX : entrance exit (interlocking operation mode)

O
ÖBB : Österreichische Bundes Bahnen (Austria)
OCU : operators control unit
OSP : optical synoptic panel

P
PC : personal computer
PIDN : process information data network
PPM : panel processor module
PRCI : *Poste tout Relais à Commande Informatique* (all relay interlocking with processor control)
PRG : *Poste tout Relais Géographique* (geographical all relay interlocking)
PROM : programmed read out memory
PRS : *Poste tout Relais à transit Souple* (all relay interlocking with sectional route release)

R
RCC : remote control centre
RENFE : Rede Nacional de los Ferrocarriles Espanõles (Spain)
RETB : radio electronic token block
RRI : remote relay interlocking interface
RZÜ : *Rechnergestützte Zugs Überwachung* (train supervision)

S
SB : signalbox
SBB : as SBB/CFF/FFS Schweizerische Budes Bahnen (Switzerland)
SDS : signalman's display system
Sh TC : short jointless track circuit
SNCF : Société Nationale de Chemins de fer Français (France)
SPAD : signal passed at danger
SSI : solid state interlocking

T
TDM : time division multiflex
TDS : train describer system
TFM : trackside functional module
TGV : *Train à Grande Vitesse* (high speed train)
TR : track relay
TTP : timetable processor
TVM : *Transmission Voie Machine* (track to train transmission)

U
UIC : *Union Internationale de Chemins de Fer*

V
VDU : video display unit
VPI : vital processor interlocking

1 Railway Signalling Principles

D. HOTCHKISS

Contents

Glossary	3
1.1 Introduction	5
1.2 Basis of signalling systems	5
1.2.1 Key elements	5
1.2.2 Application of system	5
1.2.3 Two and three aspect systems	6
1.2.4 Four aspect system	6
1.2.5 Moving block	6
1.3 Speed signalling and route signalling	8
1.3.1 Speed signalling	8
1.3.2 Route signalling	9
1.3.3 Comparative merits	10
1.4 Signal profiles	10
1.4.1 Running signals	10
1.4.2 Shunting signals	15
1.4.3 Other signals	18
1.4.4 Speed restriction signs	20
1.5 Signal aspects and aspect sequences	21
1.5.1 UIC 732 Recommendations	21
1.5.2 Aspects in use in Europe	21
1.5.3 Aspect sequences	35
1.5.4 Signal aspect and aspect sequence diagrams	37
1.6 Bidirectional working	105
1.7 Principles of interlocking and controls	105
1.7.1 Type of system	105
1.7.2 Method of operation	106
1.7.3 Route calling	107
1.7.4 Route setting	107
1.7.5 Route locking	107
1.7.6 Approach locking	107
1.7.7 Aspect controls – stop signals	108
1.7.8 Aspect controls – shunting signals	111
1.7.9 Aspect controls – warning signals	112
1.7.10 Aspect controls – other signals	113
1.7.11 Aspect replacement	113
1.7.12 Route release	113
1.7.13 Point operation	115
1.7.14 Level crossing controls	116
1.7.15 Bidirectional and single line controls	117
1.7.16 Remote control of interlockings	117
1.7.17 Special controls	118
1.7.18 Controls for staff protection systems	120

Illustrations

Fig. 1.1	Parameters of European signalling systems	7
Fig. 1.2	Advantages and disadvantages of speed and route signalling	10
Fig. 1.3	Maximum speed for lineside signals	11
Fig. 1.4	Warning signals	12
Fig. 1.5	Repeating signals	13
Fig. 1.6	Contraflow signals	14
Fig. 1.7	Preceding shunt	16
Fig. 1.8	Shunting signal types	17
Fig. 1.9	Examples of level crossing rail signals	19
Fig. 1.10	Single green	23
Fig. 1.11	Double green	24
Fig. 1.12	Single yellow	25
Fig. 1.13	Double yellow	26
Fig. 1.14	Green & yellow together	28
Fig. 1.15	Single red	29
Fig. 1.16	Single white	30
Fig. 1.17	Other combinations	31
Fig. 1.18	Uses of flashing aspects	33
Fig. 1.19	Extension of three aspect systems to give improved braking distance	36
Fig. 1.20A–D	ÖBB: signal aspects	38–40
Fig. 1.21A–B	ÖBB: aspect sequences	41–42
Fig. 1.22A–E	NMBS/SNCB: signal aspects	43–47
Fig. 1.23	NMBS/SNCB: aspect sequences	48
Fig. 1.24A–D	SBB: signal aspects	49–52
Fig. 1.25A–C	SBB: aspect sequences	53–55
Fig. 1.26A–D	DB AG: signal aspects	56–59
Fig. 1.27A–B	DB AG: aspect sequences	60–61
Fig. 1.28A–B	RENFE: signal aspects	62–63
Fig. 1.29	RENFE: aspect sequences	64
Fig. 1.30A–D	SNCF: signal aspects	65–68
Fig. 1.31A–D	SNCF: aspect sequences	69–72
Fig. 1.32A–C	BR: signal aspects	73–75
Fig. 1.33A–D	BR: aspect sequences	76–79
Fig. 1.34A–D	FS: signal aspects	80–83
Fig. 1.35A–C	FS: aspect sequences	84–86
Fig. 1.36A–C	CFL: signal aspects	87–89
Fig. 1.37	CFL: aspect sequences	90
Fig. 1.38A–B	NSB: signal aspects	91–92
Fig. 1.39	NSB: aspect sequences	93
Fig. 1.40A–F	NS: signal aspects	94–99
Fig. 1.41	NS: aspect sequences	100
Fig. 1.42A–B	BV: signal aspects	101–102
Fig. 1.43A–B	BV: aspect sequences	103–104
Fig. 1.44	Method of route release	108
Fig. 1.45	Overlap controls	110
Fig. 1.46	Shunting signal controls	112
Fig. 1.47	Opposing locking omitted	111
Fig. 1.48	Sectional release of route locking	115
Fig. 1.49	Tunnel controls (DR)	118
Fig. 1.50	Freight train controls – rising gradient (NS)	119
Fig. 1.51	Freight train tunnel controls (NS)	119

Glossary

Signals

running signal — (or main signal) is a signal controlling trains en route along a running line

stop signal — is a signal at which trains are required to stop if it is at 'danger'

warning signal — is a signal without a 'stop' aspect which indicates the aspect of a stop signal at braking distance in advance (otherwise known as a distant, advance or *avertissement* signal)

repeating signal — is a signal which indicates the aspect of a stop signal at less than braking distance in advance

shunting signal — is a signal, normally close to ground level, which authorizes shunting movements

substitution signal — is a signal aspect exhibited when the proper signalling is out of order due to engineering work or failure. It usually authorises passage at restricted speed

subsidiary signal — on BR, this term is applied to an aspect displayed alongside a 'stop' aspect to authorize passage (for example) into an occupied platform

aspect — is the display given by a signal, either at lineside or in the cab. It can be described either in visual terms, for example 'flashing yellow' or by its meaning, for example 'next signal is at steady yellow and is at reduced distance to the stop signal' (SNCF)

stop aspect — (of a signal) is the aspect of a main or shunting signal that requires a train to stop

proceed aspect — (of a signal) is any aspect other than the most restrictive that the signal can display. It may be called the OFF aspect. It allows a train to proceed on its journey or to execute shunting movements

advance — is the area through which a train travels after passing a given location

rear — is the area through which a train has travelled before reaching a given location

Interlocking

interlocking (level) — is the logic by which routes which conflict are prevented from being set at the same time

selection (level) — is the logic by which signals are allowed to show a 'proceed' aspect only when all conditions have been satisfied

approach locking — is provided to prevent a route being altered once a proceed signal could have been seen by a driver or ATP system. It is normally possible to cancel the route after a time delay, or by operation of a special procedure

route locking — is provided to prevent a route being cancelled or altered once a train has passed the protecting signal

sectional release (of route locking) — is a method of releasing the route locking in sections after the passage of a train through the route

backlocking	– is provided to prevent the restoration of a signalled route or ground frame release once it has been operated until certain conditions have been fulfilled	reverse	– (as applied to points) is the opposite position to normal – usually set for a diverging route

Level crossings — controlled – is operated under supervision and interlocked with the protecting signals. May be monitored by closed circuit television (CCTV)

release — (as applied to a ground frame) is the condition where the levers (or mechanism) is free to operate. Once the levers have been operated, the release is backlocked until they have been restored to the normal position

— (as applied to locking) is when all the conditions which caused the locking to be applied have been satisfied, for example, in the case of approach locking, the train has either:

- passed into the route or
- stopped at the protecting signal
- not occupied any of the approach locking track circuits

normal — (as applied to points) is the position in which the points are shown on the signalling plan – usually set for the straight route or to provide maximum flank protection for other routes

Note Other definitions are in use:

- '+' and '−'
- left hand (or right hand) switch closed
- left or right – direction of train at facing points (This is the exact opposite of which switch is closed!)

— (as applied to ground frame) is the state of the ground frame where the points are set for the main line and the levers (or mechanism) are locked

— automatic – is operated by approaching trains and is not normally interlocked with the signalling system.

For example on BR:

- Automatic Half Barrier – AHB
- Automatic Barrier, Locally Monitored – ABCL
- Automatic Open Crossing, Locally Monitored – AOCL
- Miniature Warning Light – MWL

1.1 Introduction

The aim of this chapter is to:

- contrast and compare the principles of the signalling systems in use on the railways of Europe;
- describe the aspects and aspect sequences in common use and their interpretation to drivers;
- set out the principles of interlocking and selection employed by the various railways.

1.2 Basis of signalling systems

The aim of all railway signalling systems is to prevent collisions and derailments whilst allowing trains to travel at whatever maximum speed is allowable by the characteristics of the line and the vehicles that run on it. This can be achieved either by giving orders to drivers by visual means, typically coloured lights or semaphore arms, or by a visual and/or audible display in the driving cab. In the latter case, the signalling information, having been transmitted to the train, can also be used to enforce compliance should the driver fail to respond to cautionary or danger signal aspects.

It thus follows that signalling systems must be designed to be fail-safe. This means that the failure of any component or subsystem must result in a default state which ensures safety in all circumstances. Systems and equipment are therefore designed, manufactured, installed and maintained with safety as the foremost consideration, as is required by the various railway administrations and their respective governments.

Other facilities and management systems are linked with the signalling, for example train identification, automatic route setting and train recording, but these must never be allowed to interfere with the safety embodied in the signalling system.

Anything less sophisticated than a fully automatic railway needs people to run it and human failures must also be taken into account in designing a safe signalling system. Of particular concern is drivers passing signals at danger and this is being addressed through various forms of Automatic Train Protection (ATP). The system must also be designed as far as possible to eliminate or reduce the effects of failures or malpractices of other members of staff, such as signalmen and signal maintainers.

However, when the signalling system does fail, trains must still be enabled to move, albeit at reduced speed. The effects of a failure and the process of restoration to normal working need careful consideration, together with adequate training of all relevant personnel.

1.2.1 Key elements

There are a number of key elements in the design of signalling systems which are vital to the manner in which they achieve the required degree of safety. They are:

- train detection, by track circuits and axle counters;
- control and detection of track elements, such as points, swing nose crossings and switch diamonds;
- method of interlocking points and signals;
- indications to the driver by lineside signals or cab signalling;
- enforcement of signal controls on the train with train stops or ATP;
- braking characteristics of the rolling stock using the line.

1.2.2 Application of system

The braking characteristics of trains are not usually determined by the signal engineer. However, the application of the signalling system to a particular line is dependent on him first knowing the distance required to stop by a train travelling at the maximum allowed speed. This distance depends on:

- the gradient between the warning and stop signals;
- the brake force available on the train, which varies according to the coefficient of friction between wheel and rail;
- the loading of the train;
- the state of wear of the brake pads and the air pressure available in the brake cylinders.

An allowance of 15–20% is usually added to the basic distance to take account of these variable factors.

Other basic parameters which the signal engineer requires to know are the frequencies with which trains are required to run, the headways, both for non-stopping trains and for trains stopping at intermediate stations, the maximum train length and the acceleration rate.

On mixed traffic lines, some of these train characteristics may vary widely. The signal engineer then has to take a view on which of them become critical for each part of the line and then design the signalling accordingly.

1.2.3 Two and three aspect systems

On U–bahn lines with trains of uniform consist travelling at moderate speeds, a signalling system having only two aspects – proceed and stop – may be possible, but generally the braking distance is greater than the sighting distance to the stop signal and some form of warning aspect becomes necessary. As will be seen later, it is the precise form and meaning of this warning aspect which is responsible for most of the visible diversity between the signalling systems of the European railways.

The warning or distant signal is placed at full service braking distance from its associated stop signal and except on lightly used lines is usually combined with the previous stop signal. This then becomes a three aspect system, each signal being capable of showing proceed, warning and stop.

1.2.4 Four aspect system

Where train speeds are limited to approximately 120km/h or where the number of trains is limited to, say one every five minutes, one warning signal so placed is sufficient. However, when a combination of high speeds and frequent trains is to be signalled, it may be appropriate to space the signals closer together, so that the warning (or braking) distance is spread over two sections or blocks, thus giving a four aspect system. **Fig. 1.33A** shows how this is done in the BR system, and how the headway is maintained with a four aspect system for trains travelling at higher speed and therefore requiring a greater braking distance.

1.2.5 Moving block

If the required headway cannot be accommodated with four aspect signalling, then five or more aspects could be employed – at the cost of greater complexity in the control equipment. Higher than four aspect systems are uncommon in lineside signalling systems, although a stretch of the BR East Coast Main Line between Werrington Junction and Stoke Tunnel, a distance of about 36km, has been equipped with five aspect signalling to enable high speed test runs up to 225km/h to be undertaken.

Ultimately the number of aspects could be increased to infinity, giving a moving block system, but although this would yield the best possible headway, it would be completely impractical with lineside signals, although theoretically possible with some concepts of cab signalling or automatic train control such as LZB or SACEM. Economically, any system of more than four aspects would suffer from the law of diminishing returns.

However, with cab signalling systems, the additional outlay in providing additional aspects is small compared with the basic cost of the equipment. It is therefore possible to extend cab signalling systems to give a more optimum headway – for example, the arrangement adopted in the Channel Tunnel is effectively a five aspect system, giving a braking zone from 160km/h of up to four x 500m blocks with a fifth block as overlap.

(It may be noted in passing that high speed moving block is in operation on motorways – but it is not a fail-safe system!)

The layout of signals is therefore based on the service braking distance calculated from the line speed or local maximum speed with adjustment for gradient, the sighting distance and the overlap. The various practices of European railways are shown in **Fig. 1.1**.

Fig. 1.1 Parameters of European signalling systems

Railway	Braking Distance	Spacing of Signals	Minimum Sighting Distance	Length of Overlap	Maximum Train Length
ÖBB	Dependent on speed & gradient 400–1500m	Minimum 400m	2.5 x v.max. (km/h in metres)	≤40km/h: 0m otherwise: 50m	700m
NMBS/ SNCB	120km/h: 900m 140km/h: 1200m 160km/h: 1500m	Minimum: 370m Maximum: 2000m	≤60km/h: 150m >60km/h: 300m	minimum: 50m (but for junctions between two lines minimum 100m)	
SBB				<45km/h: 40m Up to 165km/h: 100m Increased for falling gradient	
DB AG	Dependent on several variables. Minimum distances 80km/h: 400m 120km/h: 700m 160km/h: 1000m	160km/h: 1000m For high speed lines: up to 1300m For minor lines: 400–700m	For stop signals: 500m For warning signals: 250m	>60km/h: 200m ≤60km/h: 100m ≤40km/h: 50m ≤30km/h: <50m	750m
RENFE	Dependent on speed, gradient and characteristics of rolling stock 200km/h maximum	1500m	300m	50m	
SNCF	≤120km/h: 1100m 140km/h: 1400m 160km/h: 1500m Minimum braking distances. Can be compensated for gradient	Minimum: 500m	≤60km/h: 100m ≤120km/h: 200m ≤160km/h: 300m	100m min. distance from trailing switch. No overlap locking	
BR	Dependent on speed, gradient and characteristics of rolling stock (200km/h maximum)	150% braking distance maximum	Equivalent to 7s running time. (eg 300m @ 160km/h)	>96km/h: 183m reducing to ≤24km/h: 46m	Passenger Trains: 300m Freight Trains: 500m (International Trains: Passenger 400m Freight: 700m)
FS	–	>120km/h: 1200m (1000m if >1:100 rising) >110km/h: 1000m (800m if >1:100 rising) When cab signalling installed: 1350m	>90km/h: 200m <90km/h: 150m	100m (50m for platform starting signals)	650m generally (Rules allow for up to 1100m)

Railway	Braking Distance	Spacing of Signals	Minimum Sighting Distance	Length of Overlap	Maximum Train Length
CFL	400m Dependent 700m on speed 1000m and 1200m gradient		200–300m otherwise repeating signal is required	>60km/h: 100–200m ≤60km/h: 50m for >2.5% falling; 0m for <2.5% falling or rising gradient	
NSB	–	≤100km/h: 1000m ≤130km/h: 1200m ≤160km/h: 1800m ≤200km/h: 3000m	8s or minimum of 200m	0–400m	Local Trains: 250m IC & Reg.: 400m Freight: 1000m
NS	Dependent on gradient: 40km/h: 400m 60km/h: 500m 80km/h: 800m 130km/h: 1000m 160km/h: 1150m	400m min. 1800m max. In station areas, may be less than 400m if required for operational reasons	9s (in clear weather)	Not applicable	Passenger Trains: 400m Freight Trains: 700m
BV	130km/h: 1000m supplemented by ATC for high speed trains	1000m Up to 150m additional for falling gradient	200–350m Dependent on line speed	200m between train routes. May be reduced for reduced train speed and type of obstruction	Local Trains: 200m IC & Reg.: 400m Freight: 630m

1.3 Speed signalling and route signalling

Signalling systems can be divided into two types, by consideration of how the information is presented to the driver.

1.3.1 Speed signalling

Of the 12 European railways being considered, all except RENFE, BR and NSB employ a speed signalling system. Drivers are ordered to proceed at a given speed by the signal aspect.

Reductions in speed, due for instance, to:

- permanent speed restrictions,
- diverging routes being set,
- reduced braking distance beyond the next signal,
- signal ahead at danger,
- signal ahead at danger with a reduced overlap,

are all indicated to the driver as an order to reduce to a specified limit or target speed. Where that reduced speed must be maintained, this is also indicated as a speed limit. The driver is not told the reason for the reduction in speed, although if he is familiar with the route and the working of his train he will be able to deduce this; he is merely told to comply with the maximum speed indicated by the signal aspect.

The target or actual speed limits are indicated by various signal aspects; on some railways these are supplemented by an illuminated numeral which indicates the speed in km/h divided by 10; for example '13' means 130km/h maximum.

ÖBB
Colourlights to indicate 40 and 60km/h; numeral* for higher speeds.

NMBS/SNCB
Colourlights with numeral* to indicate speeds.

SBB
System L (Luminous) : colourlights to indicate 40, 65 and 95km/h.
System N (Numeric): colourlight plus numeral* to indicate speed.

DB AG
System HV : Double aspect to indicate 40km/h. The addition of a numeral* indicates a speed of 20, 30, 50 or 60km/h. One green light plus numeral* indicates a speed above 60km/h.
System KS : colourlight plus numeral* to indicate speed.

RENFE
Colourlights to indicate 30km/h; numeral* for higher speeds.

SNCF
Colourlights to indicate 30 or 60km/h; higher speeds indicated by numerals which show exit speed in km/h.

FS
Colourlights to indicate 30, 60 or 100km/h speeds for diverging routes.

CFL
Colourlights to indicate 30km/h; numeral* for higher speeds.

NSB
(Eventual Future System) Colourlights to indicate 40 and 100km/h; cab signals permit higher speeds.

NS
Colourlights to indicate 30 or 40km/h or maximum speed; numeral* for lower speeds than maximum.

BV
Colourlights to indicate 40 or 80km/h; no lineside numerals, but trains fitted with ATC may proceed at higher speeds as displayed by the cab equipment.

Notes
1. *Numeral x 10 indicates permitted speed in km/h.
2. Typical aspect profiles and sequences are given in the diagrams applicable to each railway in **1.5**.

1.3.2 Route signalling
The route signalling philosophy is entirely different. The signal aspects do not tell the driver at what speed he should drive, but they do indicate precisely what route he is to take, both at stations and junctions. He is expected to know, as part of his training, all the permanent speed restrictions on the line and also all the maximum permitted speed over all diverging junctions. However, to assist him in driving the train in the safest and most economical manner, the basic signalling system is supplemented on BR in the following manner.

1.3.2.1 Permanent speed restrictions
- Advance warning indicators, triangular with yellow border, at braking distance;
- Commencement indicators, circular with red border;
- Termination of restriction – circular, with red border, indicating line speed.

These signs are placed to aid the route learning process by drivers.

1.3.2.2 Diverging junctions
- Junction signal released from stop to a proceed aspect by the approach of the train, depending on the speed restriction over the diverging route relative to the allowable speed over the straight route;
- For higher speed junction, junction signal released from caution (1 yellow) by the approach of the train. This aspect may be preceded by the signal in rear of the junction showing flashing single yellow, and in four aspect systems the second signal is rear showing flashing double yellow;
- As an alternative to flashing yellows, two heads may be provided on the signal in rear of the junction signal, one for each route. This is called a splitting distant signal;
- Where the speed of the diverging route is not more than 16km/h less than that of the straight route, no additional controls are provided;
- In all cases, the diverging route is indicated by a junction indicator (bar of white lights) or, for divergences less than 96km/h, a cipher (letter or numeral) which is proved to be illuminated before the signal can show a proceed aspect;
- Where all routes are low speed – for instance when approaching large stations, a cipher is provided for the straight route as well.

	Route Signalling	**Speed Signalling**
Advantages	• Generally lower number of aspects • Drivers' action directed by implication	• Drivers action can be exactly signalled • Regulation of trains possible but rarely used
Disadvantages	• Route knowledge essential • Not easily used for permanent speed restrictions	• Higher number of aspects with attendant disadvantages: (1) More complex signal equipment (2) More scope for driver error

Fig. 1.2 Advantages and disadvantages of speed and route signalling

1.3.3 Comparative merits

The relative merits of speed and route signalling have been debated many times at Institution meetings; a detailed study of the subject was made some years ago (R.S. Wyatt, *Speed & Route Signalling Proc. IRSE 1981/2*) which included a table of advantages and disadvantages of each **(Fig. 1.2)**.

However, the distinction between speed and route signalling systems is not absolute. The approach control of signal aspects for trains approaching diverging junctions could be considered a form of speed signalling. Conversely, some speed signalled railways employ a form of indicator to advise drivers to which line they are routed.

Indeed, it is essential to provide the driver with routing information where for instance, the signal next in advance of the junction is one of a group and he must correctly observe the aspect which applies to his train.

- NMBS/SNCB: white 'V' above signal aspect indicates that train is routed to the right-hand line
- FS & CFL: sometimes indicate the route, for the information of drivers
- SNCF: one, two or three horizontal white lights above signal aspect indicates routes progressively from left
- NS: *Richtingaanwijzer* : 'V' with left or right-hand arm illuminated indicates route set to left or right. This signal has no safety significance; it is provided for operational reasons only.

1.4 Signal profiles

This section describes the appearance of signals and how differences in profile are used to convey fixed information to the driver. Some of the profiles are illustrated in the diagrams in **1.5**, which also discusses the variations in aspects and aspect sequences in current use. Shunting and other types of signal aspects are dealt with in this section since they are, in many cases, qualified by the profile of the signal.

1.4.1 Running signals

Running signals control train movements from block to block and are therefore designed to be seen at a distance of several hundred metres (see **Fig. 1.1** for minimum sighting distances of the various railways). They are normally elevated to about the level of a driver's eye if mounted on a post, or located on gantries above the track to which they refer. Individual light units are generally shielded from incident sunlight by hoods and the whole assembly is provided with a backboard painted matt black, to aid sighting. Some railways trim the edges of the backboard in white to give even greater contrast with the background scenery. A more comprehensive description of the engineering of signals and signal structures is given in Chapter 4.

Such signals are considered suitable for speeds up to 200km/h, supplemented in most countries by a track to train warning system. Above that speed, cab signalling is usually required (**Fig. 1.3**).

1.4.1.1 Stop signals

Stop signals are capable of displaying a danger of stop aspect which requires trains to halt at or before passing the signal post. The driver must then remain at the signal until authorized to move unless the aspect displayed gives him discretion to proceed under 'stop on sight' arrangements.

A signal which may be passed at danger under these circumstances is indicated as follows:

Railway	Maximum speed allowable with lineside signals	Track to Track Warning System
ÖBB	160km/h	LZB
NMBS/SNCB	160km/h	
SBB	160km/h	Above 160km/h, cab signalling system. Not yet defined
DB AG	160km/h	LZB
RENFE	200km/h	ASFA
SNCF	320km/h	KVB
BR	200km/h (225km/h for test purposes)	AWS
FS	150km/h	BACC and/or discontinuous cab signalling
CFL	160km/h	*Crocodile*
NSB	160km/h	
NS	Not applicable	
BV	80km/h on ATC fitted lines 110km/h on non ATC fitted lines	ATC (The optical system is capable of 140km/h but 80km/h is used to enforce the use of ATC where possible)

Fig. 1.3 Maximum speed for lineside signals

ÖBB
Flashing Red aspect.

NMBS/SNCB
Permissive signals are labelled 'A' or 'B'.

SBB
None. Only possible when calling-on aspect exhibited at home signals.

DB AG
Yellow/White/Yellow/White/Yellow panel at automatic block signals.

RENFE
'P' sign on signal post; 1 minute stop required.

SNCF
*'F' sign on signal post plus one red light only.

BR
'Automatic' plate (horizontal black line on white background) plus failure of the signal post telephone.

FS
'P' sign on signal post; 3 minute stop required.

CFL
None; a written order is necessary.

NSB
Flashing Red aspect in block signals and home signals. A minimum stop of 20 minutes is required.

NS
'P' sign on signal post, plus special authority.

BV
(Intermediate Block Signals only). Failure of contact with dispatcher; 15 minute stop required.

Notes
*'F' means *franchissable*; stop signals which must not be passed at danger display two red lights and are plated 'Nf' – *non-franchissable*.

1.4.1.2 Warning signals (Fig. 1.4)

Warning signals do not themselves carry a stop aspect, their most restrictive indication meaning 'be prepared to find the next signal at stop'. They may appear on their own, located at braking distance from their associated stop signal, or they may be combined with the previous stop signal, to give a three aspect system. Some railways call them distant signals, others advance or *avertissement*. When combined with a stop signal they may retain their separate identity or be merged as one signal. When on their own, most railways differentiate between a stop and a warning signal, but do it in a variety of ways.

Railway	Name		Differentiation with Stop Signal	Combined Stop and Warning
ÖBB	Distant*		Square background (Stop signals have rectangular backboard)	Located below stop signal
NMBS/SNCB	Warning*		Yellow number plate on post (stop signals have white number plate)	Combined within same backboard
SBB	Distant*	System L	Square background (stop signals have rectangular backboard)	Located below stop signal
		System N	Rectangular white border (Stop signals have circular border) Number plate shows one star	Combined within same backboard
DB AG	Distant*	System HV	'X' identification panel	Located below stop signal
		System KS	Yellow '∨' identification panel (Stop signals have red '∧' panel)	Combined within same backboard Post carries both red '∧' and yellow '∨' panels
RENFE	Warning*		No number plate	Combined within same board
SNCF	*Avertissment*		Circular backboard (Stop signals have irregular shaped backboard)	Combined within same backboard
BR	Distant		Number plate carries suffix 'R' (or 'RR' for YY aspect) No telephone provided	Combined within same backboard
FS	*Avviso*		Horizontal black stripes on signal post and countdown markers (Stop signals have plain light grey post and diagonal stripes on countdown markers)	Combined with stop signal aspect (searchlight mechanisms)
CFL	Distant*		Square backboard (Stop signals have rectangular backboard)	Located below stop signal
NSB	Distant*		Always displays flashing aspects	Located below stop signal
NS	*Voorsein*		Rectangular backboard (Stop signals have circular/oval backboard)	Combined within same backboard
BV	Distant*		Circular backboard with distinctive pattern. Always displays flashing aspects	Combined within same backboard. Warning aspects flash

* English translation

Fig. 1.4 Warning signals

1.4.1.3 Repeating signals (Fig. 1.5)

When signals cannot be sighted at a sufficient distance to enable a driver to approach at a reasonable speed but still observe and act on the signal indication, a repeating signal is provided. Whilst on most railways, a variation on the warning aspect is used, the BR practice is to use an entirely different indication, based on a semaphore arm. The 'banner' signal consists of a black arm rotating against a white background, although a more modern equivalent is now in use which uses fibre optic techniques to simulate the semaphore indication. NMBS/SNCB and NS use a white bar of light on a black background whilst ÖBB take the principle a stage further, by using an upwards diagonal row of white lights as a representation of a semaphore arm to repeat an

Railway	Separate Signals		Main signal aspect modified to indicate reduced braking distance
	ON Aspect	OFF Aspect	
ÖBB	Vertical/horizontal light bar	Vertical/inclined light bar	No
NMBS/SNCB	Horizontal row of white lights	Diagonal (\) row of white lights	Repeater shows 'ON' aspect for a running signal carrying a speed restriction
SBB	Repeat aspect of signal in advance	Repeat aspect of signal in advance	Number plate shows two stars
DB AG	None	None	Small white light at top left-hand corner of warning signal
RENFE	Vertical (rectangular) white light bar	Light bar extinguished	No
SNCF	None	None	No, but flashing single yellow at signal in rear
BR	Horizontal black arm on circular white background	Inclined black arm on circular white background	No
FS	Two white lights on the platform in rear of a departure signal	White lights extinguished	No
CFL	None	None	White aspect at bottom right-hand corner of warning signal
NSB	None	None	None
NS	Horizontal row of white lights	Diagonal (/) row of white lights	No
BV	Normal Warning Signal, plus target distance on marker plate		Two or three green lights

Fig. 1.5 Repeating signals

unrestricted running signal, and a downwards row of white lights to repeat a restricted speed running signal; a horizontal row of white lights repeats the stop aspect.

1.4.1.4 Contraflow signals (Fig. 1.6)

On NMBS/SNCB, SBB, most of SNCF, BR, FS, part of CFL and BV, trains run normally on the left-hand track; on the remaining railways right-hand running is practised. However, widespread use is now made of reversible or bidirectional running, both for normal use and during engineering works or emergencies. Some railways distinguish signals applicable to trains routed on the contraflow line.

Railway	Distinction
ÖBB	No distinction
NMBS/SNCB	Distinctive backboard is opposite handed. Aspects always flash. Change to and from contraflow track is indicated by a white 'V' above aspect
SBB	No distinction
DB AG	Change to and from contraflow track is indicated by a zig-zag fibre optic sign
RENFE	No distinction
SNCF	Signals controlling entrance and exit contraflow routes are equipped with luminous symbols showing the beginning and the end of the contraflow route TECS (*Tableau d'entrée à contresens*) – and TSCS (*tableau de sortie à contresens*) – contraflow entrance or exit luminous board
BR	Signal number plates carry suffix 'X' (applies to parts of BR only)
FS	Searchlight signal heads have square backboards for right-hand track
CFL	Change to and from contraflow track is indicated by a fibre optic sign
NSB	No distinction
NS	No distinction
BV	No distinction; all main tracks have equal status in both directions

Fig. 1.6 Contraflow signals

1.4.1.5 Number plates and signs

Obviously, it is of utmost importance that a signal should be seen; a black signal, due to lamp or power failure is not much use in the dark! A solution practised by railways which do not have a policy of proving lamps alight is to have a retro-reflective sign on the signal post or to paint the latter in stripes of retro-reflecting paint. This is the case on ÖBB, DB AG and CFL. The BV practice is to trim the backboards with high intensity retro-reflecting tape. Other railways paint their posts in a light colour (SNCF, BR, FS (stop signals)) or striped (FS (warning signals), NS); this also helps to locate signals in conditions of poor visibility.

Most signals carry a number plate for identification when, for instance, a defect has to be reported or a driver is being instructed to pass a signal at danger. In interlocking areas, the numbering is related to the sequence used for the interlocking but automatic block signals may have their identification related to distance (some parts of BR, RENFE). The number plate may carry a reference to the controlling signalbox or to the line, for example EN for Euston or DM for down main (BR). Other features defined by the signal number plate are warning signals (NMBS/SNCB, DB AG, BR) and permissive stop signals (RENFE, SNCF, BR, FS). NS define permissive stop signals by a separate plate, a black 'P' on white background.

1.4.1.6 Countdown boards

Countdown marker boards are used extensively on SNCF (lines above 120km/h), FS, CFL and NS. For instance, SNCF place three boards at 300m (3 diagonal black stripes), 200m (2 stripes) and 100m (1 stripe) in rear of stop signals at locations where the sighting of the signal is poor, whilst DB AG and CFL locate similar boards at 250m, 175m and 100m in rear of warning signals not combined with a stop signal. RENFE place markers at 300m, and BR have provided countdown boards at a small number of signals which have a poor record of being passed at danger. FS differentiate between warning signals, which have countdown boards with horizontal black and white stripes, and those for stop or combined signals which carry diagonal stripes. On electrified lines, four such boards are carried or painted on the catenary structures.

NS place markers in rear of all signals on the open line as well as for main track signals in small stations. These markers consist of three reflective plates at 180m, two at 120m and one at 60m (Type A). Different types of markers are used to mark:

- the last permissive signal on a line (Type B);
- warning signal (Type C);
- a signal at braking distance in rear of junction signals and signal protecting movable bridges (Type D).

Fig. 1.40F shows markers of Types A to D inclusive.

1.4.2 Shunting signals

Whilst the primary object of running signals is to keep the trains apart, shunting signals are necessary to control the bringing of trains together, for attaching, detaching or reforming. The profiles and aspects displayed in shunting signals are therefore different, both in shape and meaning.

The main difference in profile is that, being designed for low speed moves, they can be short range, with simpler optical systems. Position light aspects, rather than coloured lights, are used on some railways (ÖBB, SBB, DB AG, BR, FS, NSB, BV), to give a quite different indication from those of running signals. This particularly applies to the proceed aspects.

1.4.2.1 Shunting aspects on running lines

A shunting move may be required from a running signal, either to admit the train to an occupied platform (a calling-on route, for example yellow over yellow on FS), or to allow access to a siding or marshalling yard where main line signalling regulations do not apply. A separate ON aspect for a shunting signal located on the same post as a running signal is not usually provided, since the latter's main stop aspect suffices. (CFL do, however provide a separate ON aspect at such signals, giving a combined ON aspect of red and blue.) When a proceed aspect is displayed, it generally requires 'drive on sight' and a maximum speed is prescribed, although the controls sometimes include line of route tracks in the yard switching area, but this is not apparent to the driver.

Some railways differentiate between calling-on routes and those for ordinary shunting (SBB, CFL). NS do not use shunting signals at all – but its flashing yellow aspect in either a high or low signal has the same meaning – drive on sight, max. 30km/h – as do most railways' shunting signals.

Where shunting moves are required other than from running signals, for example setting back where contraflow signals are not provided, the shunting signals are provided with ON aspects and can be either elevated – principally to aid visibility when propelling moves are being made – or ground mounted. In some places where tracks are close together, elevated post mounted shunting signals are not possible, the only alternative to ground signals being to place them on gantries, an expensive option. When siting ground shunting signals it is necessary to consider their visibility from the cab, since some modern trains have long noses which require a driver to stop several metres short of such a signal, to keep it in view. This may affect the standage of trains in loops or between turnouts. Another consideration is that ground mounted signals tend to become dirty quite quickly, which affects their visibility. They require frequent cleaning, the cost of which may equate, over the lifetime of the installation, to the additional first cost of providing elevated signals.

1.4.2.2 Preceding shunts

To avoid shunting movements having to run to the extremity of the switching area to get behind a running signal, shunting signals may be located at convenient places within the pointwork. If these have to be passed by trains proceeding under the authority of a running signal, the former are also required to show a proceed aspect when the main route is set. Whilst this may be done automatically, as part of the main route setting operation, the operator is enabled to restore such signals to danger in an emergency after the train has passed the running signal. On BR, such signals are called preceding shunts (see **Fig. 1.7**). An alternative is to extinguish the shunting signals when a main route is set.

1.4.2.3 Shunting to and from sidings

Shunting to and from sidings may either be controlled by the signalman setting routes individually, or the whole area may be given over to shunting.

Where the signalman retains control, the shunting signals are of the types already described; where however continuous shunting is permitted, the fixed signals may be switched off altogether, or show a distinctive aspect (see **Fig. 1.8**) with movements controlled by hand signal.

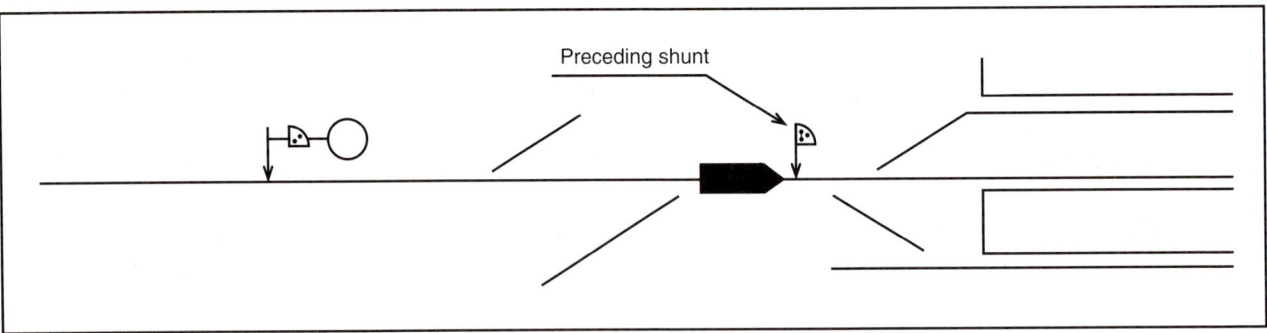

Fig. 1.7 Preceding shunt

Railway	Proceed Aspect when Combined with Main Signal	Independent Shunting Signals Stop (ON) Aspect	Proceed (OFF) Aspect
ÖBB	Two white lights (/)	Two white lights (–)	Two white lights (/)
NMBS/ SNCB	Red over white	Two white lights (–) or One red light	Two white lights (\) or One yellow light
SBB	System L (calling-on) Bar of yellow light (–) or (/)	Bar of white light (– or x)	Bar of white light (/) or (l)
	System N (calling-on) Flashing red	Two white lights (–)	Two white lights (\) or (l)
DB AG	Two white lights (/)	Two red lights (–)	Two white lights (/)
RENFE	Red over flashing white	One red light	Straight Route: Two white lights (l) Diverging Route: Two white lights (–) Local shunting: Red over white
SNCF	One white light (Steady or flashing)	One violet light	One white light (Steady or flashing)
BR	Main red aspect plus two small white lights (\)	One red and one white light	Two white lights (\)
FS	Yellow over yellow	Tall signals: Bar of white light (–) Ground signals: Two white lights (–)	Bar of white light (l or ll or ⋈) Two white lights (l)
CFL	One white light	One blue light	One white light
NSB	Main dwarf signal Two white lights (l)	Tall Signals: Bar of white light (–) Ground Signals: Two white lights (–)	Bar of white light (/) Two white lights (l) or (\) or (/)
NS	Not applicable	Not applicable	Not applicable
BV	(Elevated main signal with ground shunt signal or main dwarf signal): Two white lights (l)	Two white lights (–)	Two white lights (l) or (/) or (\)

Fig. 1.8 Shunting signal types

1.4.2.4 Trains starting from sidings

Whilst shunting movements from sidings on to running lines are controlled by shunting signals, the departure of a train into the block section after shunting has been completed may be differently signalled. If the driver, once on the main line, has to travel some distance before reaching the first running signal, the drive on sight aspect given by the shunting signal may be inappropriate. In such cases, BR consider placing a running signal at the siding exit, whilst SNCF's flashing white aspect specifically forbids departure to the main line. NMBS/SNCB use the colour light version of their dwarf signal to control such train movements.

1.4.3 Other signals

1.4.3.1 Protection signals

Yard exits are not normally equipped with a separate running signal for each road. Practices vary between railways; ÖBB and DB AG locate a protection signal at the stopping or fouling point of each road. On ÖBB, its red aspect is the same as for normal running signals, (with white/red/white reflecting bar) and the OFF aspect is two white vertical lights; DB AG use normal shunting signal aspects. SBB indicate the permissible speed of their group starting signals by taking into account the speed restrictions over the yard points in rear of such signals. This is an exception to the rule that the indicated speed commences at a running signal.

BR normally provide fixed signs 'stop and await instructions'; trains are then called forward by handsignals from the shunter.

1.4.3.2 Substitution signals

It is recognized that, whilst the signalling system is designed to fail-safe, the act of running trains during a signal failure is very considerably less safe than normal working. Instead of the interlocking logic of the control system, safe movement has to rely on the human observations and actions of signalmen, handsignalmen and drivers – not to mention inspectors and supervisors, whose main objective may be to get trains moving.

Rules and procedures are therefore framed to avoid an incident due to the lack of the usual signalling facilities, or to reduce the effects to a minimum if one should occur:

- Avoidance – formal procedures to govern train movements
  - operation of releases logged on a counter or otherwise recorded
  - proposed movements checked by a second operator
  - points to be secured on the panel or on the ground
- Reduction – speed limits on movements past red signals
  - limit of one (or two) simultaneous movements

One of the principal causes of incidents during handsignalling remains the mistransmission of messages between signalman/handsignalman/driver and other communication difficulties. Even the advent of driver-shore radio has not eliminated the problem – it has been known for drivers to be able to speak to a signalbox on an entirely different line! For this reason, ÖBB, SBB and FS have adopted procedures which involve signal aspects rather than verbal or written messages to order drivers to proceed.

- ÖBB: Substitution signal :
  - : White Flashing aspect: max. 40km/h
  - : Red Flashing aspect: stop on sight
- SBB: Main aspect, enabled by special procedures to be adopted by the signalman (see **1.7.10.3**).
- DB AG: Substitution signal
  - : 3 white lights (triangle): Proceed
  - : 3 yellow lights (inverted triangle): Caution
- FS : for departure signals: two horizontal yellow lights, steady or flashing, depending on whether or not the warning signal of the next main signal is combined with the departure signal
  - : for entry signals: two horizontal flashing milky white lights

ÖBB and DB AG display the substitution aspect when the signal cannot be cleared in the proper manner due to point or track circuit failure or when a signal lamp failure occurs. On SBB the signal can be specially operated under track circuit, block or barrier failure conditions, provided that the interlocking is in order and a route can be set.

On FS the substitution signal can be displayed for all types of failures for entry routes. For departure routes the block conditions are required to be satisfied.

1.4.3.3 Platform duties signals

Platform duties signals are not, in general, considered to be part of the hierarchy of running and shunting signals although Right Away indicators may, for convenience, be mounted on the post which carries the starting signal.

On ÖBB the Right Away indication is given by a flashing green aspect which complements the proceed aspect of the starting signal, elsewhere a green ring may be used (this being a derivative of the station master's green baton); on BR the cipher 'R' or 'RA' is used.

NMBS/SNCB use platform duties signals which are nicknamed 'marguerites'. They normally show no light; when the platform inspector operates the Right Away plunger, there is a short period of 7–10 seconds during which a single red light is shown. The aspect then changes to a ring of white lights; this is the signal for the train to depart.

Other platform duties signals are Brake Test (ÖBB, FS) and Close Doors (BR). The latter is a recent innovation, brought about by the advent of driver only operation (DOO) of multiple unit trains with sliding or plug doors. At unstaffed stations, a mirror or closed circuit TV is provided, but at staffed stations, the person in charge is required to signal to the driver to close the doors (CD indication), then to check that the doors are properly closed and that no passengers or clothing are trapped, before operating the Right Away indicator.

1.4.3.4 Level crossing rail signals

At level crossings where the barriers are interlocked with the signalling, protection for rail traffic is afforded by the normal running signals. Automatically operated level crossings may have no rail signals at all on NMBS/SNCB, SNCF, BR (AHB or MWL crossings), FS, CFL and NS, reliance being placed on the road user not to obstruct the crossing when the road signals and/or barriers are operating.

However, some railways have automatic level crossings where the train driver is advised by signal that the road signals and/or barriers have commenced to operate. If the proceed rail signal is not shown by the time that the train has reached braking distance from the crossing, its driver is required to stop and

Railway	Stop Aspect	Proceed Aspect
DB AG	1 yellow (yellow border to backboard)	Flashing white over yellow
BR (ABCL & AOCL crossings only)	Flashing red	Flashing white
BV (See **Fig. 1.43B**)	3 flashing yellows (∇) at >300m and visible at b.d Red at crossing	3 steady yellows (∇) at >300m White at crossing

Fig. 1.9 Examples of level crossing rail signals

not proceed over the crossing until he can satisfy himself that it is clear. A similar course of action is necessary if the train driver cannot see, from the braking point, that the crossing is unobstructed.

Examples of rail signals are given in **Fig. 1.9**:

It will be appreciated that if the crossing or the warning signal cannot be seen from the braking point, then the provision of signals is pointless. On BR such crossings therefore tend to be on low speed lines, or carry a local speed restriction, but BV use this method on all lines where it is not possible to afford protection by fixed signals. Where ATC is provided, the crossing signals are enforced by the ATC.

1.4.3.5 Signalling for freight trains

For the most part, freight trains are required to observe the same signals as passenger trains. This is generally appropriate since, although they may be heavier and have different braking characteristics to passenger trains, they tend to travel at lower maximum speeds.

NS, however, have identified two circumstances in which differential signalling is necessary. These are described in detail in **1.7.17.2** and the associated special signals are illustrated in **Fig. 1.40D**.

On the former Southern Region of BR, the lines are signalled to braking data which is applicable to multiple unit electric trains, due to the intense suburban services operated by these units. Freight trains running on these lines are therefore required to observe a limit of two-thirds (66%) of the line speed for passenger trains. On other parts of BR, trains with enhanced brake power of up to 9%g are allowed to attain a proportionately higher maximum than the line speed.

1.4.3.6 Fixed signs

Although much less dramatic and impressive than signals which light up, flash and change colour, fixed signs play an important part in the operation of most railways. Mention has already been made of signs on signal posts and countdown markers; permanent and temporary speed restriction indicators are surveyed in **1.4.4**. However, there remain other lineside signs to consider.

BR, FS and BV use fixed signs to indicate the presence of certain level crossings and BR use them extensively in connection with RETB installations. Indeed, with the exception of the Points Set indicator (single yellow), only fixed signs are used trackside to complement the radio token equipment. (See **Fig. 1.33D** for details.)

Other fixed signs are provided to instruct drivers of electric trains when to shut off and apply power at neutral sections, when to lower and raise pantographs and, on NMBS/SNCB, CFL and NS, when to change traction voltage.

Not quite in the category of fixed signs are mechanical points indicators. These are provided on some railways to assist in local shunting movements. Apart from being operated direct from the points, they have no connection with the signalling system.

1.4.4 Speed restriction signs

1.4.4.1 Permanent speed restrictions

With speed signalling it would theoretically be possible to include all permanent speed restrictions in the signalling system, without any additional indicators. Although this is done at facing junctions, a separate system of fixed indicators is often adopted in plain line areas to avoid additional running signals, use of which might upset the braking and headway requirements and would in any case be more expensive.

The signs adopted by some of the European railways are shown in **1.5.4**. In most cases there is an indicator at braking distance to the restriction, with the permitted speed in km/h shown as a numeral, divided by 10 (NMBS/SNCB, NSB, NS) or actual (RENFE, SNCF, FS, CFL). Where ATC is in general use a beacon giving commensurate information in the cab is provided (FS, BV). The commencement of the restriction is indicated by a sign giving similar speed information (NMBS/SNCB, NS), a common cipher, for example 'Z' on SNCF and CFL or a small yellow triangle (NSB). The end of the restriction is also indicated either by a sign giving the new maximum speed – line speed, or another restriction – (NMBS/SNCB, NSB, NS) or a different common cipher (R on SNCF and CFL). All these signs are reflectorized. On BR, the speeds are shown in mile/h, but the same principle applies, with warning and commencement signs showing the restricted speed and the termination sign the speed that applies beyond the restriction. An AWS indicator is provided on the approach to the warning sign.

The indication of permanent speed restrictions over diverging junctions is extended on RENFE, SNCF, BR, FS (up to 200km/h) and CFL by the addition of arrows showing to which direction the restriction applies. This complements the indications given by the signalling system.

1.4.4.2 Temporary speed restrictions

Whilst indicators for permanent speed restrictions could, at a pinch, be considered almost a luxury – because they are always there and are provided primarily for drivers' route learning purposes – the same cannot be said of indicators for temporary speed restrictions. Although booklets listing the places where they have been imposed are published regularly and drivers on most railways are required to acknowledge their receipt, the precise location of a temporary speed restriction remains to be defined by the trackside signs.

As with permanent speed restrictions, it is usual to provide warning, commencement and termination indicators. However, additional considerations apply:

- the signs must be portable, not too heavy to carry but stable enough not to be blown over by the wind (or stolen);
- they must be visually striking, since they appear in a variety of unusual places. This has led most railways to retain some form of illumination, not relying entirely on reflective surfaces although some do rely on high intensity retro-reflective backgrounds.

NMBS/SNCB use the same design of indicator as for permanent speed restrictions, with the addition of a pair of flashing lights to ensure visibility, whilst SNCF provide circular boards showing the restricted speed for warning and commencement, with a plain white circular board at the termination point. FS use shaped yellow and black boards with yellow lights for warning and commencement, with a rectangular green and white board plus green light for the termination. CFL provide a circular yellow board with two flashing yellow lights for warning, a rectangular board for commencement both with speed indicated, and a plain rectangular white board for termination. NS provide rectangular yellow boards lettered L with two yellow lights – followed at a distance of 50m by an eight-sided speed indicator, a yellow board lettered A with one yellow light and a green board lettered E with one green light for commencement and termination respectively.

On BR, besides the yellow warning board with two flashing white lights and speed indicator in mile/h, commencement speed indicator and termination 'T' board, an additional indicator is provided for emergency speed restrictions which it has been necessary to impose at short notice. This consists of black and white chevrons with two brilliant white strobe lights which are required to flash at all times of day or night. Like the temporary speed warning indicator, it is preceded at 180m by an AWS inductor.

On some railways (BR, NS) differential temporary speed restrictions are permitted; the higher speed applies to passenger trains, the lower to freight trains.

A dilemma surrounds the imposition of a severe temporary speed restriction on the approach to an automatic level crossing. Should the level crossing controls be modified to compensate for the longer road warning time that the lower approach speed will impose, since it is a tenet of automatic level crossings that the road is closed for the minimum time before each train arrives? But if the controls are modified, what happens if the temporary speed restriction is lifted without the level crossing controls being reinstated?

1.5 Signal aspects and aspect sequences

There are two signal aspects that are almost universal amongst the European railways and indeed, worldwide:

- Red means STOP
- Green means PROCEED, except for flashing green on BV

However, these two aspects give only the most basic information to the driver; the efficient operation of a main line railway requires a degree of advance warning of conditions ahead. Herein lies the railway equivalent of the Tower of Babel – the multiplicity of aspects which mean slow down to, or proceed at, a certain speed.

Each railway must, of necessity, claim that its own choice of aspects and aspect sequences is optimum! However, there are some basic factors which can be traced through the various railways systems:

- the aspect sequences must be logical;
- account must be taken of lamp failure – it must not produce a less restrictive aspect;
- mode of failure of flashing aspect;
- attempts to create a four aspect system to achieve an improved headway.

1.5.1 UIC 732 Recommendations

UIC 732 Committee has published a model set of aspects for speed signalling systems:

- Green : proceed at line speed
- Yellow : warning : the next signal is at red
- Red : stop before passing the signal

This is a basic three aspect system; as such it does not permit signals to be spaced at less than braking distance and consequently restricts headways on heavily used lines. However, four intermediate speed levels of 30–40km/h, 60–70km/h, 90–100km/h and 120km/h can be used for control of trains through diverging junctions and this is one method that some railways use additionally for obtaining an improved headway.

The UIC 732 Recommendations also permit the qualification of the red aspect for permissive working; this is accomplished in various ways as detailed in **1.4.1.1**.

The various strategies adopted by European railways in interpreting and building upon UIC 732 Committee's Recommendations bear some consideration and analysis.

1.5.2 Aspects in use in Europe

Diagrams of aspects used in running signals by the various railways are shown in **1.5.4** together with brief explanations. The following sections discuss the meaning, on each railway, of the principal aspects, grouped by colour.

1.5.2.1 Single green (Fig. 1.10)
Although precise definitions vary slightly, on most railways one steady green light means line clear, proceed at maximum permitted speed. An exception is NSB, where steady green means line clear, proceed to route with reduced speed, but flashing green in a warning signal means line clear – maximum speed. On DB AG (KS) and NS, those meanings are totally reversed, since flashing green means proceed at 40km/h maximum – or as indicated by numeral.

A more significant exception is to be found on BV where, although a green aspect in a stop signal means line clear, a flashing green aspect in a warning signal indicates next signal at stop.

1.5.2.2 Double green (Fig. 1.11)
The principal use of two greens lies with those railways which use separate warning aspects. On ÖBB, SBB (L) and DB AG (HV), an inclined pair of green lights indicates line clear – an analogy of the former lowered semaphore arm.

ÖBB, SBB (L) and BV use two greens vertically displayed in stop signals to indicate a reduced speed route, whilst on NSB this aspect indicates line clear – line speed.

The only use of flashing double green, vertically displayed is on BV where it is the warning signal 'clear – reduced speed' indication.

1.5.2.3 Single yellow (Fig. 1.12)
Steady single yellow is used by all except ÖBB, NMBS/SNCB, NSB and BV and in each case it means prepare to stop at next signal. On NSB flashing single yellow is used for this purpose.

SBB and NS may qualify the single yellow aspect by a numeral or cipher and on BR RETB systems using train operated points it has an altered meaning, indicating that the facing points are correctly detected and the train may proceed at 20km/h into the loop.

Flashing single yellow has a variety of meanings, but whereas on RENFE it means reduced braking distance to the next signal, SNCF and FS use the same aspect to indicate reduced braking distance between the next signal and the signal beyond. On NS, it is the nearest aspect to a shunting signal – drive on sight; 30km/h maximum.

Railway			Steady	Flashing
ÖBB		Stop Signal	Proceed at maximum local speed	Departure signal (displayed with proceed aspect)
NMBS/SNCB		Combined Signal	Proceed; next signal is at a proceed aspect	Proceed – applicable to RH track for contraflow moves
SBB	System L	Stop Signal	Line Clear; proceed at line speed	None
	System N		Line Clear; proceed at line speed (Qualified by numerical speed indicator)	None
DB AG	System HV	Stop Signal	Proceed at line speed unless speed restriction indicated	None
	System KS		Proceed at line speed unless speed restriction indicated	Proceed; reduce to speed indicated (Yellow numeral x 10km/h)
RENFE			Proceed at line speed	None
SNCF			Proceed at line speed (up to 160km/h)	Reduce speed to 160km/h by next signal
BR			Proceed at maximum permissible speed	Proceed: next signal is showing green or flashing green Test trains may run at up to 225km/h (Experimental: installed for high speed testing of rolling stock)
FS		Stop Signal	Line Clear	None
		Warning Signal	Line Clear at maximum permissible speed	None
CFL		Stop Signal	Line Clear	None
		Warning Signal	Line Clear	None
NSB		Stop Signal	Line Clear; proceed to route with reduced speed	None
		Warning Signal	None	Line clear at line speed
NS			Permission to proceed at the speed indicated by the fixed speed signs along the line	Unqualified: permission to proceed at 40km/h Qualified by indicator: permission to proceed at speed indicated (number x 10km/h)
BV		Stop Signal	Line Clear	None
		Warning Signal	None	Next signal is at stop

Fig. 1.10 Single green

Railway		Steady	Flashing
ÖBB	Stop Signal	(I) Proceed at 60km/h	None
	Warning Signal	(/) Next stop signal is at 'proceed'	None
NMBS/SNCB		None	None
SBB	System L Stop Signal	(I) Clear to proceed at 65km/h	None
	Warning Signal	(/) Line clear	None
	System N	None	None
DB AG	System HV		
	Warning Signal	(/) Next main signal shows clear	None
	System KS	None	None
RENFE		None	None
SNCF		None	None
FS		None	None
CFL		None	None
NSB		(I) Line clear at line speed	None
NS		None	None
BV	Stop Signal	(I) Proceed at 40km/h (40–80km/h with ATC); expect to stop at next signal	None
	Warning Signal	None	(I) Next signal shows 2 or 3 green lights

Fig. 1.11 Double green

Railway		Steady	Flashing
ÖBB		None	None
NMBS/SNCB		None	None
SBB	System L	None	None
	System N	Unqualified: caution, next signal shows stop.	None
		Qualified by numeral: proceed at speed indicated. Can also be qualified by horizontal yellow bar or V See **Fig. 1.24B & C**	
DB AG	System HV Warning Signal	Braking distance from level crossing	None
	System KS	Caution: next signal shows STOP	None
RENFE		Caution: be prepared to stop at next signal	Caution; reduced bd to next signal
SNCF		Prepare to stop at next signal	The next signal is at single yellow (steady) and is at reduced bd to the succeeding stop signal
BR		Caution: Prepare to stop at next signal (On RETB lines: facing points are detected in normal position)	Next signal is showing one yellow with junction indicator for diverging route
FS	Warning Signal	Prepare to stop at next signal	The next signal displays a proceed
NSB	Warning Signal	None	Next signal at STOP
NS		Unqualified: Reduce speed to 40km/h, prepare to stop at red signal. Qualified by numeral: reduce speed to speed indicated before passing next signal. Qualified by flashing numeral: reduce to or maintain speed indicated. Distance to next signal is <bd	Permission to drive on sight at max. speed of 30km/h
BV		None	None

Fig. 1.12 Single yellow

1.5.2.4 Double yellow (Fig. 1.13)

Steady double yellow is especially popular amongst the railways which use dedicated warning signal aspects and it is displayed vertically, horizontally and inclined (∕ and ∖). ÖBB, NMBS/SNCB, SBB (L) and DB AG (HV) all use it to indicate 'be prepared to stop at next signal'. On CFL at a warning signal (horizontal) or a stop signal (vertical) it means respectively reduce to, or proceed at, 30km/h; it can be qualified by a numeral to indicate higher speeds or by a 'U' to indicate a calling-on aspect. CFL is the only railway which has a warning aspect for a calling-on route.

SNCF use horizontal double yellow to warn of a 30km/h turnout; vertical for the immediate approach; flashing indicates similarly for a 60km/h turnout.

Fig. 1.13 Double yellow

Railway		Steady	Flashing
ÖBB	Warning Signal	(–) Next stop signal is at stop	None
NMBS/SNCB		(∕) Be prepared to stop at next signal	(∖) Be prepared to stop at next signal – applicable to RH track for contraflow moves
SBB	System L Stop Signal	(l) Caution, next signal at stop with bd for <40km/h	None
	Warning Signal	(–) Expect next signal at stop	None
	System N	None	None
DB AG	System HV Warning Signal	(∕) Caution, next signal shows stop	None
	System KS	None	None
RENFE		None	None
SNCF		(–) Reduce speed to 30km/h when traversing the junctions ahead of the next signal (l) Do not exceed 30km/h when traversing the junction ahead of this signal	(–) Reduce speed to 60km/h when traversing the junctions ahead of the next signal (l) Do not exceed 60km/h when traversing the junctions ahead of this signal
		Note: the above aspects may be qualified by the single yellow (steady or flashing) in main signal (see **Fig. 1.31C**)	
BR		(l) Preliminary caution: be prepared to find the next signal at single yellow	(l) Next signal is showing flashing single yellow
FS	Warning Signal	(l) Warning of occupied track at abnormally reduced bd or occupied or short track	None
	Substitution Signal	(–) For departure signal (includes warning signal)	(–) For departure signal (not including warning signal)

Railway		Steady	Flashing
CFL	Stop Signal	(l) Unqualified: Proceed at 30km/h Qualified by numeral e.g. '6': Proceed at 60km/h Qualified by white ⊔: Proceed with caution to platform	None
	Warning Signal	(−) Unqualified: Reduce speed to 30km/h Qualified by numeral e.g. '6': Reduce speed to 60km/h Qualified by yellow ⊔: Reduce speed to 30km/h	None
NSB		None	None
NS		None	None
BV		None	None

These *Ralentissement* and *Rappel* aspects may be qualified by the main aspect, for example steady or flashing yellow.

BR use steady double yellow as the preliminary caution aspect; very similar to the flashing single yellow of SNCF and FS, but the latter uses steady double yellow to signify an abnormally reduced distance to the stop signal – a meaning totally opposite to that of BR and needing specific lamp proving to avoid a wrong side failure due to lamp burn out.

1.5.2.5 Green and yellow together (Fig. 1.14)

Since all railways being considered here except BV use both yellow and green aspects, it is not surprising that, when an additional aspect was called for, simultaneous display of yellow and green came to be considered on economic grounds. However, both the meaning and the relative disposition of the aspects vary.

With regard to the former, on most railways a yellow and green aspect requires a reduction of speed to be made or maintained. But the less restrictive meaning of a single green aspect (**1.5.2.1**) means that precautions are necessary against failure of the yellow light. Most yellow and green aspects are displayed vertically; Green over Yellow on ÖBB, NMBS/SNCB, SBB (L), DB AG (HV) and RENFE; and Yellow over Green on FS and NSB.

NMBS/SNCB also use a horizontal display, whilst warning signals on SBB (L) and DB AG (HV) show Yellow over Green (inclined) and Green over Yellow (inclined the other way) respectively. Only SBB (L) uses both Green over Yellow – vertically, at stop signals and Yellow over Green – inclined, at warning signals.

Apart from the standard policies on flashing aspects adopted by NMBS/SNCB and NSB, only FS uses flashing combinations of Yellow over Green – flashing in phase to indicate a 60km/h restriction and flashing alternately to indicate 100km/h.

Railway			Steady	Flashing
ÖBB		Stop Signal	(l: G over Y) Proceed at 40km/h	None
NMBS/SNCB			(– G, Y) Proceed, next stop signal	(– Y, G) Proceed, next signal carries a carries a speed restriction speed restriction. Applicable to RH track for contraflow moves
			(l: G over Y) Proceed; be prepared to stop at second signal. Next signal is at <bd	(l: G over Y) Proceed; be prepared to stop at second signal. Next signal is at <bd Applicable to RH track for contraflow moves
SBB	System L	Stop Signal	(l: G over Y) Proceed; max. speed 40km/h	None
		Warning Signal	(\: Y over G) Reduce speed to 40km/h	None
	System N		None	None
DB AG	System HV	Stop Signal	(l: G over Y) Pass at 40km/h unless otherwise indicated (e.g. by numeral)	None
		Warning Signal	(/: G over Y) Next stop signal at (G over Y)	None
	System KS		None	None
RENFE			(l: G over Y) Pass next signal at restricted speed	None
SNCF			None	None
BR			None	None
FS	Warning Signal		(l: Y over G) Reduce speed to 30km/h by next signal	(l: Y over G: flashing in phase): reduce speed to 60km/h by next signal (l: Y over G: flashing alternately): reduce speed to 100km/h by next signal
CFL			None	None
NSB	Warning Signal		None	(l: Y over G) next signal at clear to route with reduced speed
NS			None	None
BV			None	None

Fig. 1.14 Green & yellow together

1.5.2.6 Single red (Fig. 1.15)

Steady single red means stop in all signalling systems. The only qualification is on SNCF, where the absolute stop signal is double red; after bringing his train to a stand at a single red the driver may proceed – prepared to stop on sight. On other railways, derogation to pass a signal at red is given by other means (see **1.4.1.1**).

Flashing red has a number of meanings. On ÖBB it is the caution aspect of the substitution signal (see **1.4.3.2**); on SBB (N) it is the call-on aspect (maximum 20km/h), whilst on SNCF it fulfils a similar role (maximum 15km/h). BR use it to indicate that the road traffic signals at a locally monitored level crossing have not started to operate or that the barriers, where provided, have not commenced to descend; NSB flash the red stop aspect in their home and block signals.

A single red light, or very often two, carried on the rear vehicle of all trains and on buffer stops of bay and terminal lines is, of course, the ultimate stop aspect. A flashing red tail lamp is used on some railways when built-in vehicle power is not available or reliable.

Railway			Steady	Flashing
ÖBB		Stop Signal	Stop	Caution signal
NMBS/SNCB			Stop	Stop: Applicable to RH track for contraflow moves
SBB	System L	Stop Signal	Stop	None
	System N		Stop	Calling-on signal: max. speed 20km/h
DB AG	System HV	Stop Signal	Stop	None
	System KS		Stop	None
RENFE			Stop	None
SNCF			Stop & proceed. Be prepared to stop on sight (Plated F)	May be passed without stopping but speed not to exceed 15km/h. Be prepared to stop on sight
BR			Stop	(At locally monitored level crossing): Road signals/barriers not operating. Stop before crossing; proceed when safe to do so
FS		Stop Signal	Stop	None
CFL		Stop Signal	Stop	None
NSB		Stop Signal	Stop (Starting signals)	Stop (home and block signals)
NS			Stop	None
BV		Stop Signal	Absolute stop	None

Fig. 1.15 Single red

1.5.2.7 Single white (Fig. 1.16)

Not much use is made of white as a running signal aspect, except on BV, who do not use yellow and who regard green as the caution aspect in warning signals. Flashing white on BV therefore means that the next stop signal is at clear – green; this anomaly has been discussed earlier. Otherwise flashing white is used on ÖBB and DB AG (System KS) as a substitution aspect; on SNCF, both steady and flashing, to control shunting and on BR and FS (flashing) and BV (steady) to indicate that a level crossing may be traversed.

On DB AG and NS a single white aspect is used to show that the signal is otherwise switched off. This is to indicate that the signal has not failed, but is out of use for a specific reason for example:

- DB AG – main signal at half braking distance
- NS – shunting is taking place

The relative unpopularity of white as a railway signalling colour probably stems from the desire to avoid confusion with sundry lights outside the railway. However, the almost universal employment at present of sodium vapour lamps for street and motorway lighting means that steady single yellow aspects are also at risk from misinterpretation. Perhaps BV is right after all!

Railway		Steady	Flashing
ÖBB	Stop Signal	None	Substitution Signal
NMBS/SNCB		None	None
SBB		None	None
DB AG	Stop, Warning and Shunting Signals	Signal switched off for shunting	Substitution Signal (System KS)
RENFE	Warning Signal	Movement authorized up to next signal	None
SNCF		Proceed as shunting movement	Restricted (short) shunt; departure to main line prohibited
BR		None	(At locally monitored level crossing): Road signals/barriers are operating. Proceed if crossing clear
FS		None	Automatic open level crossing with road signals: road signals operating
CFL		None	None
NSB		None	None
NS		Signal switched off for shunting (local control)	None
BV	Stop Signal	Level crossing barriers closed (see **Fig. 1.43B**)	None
	Warning Signal	None	Next signal shows 1 green light

Fig. 1.16 Single white

1.5.2.8 Other combinations (Fig. 1.17)

Most railways use other combinations of colours and/or patterns for specific meanings. Some of these are related to more precise indications to reduce or maintain speeds at diverging junctions or other sites – ÖBB, SBB (L), FS and BV fall into this category. NMBS/SNCB and RENFE use combinations of red and white to enable shunting moves on running lines; DB AG (HV) and BV use flashing white over steady yellow and a triangle of yellow respectively to indicate level crossing clear and DB AG (HV) and CFL use an additional white light in their warning signals to indicate less than braking distance to the associated stop signal.

Several railways use white light arrays to indicate divergences; as well as the route-signalled BR, with five light lunar white junction indicators (**Fig. 1.32B**), RENFE, SNCF and NS also make use of white lights in this manner. ÖBB use white over white as their protection signal proceed aspect and DB AG (HV) use triangles of white and yellow for substitution signals (proceed and caution respectively).

BR use a position light signal to control mechanized loading/unloading operations, for instance at collieries and power stations. It is illustrated in **Fig. 1.32C**.

The SNCF 'Nf' double red aspect has already been mentioned; this can be shown either vertically or horizontally. At the exit from yards a single violet light may be used instead. Although rather difficult to see in strong daylight, it is used only for low speed moves and has the advantage of not distracting drivers on adjacent running lines.

Although FS – using searchlight units – can generate impressive displays of various colours; up to three lights, plus qualifying ciphers; the record for aspect complexity must go to SBB (L) where, with a stop and warning signal on the same post, up to six lights can be shown!

Fig. 1.17 Other combinations

Railway		Combination	Meaning
ÖBB	Stop Signal	(I: White over White)	: Protection signal: 'Proceed'
	Warning Signal	($>:_G^YG$)	: Next stop signal shows restriction to 60km/h
		($\triangledown:_G^{YY}$)	: Next stop signal shows restriction to 40km/h
NMBS/SNCB		(I: Red over White)	: Shunt: from running signal
		(I: Flashing Red over Flashing White)	: Shunt: from running signal (applicable to RH track)
SBB	System L Stop signal	(I: Green over Green over Green)	: Proceed at 95km/h
	Warning signal	($\triangle:_{GY}^G$)	: Reduce speed to 95km/h
		($\triangleright:_G^YG$)	: Reduce speed to 65km/h
	System N	None	None

Railway		Combination	Meaning
DB AG	Stop Signal System HV	($\triangle$: ^W_W W)	: Substitution Signal: Proceed
		($\triangledown$: ^Y_Y Y)	: Substitution Signal: Caution
		(I: Flashing White over Steady Yellow)	: Level crossing protection operative
	Warning Signal	($\triangledown$: $^{WY}_Y$)	: Warning signal at less than bd from stop signal
	System KS	None	None
RENFE		(I: Steady Red over Flashing Yellow)	: Permissive Entry
SNCF		(I: Red over Red) (–: Red + Red): (Violet)	: Absolute Stop (Plated Nf)
		1, 2 or 3 white lights (qualifying proceed aspect)	: Route set 1st, 2nd, 3rd from left
BR		Row of 5 white lights (qualifying proceed aspect)	: Route set to left or right (see **Fig. 1.32B**)
FS		(I: Red over Green)	: Line clear and confirmation of 30, 60 or 100km/h restriction
			: (Unqualified = 30km/h; one white bar under aspect = 60km/h: two white bars = 100km/h)
		(I: Red over Yellow)	: Line clear and confirmation of 30, 60 or 100km/h restriction; next signal at stop
		(I: Steady Red over Flashing Yellow)	: Line clear and confirmation of 30, 60 or 100km/h restriction; next signal at reduced bd from signal beyond
		(–: White + White)	: Substitution signal (for entry signal)
CFL		(\: Yellow over White)	: Repeater for stop signal at red
		(–: Green + White)	: Repeater for stop signal at green
		($\triangledown$: $^{YY}_W$)	: Repeater for stop signal at double yellow
NSB		None	None
NS		None	None
BV	Stop Signal:	(I: Green over Green over Green)	: Exceptionally short route: 40km/h max. (ATC: 40–80km/h)
	Warning Signal for level crossing	($\triangledown$: ^Y_Y Y)	: Level crossing barriers open to road } see **Fig. 1.43B**
		($\triangledown$: Flashing ^Y_Y Y)	: Level crossing barriers closed to road

1.5.2.9 Uses of flashing aspects (Fig. 1.18)

All railways except CFL make some use of flashing aspects but their meanings vary tremendously. In nearly every case flashing was adopted to provide additional aspects whilst using the same hardware – a relatively easy and cheap way of increasing the versatility of the signalling system, whilst at the same time providing indications which are eye catching and distinctive.

On three railways flashing aspects have been given quite specific generic meanings. On NMBS/SNCB all aspects of signals applicable to contraflow movements on the right-hand track flash; their meanings are similar to the steady aspects in the left-hand track signals. On NSB and BV all aspects of warning signals flash, even when the warning signal is combined with a stop signal. NSB also flash the red aspect of stop signals when they function as home or block signals (starting signal red aspect is steady).

That said, there is wide diversity amongst the remaining railways in the uses to which flashing aspects are put.

Flashing green, for instance, means on:

- ÖBB: Right Away signal (qualified by steady green in starting signal)
- DB AG (System KS): reduce speed to that indicated by numeral displayed (× 10km/h)
- SNCF, RENFE: reduce speed to 160km/h by next signal
- BR: next signal is at green or flashing green (installed for rolling stock test purposes only)
- NS: proceed at 40km/h or other speed indicated

Flashing yellow is used in almost as diverse a manner, but always in a cautionary role.

- RENFE: caution, reduced braking distance to next signal
- SNCF: next signal at steady yellow with reduced braking distance to the second signal
- BR: next signal is at steady yellow with a diversion indicator for the highest speed diverging route
- FS: next signal is at reduced braking distance from the signal beyond
- NS: drive on sight; maximum 30km/h

Flashing double yellow is in use on only two railways:

- SNCF: reduce to 60km/h (horizontal) or maintain 60km/h (vertical) over junctions
- BR: (four aspect systems) next signal is showing flashing single yellow

Flashing red is used for:

- ÖBB: substitution caution signal (stop on sight)
- SBB (System N): calling-on aspect, maximum speed 20km/h
- SNCF: maximum speed 15km/h; stop on sight
- BR: level crossing road signals/barriers not operating

Fig. 1.18 Uses of flashing aspects

Railway	Aspect	Meaning	Flashing Rate/min
ÖBB	Flashing Green	'Right Away' signal (order to driver to start)	60
	Flashing White	Substitution signal	
	Flashing Red	Substitution caution signal (stop on sight)	
NMBS/SNCB	All signals applicable to contraflow routes on the RH track display flashing aspects equivalent to the steady aspects shown for the LH track		
SBB System N	Flashing Red	Calling on: max. 20km/h	
DB AG System KS	Flashing Green	Proceed; reduce to speed indicated by numeral	60
	Flashing White	Substitution signal	

Railway	Aspect	Meaning	Flashing Rate/min
RENFE	Flashing Yellow	Caution, reduced bd to next signal	
	Steady Red over Flashing Yellow	Permissive entry	
SNCF	Flashing Green	Reduce speed to 160km/h by next signal	
	Flashing Yellow	Next signal at 1 Yellow and @ < bd	
	Flashing double Yellow	(−) reduce speed to 60km/h; (I) do not exceed 60km/h	
	Flashing Red	Max. 15km/h stop on sight	
	Flashing White	Restricted short shunt	
BR	Flashing Green	Test Trains may exceed 200km/h (up to 225km/h)	60
	Flashing Yellow	Next signal at Yellow with diversion indicated	60
	Flashing double Yellow	Next signal at Flashing single Yellow	60
	Flashing Red	Level crossing road signals/barriers not operating	120
	Flashing White	Level crossing road signals/barriers operational	80
FS	Flashing Yellow	Next signal at reduced bd from signal beyond	60
	Flashing Yellow over Flashing Green (in phase)	Reduce speed to 60km/h by next signal	
	Flashing Yellow over Flashing Green (alternately)	Reduce speed to 100km/h by next signal	
	Steady Red over Flashing Yellow	Diversion; next signal at less than bd from signal beyond	
	Flashing double Yellow	(−) Substitution signal (for departure signal − not combined with warning signal)	
	Flashing double White	(−) Substitution signal (for entry signal)	
CFL	None		
NSB	Warning signals always display flashing aspects		90
	Flashing Green	Line clear at line speed	
	Flashing Yellow	Next signal at stop	
	Flashing Yellow over Flashing Green	Next signal at clear: reduced speed	
	Flashing Red	Stop: (home and block signals)	
NS	Flashing Green	Proceed at 40km/h or as indicated	75
	Flashing Yellow	Drive on sight 30km/h max.	
BV	Warning signals always display flashing aspects		80
	Flashing White	Next signal shows 1 Green light	
	Flashing double Green	Next signal shows 2 or 3 Green lights	
	Flashing Green	Next signal is at stop	

Some railways even use flashing white:

- ÖBB and DB AG (System KS): substitution signal
- SNCF: restricted speed: short shunt
- BR: level crossing road signals/barriers operational
- FS: substitution signal, and automatic open crossing road signals operational

But pride of place in the flashing aspect field must go to FS. Flashing yellow over flashing green has two meanings:
1. aspects flashing in phase: reduce speed to 60km/h by next signal
2. aspects flashing alternately: reduce speed to 100km/h by next signal

Two railways use steady red over flashing yellow:

- RENFE: permissive entry to station
- FS: diversion; next signal at less than braking distance from signal beyond

1.5.2.10 Policy on lamp failure

The precautions taken by railways against lamp failure causing a dark signal vary from very little to extensive.

At one end of the scale are NS and ÖBB. On the latter, double filament lamps are used only on high speed lines and lamp proving is restricted to returning the signal to red if a proceed aspect lamp burns out. Reliance is placed on the provision of a retro-reflective white–red–white pattern on the signal post, or on the signal backboard if gantry mounted. RENFE, on the other hand, operate a comprehensive system of aspect sequencing in which the failure of any aspect will modify either its own signal, or the one in rear, or both.

Of the remaining European railways, SNCF use single filament lamps but provide two separate light units for some of the more important aspects (C, R and RR – see **Fig. 1.30**); BR always use double filament lamps and return the signal in rear to red if both filaments of the lamp which should be lit fail, but there is no local switching to find a working aspect in the signal concerned. FS use single filament lamps but prove an aspect alight in the signal in rear.

The one thing in common to all railways is that a dark signal must be taken as its most restrictive aspect.

1.5.3 Aspect sequences

Apart from the stop indication, all signal aspects give some information about the state of the line ahead. This information may be quite basic, as with the proceed aspect of a two aspect system: proceed until you come to a stop aspect, or with the majority of shunting proceed aspects: proceed as far as the line is clear; be prepared to stop on sight of an obstruction.

However, with three and four aspect systems, a proceed signal aspect must give information, not only on the state of the line up to the next signal, but also on the aspect being displayed by that signal. In speed signalling systems, this must include the speed at which next signal must be passed if this will (or may) entail a reduction, on the basis that trains cannot change their speed instantaneously.

1.5.3.1 Signalling for following trains

The basic principles of two, three and four aspect signalling systems have been explained in **1.2** and a diagram showing BR practice is shown in **Fig. 1.33**. It should be emphasized that an aspect sequence with isolated warning signals is still a three aspect system; even though each signal shows only two aspects, the system as a whole is capable of showing three – proceed, be prepared to stop at next signal and stop. Although most railways claim to operate a three aspect system, there are at least two ways in which it can be extended to create what is effectively a four aspect system by ensuring that the warning signal is passed at a speed from which the train can be safely braked to a stand at the stop signal. This may be done by an indication at the signal in rear of the warning signal which either: (a) prescribes the maximum speed at which the warning signal may be passed (ÖBB, SBB (L), CFL, NS); or (b) indicates that the warning signal is at reduced braking distance from the stop signal:

- NMBS/SNCB-Green over Yellow
- RENFE-Flashing Yellow, for speeds between 160 and 200km/h
- SNCF-Flashing Yellow
- BR-Double Yellow
- FS-Flashing Yellow
- BV-Flashing Double Green

The former (a) (maximum speed at the warning signal) is the more flexible method, since it can be extended even further, to give a reduction of speed over more than two sections, as is done in some cab signalling systems. The 'reduced braking distance' method (b) relies to a greater extent on drivers' route knowledge, since the reduction is likely to vary between locations due to topographical features such as tunnels or stations. The BR practice is to space four aspect signals as evenly as possible, with never less than one third (33%) of braking distance from line speed between any two consecutive signals (both adjacent sections would then have to be of 67% braking distance each to preserve the full braking distance from the double yellow to the red). Such variations in signal spacing would, of course, affect the headway.

It is permissible, though undesirable, to intersperse an isolated four aspect signal into an otherwise three aspect sequence – for instance, to accommodate the reduced distance between entry and exit signals at a station, so as to avoid having to locate the entry signal too far away from the entry points. However, it is not good signalling practice to place isolated three aspect sections into an otherwise four aspect sequence, even if there is sufficient braking distance between successive signals. This would create a trap for drivers who would be used to receiving two warnings before a signal at STOP.

An alternative method of creating a better headway out of a basic three aspect system is in use on DB AG and CFL. On DB AG (HV System), each main signal showing STOP is preceded by a warning signal at full braking distance. The aspect of the stop signal at half braking distance is switched off, a white marker light being substituted. The associated warning signal then functions as a repeater, shown by the addition of a small white light. As the train moves forward, the aspect sequence steps up, the signal concerned becoming the warning signal at full braking distance. CFL operate a similar aspect sequence. SNCF and DB AG/CFL arrangements are shown in **Fig. 1.19**.

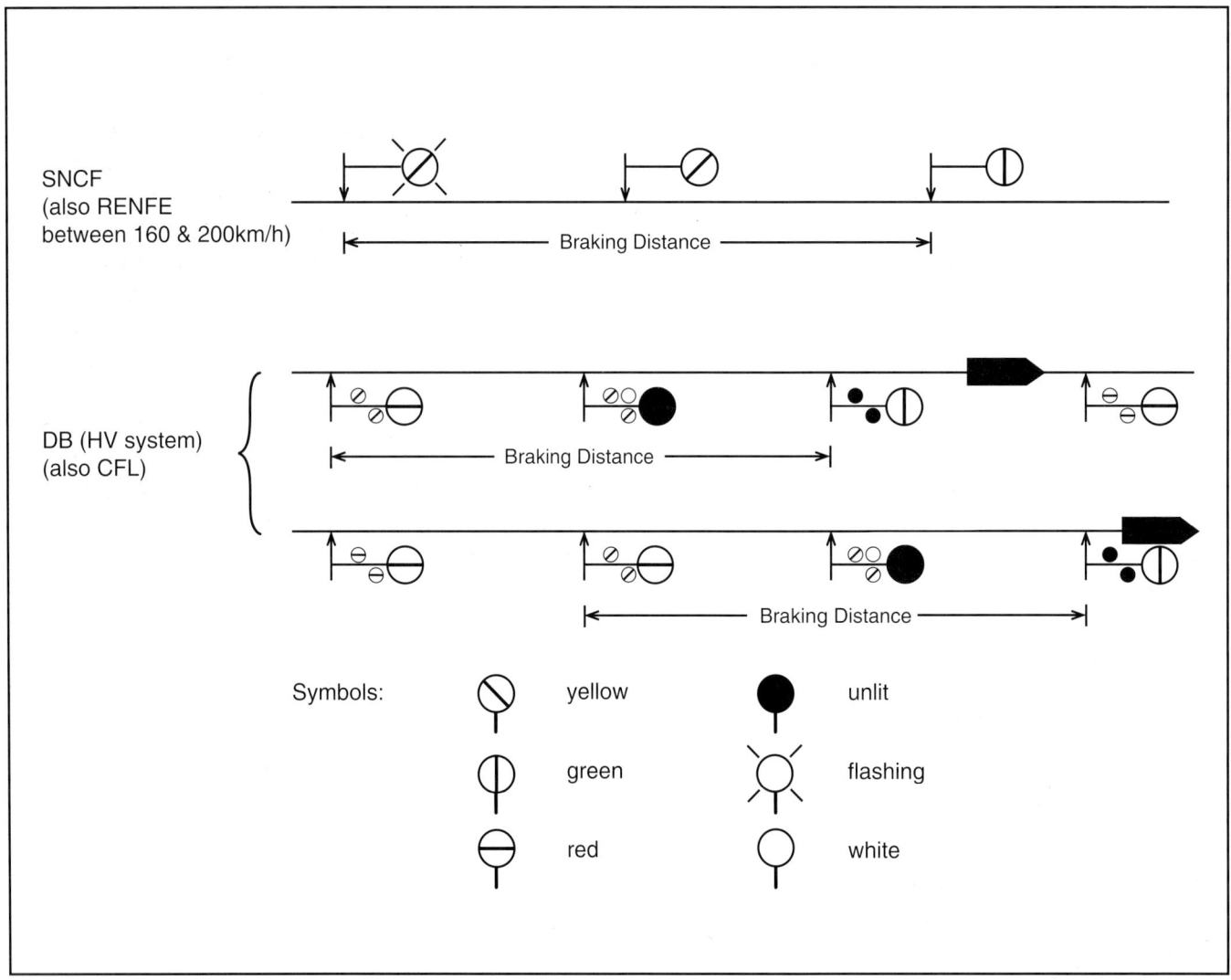

Fig. 1.19 Extension of three aspect systems to give improved braking distance

1.5.3.2 Signalling for diverging junctions

The way in which diverging junctions are signalled represents the biggest difference between speed and route signalling systems and even within these groupings, each railway has developed its own methods.

For low speed divergencies, all except SBB(N), DB AG(KS) and NS use colours – combinations of yellow or yellow and green; FS manages to include red over green as well! For higher speed junctions, illuminated numerals are used on ÖBB and SNCF (over 60km/h) and on CFL (over 30km/h). NMBS/SNCB use yellow/green (horizontal) in warning signals and a numeral to qualify the stop signal, whilst on the remaining systems, colours are used for all speeds.

On RENFE and BR, operating route signalling systems, white direction indications qualify the junction signal aspect, with appropriate caution aspects in rear. The route system used by NSB uses only the single green in the junction signal to indicate a reduced speed divergence (line speed is double green on NSB).

SNCF and FS also use direction indicators in an auxiliary mode, to supplement the speed signalling system. NMBS/SNCB and DB AG indicate that a train is being switched to or from a contraflow line to facilitate drivers sighting the correct signal in advance of the junction.

1.5.3.3 Signalling in stations

For trains running into stations with the line clear, including the overlap, where the railway provides them, the usual signal aspects apply. This is supplemented, if required, by an indicator showing the permitted speed or, on route signalled systems, by an indication of to which platform the train is routed.

Where the overlap appropriate to the approach speed is not available, clearance of the previous stop signal may be delayed or a more restricted speed limit enforced.

If the platform is partially occupied, a distinctive entry aspect is usually employed. This calling-on aspect authorizes the driver to proceed as far as the line is clear, to couple to the train in front or to enable two short trains to use the same platform for connectional purposes or to enable a train to enter a platform behind a train which is departing. Strict control of speed is enforced, since this movement breaks the golden rule that only one train must be in a section at any one time. At terminal stations, or for a bay platform at a through station, the length of the second train may be measured, for example by means of short track circuits, to ensure that it can be accommodated wholly in the platform without the rear part of it standing foul of other lines, thus causing operating inconvenience.

The aspects exhibited by calling-on signals are generally similar to those shown in **Fig. 1.8**.

The starting of trains from stations follows usual practice except that at large stations where all trains are starting from rest, short range signals may be employed. ÖBB use Protection Signals – in some cases with the warning signal applicable to the group starting signal ahead – for this purpose.

1.5.4 Signal aspect and aspect sequence diagrams

The following diagrams show the principal signal profiles, aspects and aspect sequences currently in use by the various railways.

They are grouped by railway administration, with signal aspect diagrams being followed by aspect sequences. The form, layout and meaning of the profiles and aspects are as supplied by the various railways; the title of each aspect is given in the mother tongue where possible, together with its translation into English. Likewise, the method of depicting aspect sequences varies from railway to railway; these have been reproduced as faithfully as possible so as to retain their respective national characteristics.

The diagrams represent current practice; for SBB and DB AG two systems are shown: *L* and *N* for SBB; *HV* and *KS* for DB AG. In each case the first system, though in current use, is obsolescent; the second is the present standard.

Fig. 1.20A ÖBB: signal aspects

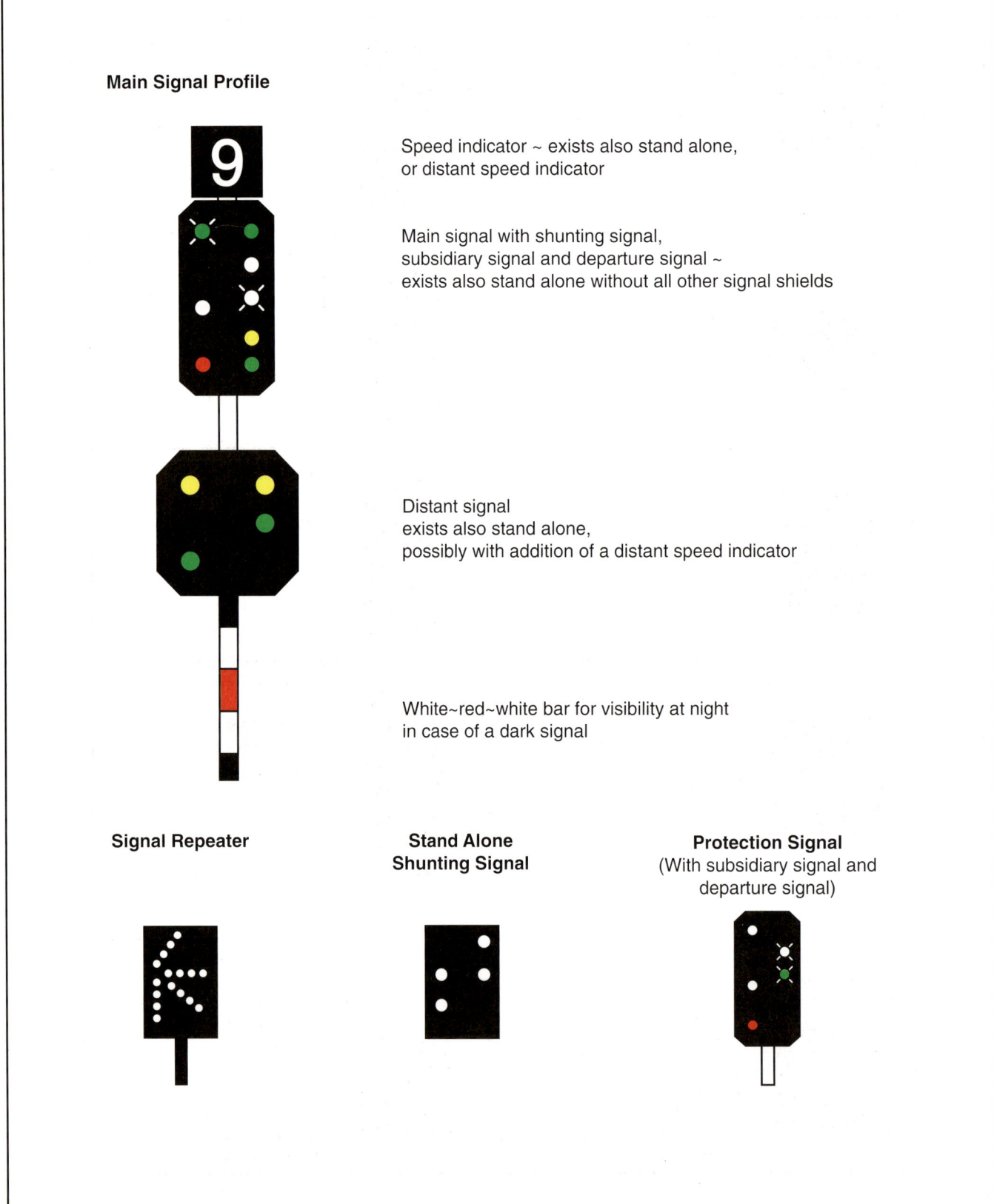

Fig. 1.20B ÖBB: signal aspects

Distant signal	Interpretation	Main signal	Interpretation
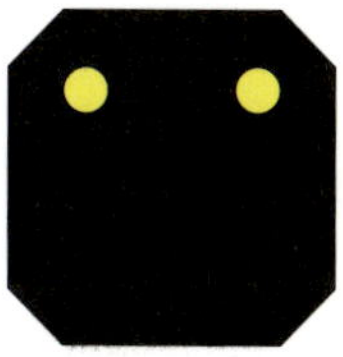 (black octagon with two yellow lights)	*Vorsicht* Warning (main signal shows stop)	(black rectangle with red light)	*Halt* Stop
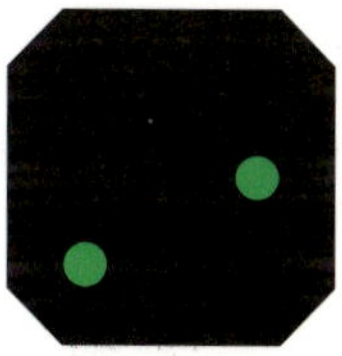 (black octagon with two green lights)	*Hauptsignal frei* Main signal shows free	(black rectangle with green light)	*Frei* Free (with maximum local speed)
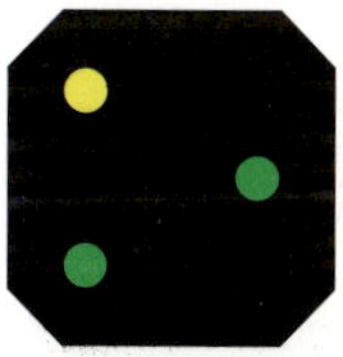 (black octagon with yellow and two green lights)	*Hauptsignal frei mit 60km/h* Main signal shows restriction to 60km/h	(black rectangle with two green lights)	*Frei mit 60km/h* Free with speed restriction to 60km/h
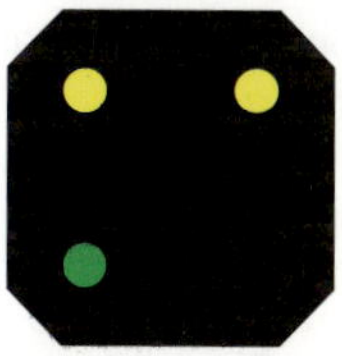 (black octagon with two yellow and one green light)	*Hauptsignal frei mit 40km/h* Main signal shows restriction to 40km/h	 (black rectangle with green and yellow light)	*Frei mit 40km/h* Free with speed restriction to 40km/h

Distant speed indicator	Interpretation	Speed indicator	Interpretation
 **9** (yellow)	*Geschwindigkeits- voranzeiger* Speed indicator shows the speed (number x 10), here 90km/h	 **9** (white)	*Geschwindigkeits - anzeiger* Maximum speed is (number x 10), here 90km/h

Fig. 1.20C ÖBB: signal aspects

Signal repeater	Interpretation
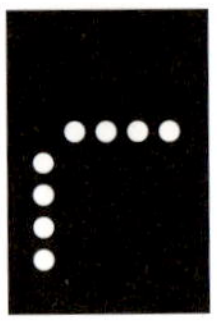 (black signal with L-shaped pattern of white dots)	*Hauptsignal zeigt halt* Main signal shows stop
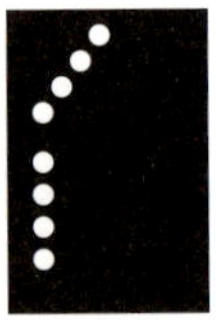 (black signal with diagonal line of white dots)	*Hauptsignal zeigt frei* Main signal shows free
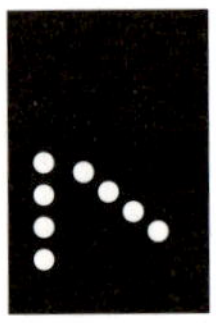 (black signal with K-shaped pattern of white dots)	*Hauptsignal zeigt frei mit Geschwingigkeitsbeschraenkung* Main signal shows speed restriction (40km/h or 60km/h)

Shunting signal	Interpretation
 (black signal with two horizontal white dots)	*Verschubverbot* Stop aspect (shunting forbidden, not relevant for a train route)
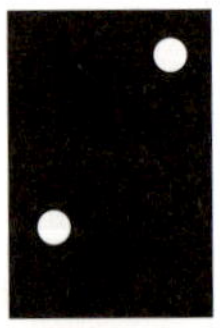 (black signal with two diagonal white dots)	*Verschubverbot aufgehoben* Free aspect (shunting admitted)

Fig. 1.20D ÖBB: signal aspects

Protection signal	Interpretation
(black signal with one red light)	*Fahrverbot* Stop
(black signal with two white lights)	*Fahrverbot aufgehoben* Free (free for a train route, or shunting is admitted)

Subsidiary signal	Interpretation
(black signal with one white light)	*Ersatzsignal* Free aspect (If the white light is replaced by a red light, it becomes the caution signal, *Vorsichtssignal*)

Departure signal	Interpretation
(black signal with white and green light)	*Abfahrt* Departure~order to the engine driver to start Only valid together with a free signal aspect

Fig. 1.21A ÖBB: aspect sequences

Examples for sequences:

M	~ Main Signal	G	~ one green light
D	~ Distant signal	2xG	~ two green lights
P	~ Protection signal	Y	~ one yellow light
SI	~ Speed Indicator	2xY	~ two yellow lights
DSI	~ Distant Speed Indicator	R	~ one red light
		2xW	~ two white lights

The distant signal may be also a stand alone signal, if the distance between the main signals is far more than the braking distance.

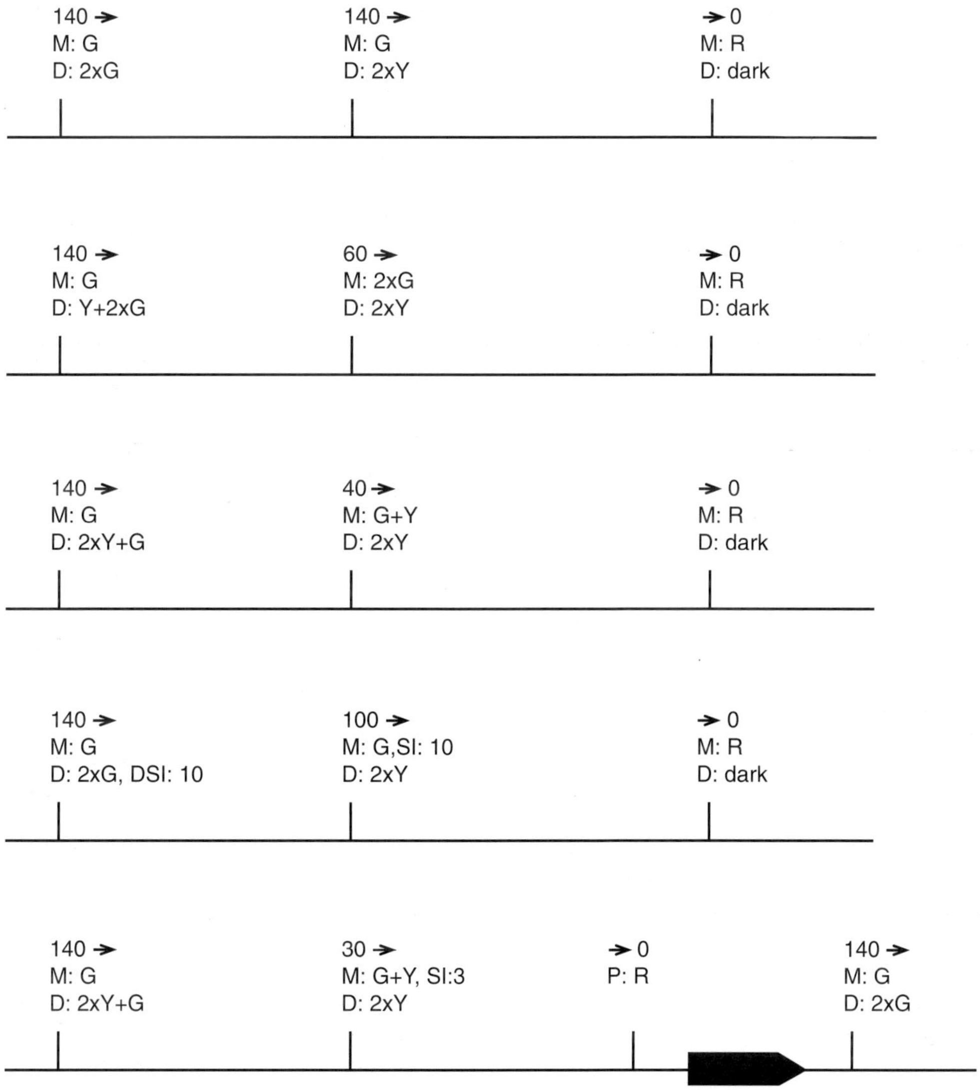

```
140 →              140 →              → 0
M: G               M: G               M: R
D: 2xG             D: 2xY             D: dark
 |                  |                  |
─┴──────────────────┴──────────────────┴────────


140 →              60 →               → 0
M: G               M: 2xG             M: R
D: Y+2xG           D: 2xY             D: dark
 |                  |                  |
─┴──────────────────┴──────────────────┴────────


140 →              40 →               → 0
M: G               M: G+Y             M: R
D: 2xY+G           D: 2xY             D: dark
 |                  |                  |
─┴──────────────────┴──────────────────┴────────


140 →              100 →              → 0
M: G               M: G, SI: 10       M: R
D: 2xG, DSI: 10    D: 2xY             D: dark
 |                  |                  |
─┴──────────────────┴──────────────────┴────────


140 →              30 →               → 0        140 →
M: G               M: G+Y, SI:3       P: R       M: G
D: 2xY+G           D: 2xY                        D: 2xG
 |                  |                  |          |
─┴──────────────────┴──────────────────┴━━━━━━▶──┴──
```

Fig. 1.21B ÖBB: aspect sequences

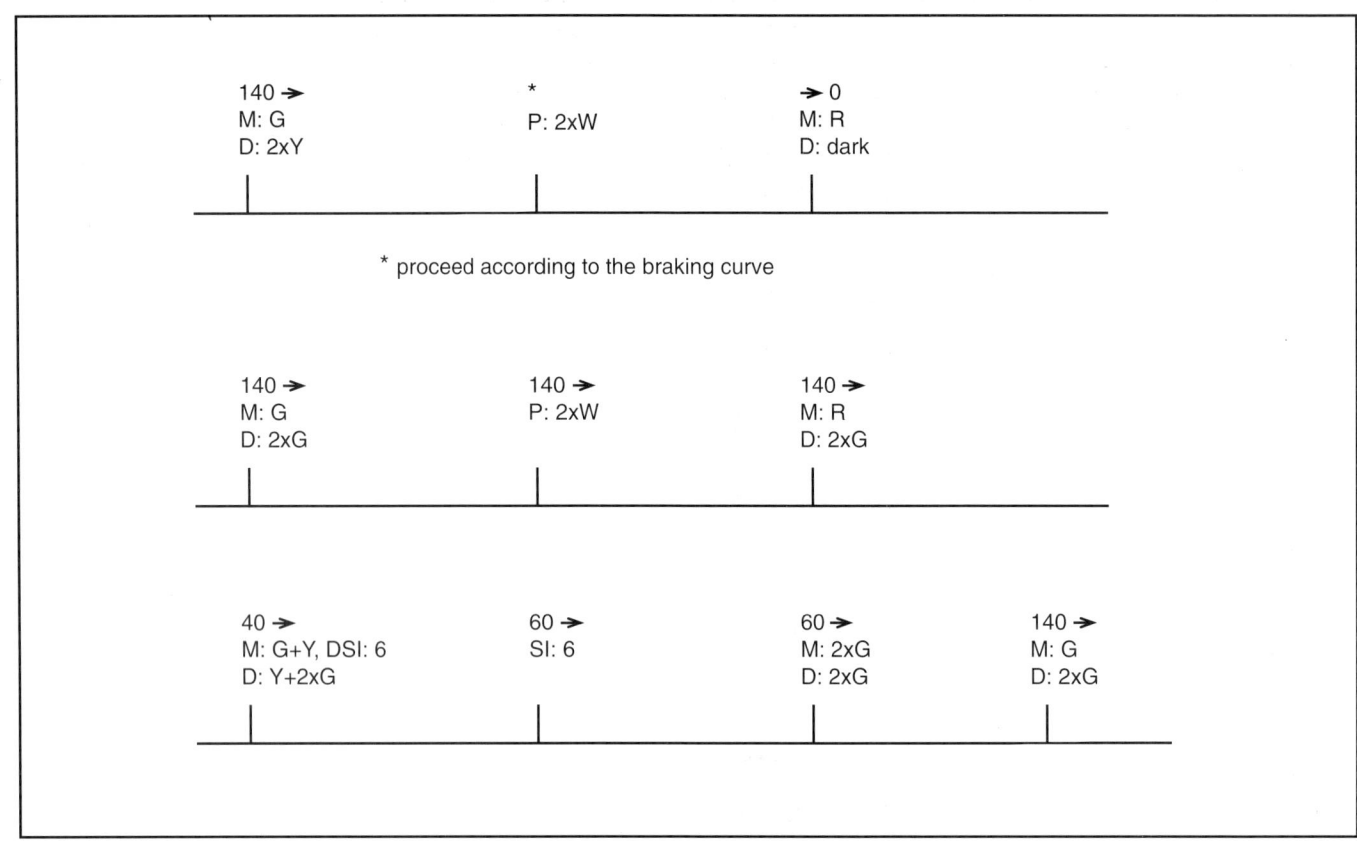

Fig. 1.22A NMBS/SNCB: signal aspects

Description	Aspect	Meaning
Combined Stop and Warning Signal **Profile**	 (green, red, yellow, white lamps)	Note: Signals not having a Stop aspect (warning only signals) are identified by a yellow number plate
Red	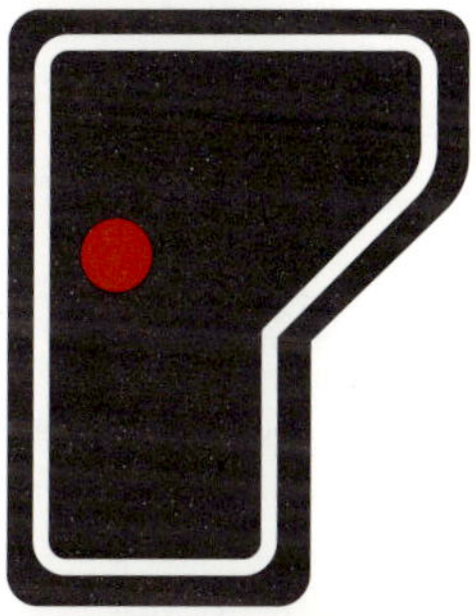 (red)	Stop
Green	 (green)	Proceed; next signal is exhibiting a proceed aspect
Double Yellow	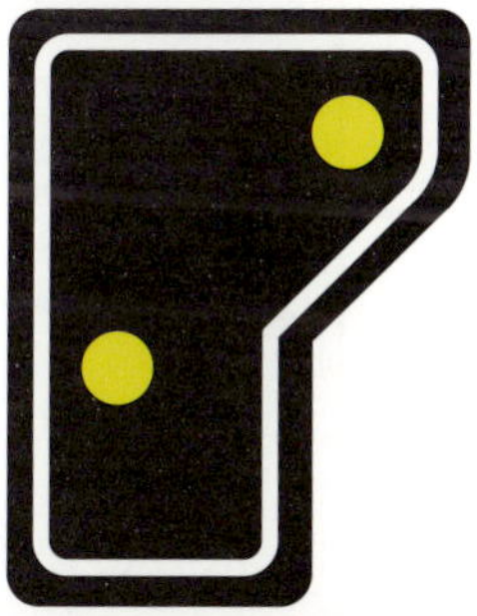 (yellow over yellow)	Be prepared to stop at the next signal
Vertical Green over Yellow	 (green over yellow)	Proceed but be prepared to stop at the second next signal (the next signal is at double yellow but is located at less than braking distance from the red). Alternatively it carries a speed restriction for which there is less than braking distance

Fig. 1.22B NMBS/SNCB: signal aspects

Description	Aspect	Meaning
Doble Yellow with speed indicator (White numeral)		Proceed at maximum speed of 40km/h. Be prepared to stop at the next signal
Horizontal Yellow~Green		Next stop signal is at clear but carries a speed restriction. Note: If variation in speed restricted routes is greater than 30km/h a speed indicator (Yellow numeral) is provided to show the actual speed restriction
Double Yellow with speed indicator and white chevron		Proceed at maximum speed of 40km/h Be prepared to stop at next signal. You are routed to the right-hand track; the next signal is located on the right and its aspect is flashing Note: Also applies vice versa
Double Yellow (flashing @ 1Hz)		Be prepared to stop at next signal. Signal at right-hand side of right-hand track for trains routed along that track in the contraflow direction. (This convention applies to all aspects)

Fig. 1.22C NMBS/SNCB: signal aspects

Description	Aspect	Meaning	
Shunting Signals Red over White aspect on running signal	 (Red over White on running signal shape)	Proceed	
Position Light shunt ON Aspect	 (Triangle with two white dots)	Stop	Applicable for shunting movements only
OFF Aspect	 (Triangle with three white dots)	Proceed	
Colour Light shunt (ground mounted) or Yard / Low speed line (elevated) ON Aspect	 (Red light)	Stop	Applicable for train movements as well as shunting
OFF Aspect	(Yellow light)	Proceed	

Fig. 1.22D NMBS/SNCB: signal aspects

Description	Aspect	Meaning
Line Speed Indicator 100km/h	 green ▲ "10"	Line speed is 100km/h
Permanent Speed Restriction Warning	 yellow ▼ "2"	Service braking distance from commencement of permanent speed restriction of 20km/h
Commencement	 ☐ ○ "2"	Commencement of permanent speed restriction of 20km/h
Termination into a second Permanent Speed Restriction	 green ▼ with yellow ▼ "5"	Termination of permanent speed restriction. Speed of 50km/h applies (lower than line speed). If line speed applies, a line speed indicator would be provided

RAILWAY SIGNALLING PRINCIPLES 47

Fig. 1.22E NMBS/SNCB: signal aspects

Description	Aspect	Meaning
Temporary Speed Restriction Warning Indicator	 Yellow downward triangle with "8" and flashing lights	Service braking distance from commencement of temporary speed restriction Note: commencement and termination indicators also similar to those for permanent speed restrictions, but with flashing lights
Change of Voltage Indicator	 Blue diamond "25000" Blue diamond "3000"	Requires pantographs to be lowered/raised. Change from 25Kv to 3Kv (Also applies vice versa)
Platform Duties Signal (Marguerite) Note: normally shows no aspect	Single red light on black background	Single red~shows for 7~10s after operation by platform inspector
	Ring of white lights on black background	After the 7~10s delay, the red aspect is replaced by a ring of white lights

Fig. 1.23 NMBS/SNCB: aspect sequences

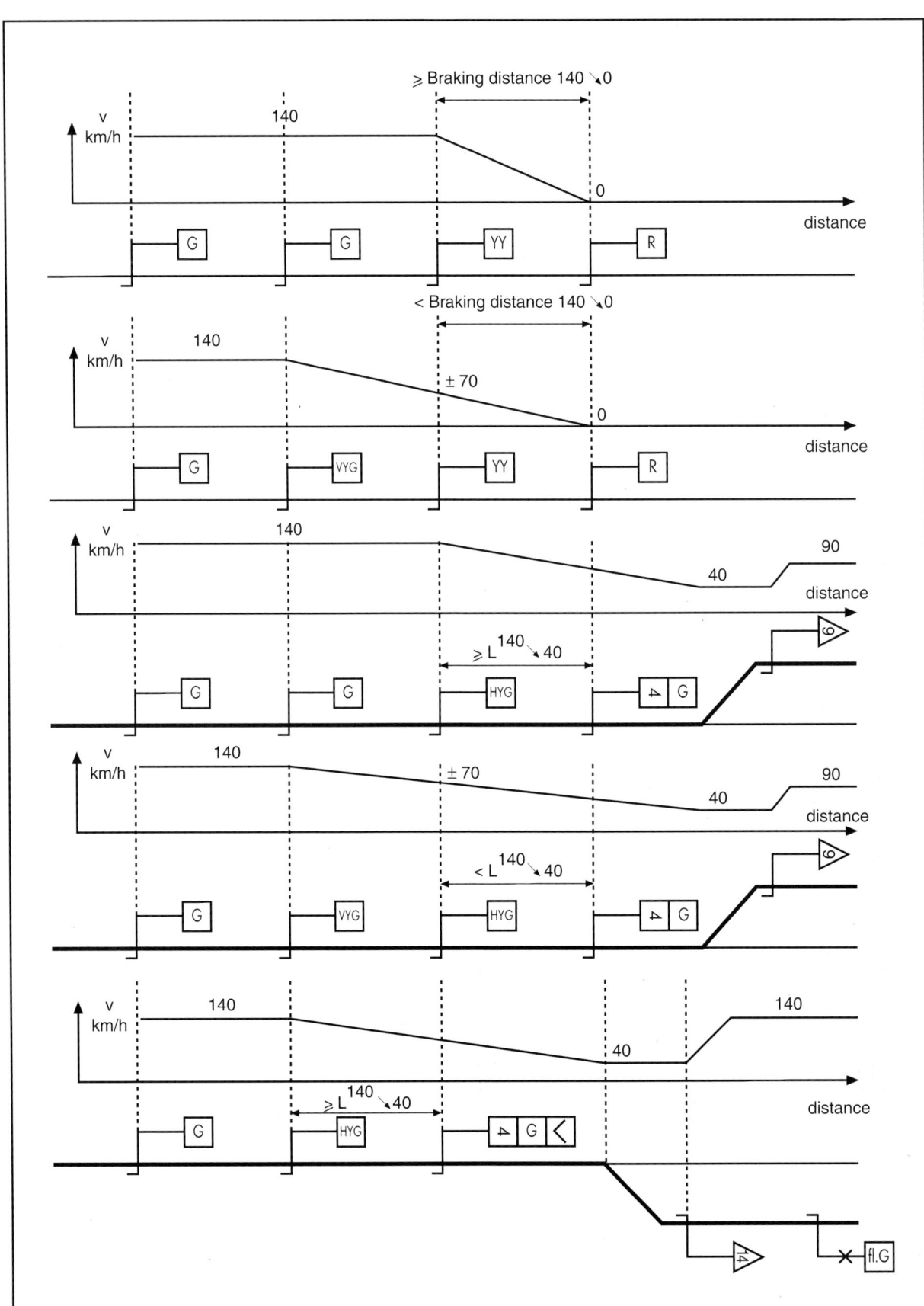

RAILWAY SIGNALLING PRINCIPLES 49

Fig. 1.24A SBB: signal aspects

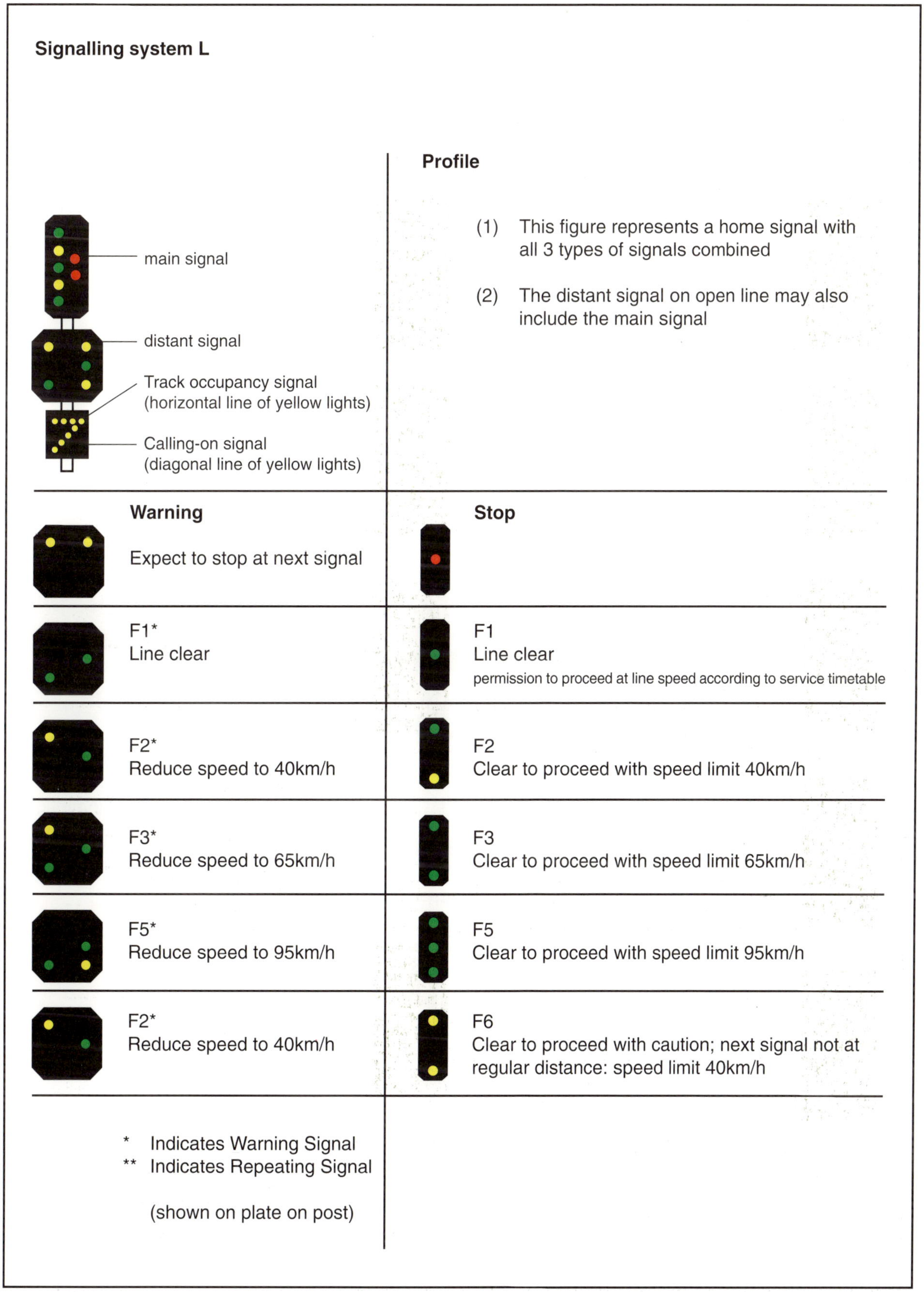

Signalling system L

main signal — distant signal — Track occupancy signal (horizontal line of yellow lights) — Calling-on signal (diagonal line of yellow lights)	**Profile** (1) This figure represents a home signal with all 3 types of signals combined (2) The distant signal on open line may also include the main signal	

Warning		Stop	
	Expect to stop at next signal		
	F1* Line clear		F1 Line clear permission to proceed at line speed according to service timetable
	F2* Reduce speed to 40km/h		F2 Clear to proceed with speed limit 40km/h
	F3* Reduce speed to 65km/h		F3 Clear to proceed with speed limit 65km/h
	F5* Reduce speed to 95km/h		F5 Clear to proceed with speed limit 95km/h
	F2* Reduce speed to 40km/h		F6 Clear to proceed with caution; next signal not at regular distance: speed limit 40km/h

* Indicates Warning Signal
** Indicates Repeating Signal

(shown on plate on post)

Fig. 1.24B SBB: signal aspects

Signalling system N

Profile	Profile
Circular, white border. This signal can display stop	Rectangular, white border. This signal cannot display stop. It is only used as a distant signal
-0 Caution, slow down; the next signal shows stop	H Stop
-4 Speed (announcement) The indicated speed (digit times 10) applies from the next signal	4- Speed (execution) The indicated speed applies from this signal
M Line clear Permissible line speed according to service timetable	Permissible line speed applies as soon as the rear of the train has cleared the last point in diverging position
=0 Short entry into a short track; expect stop at next signal. Previous signal displays permissible speed but max. 40km/h (-4)	★ ★ indicates repeating signal

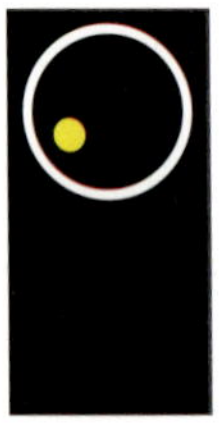

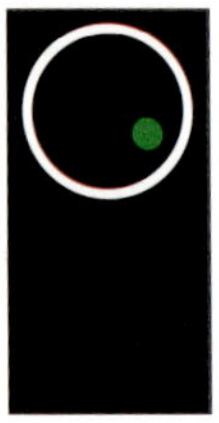

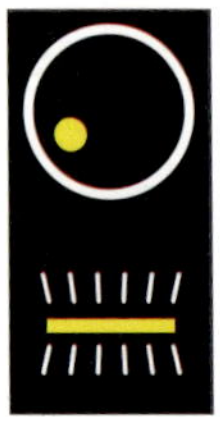

Fig. 1.24C SBB: signal aspects

Signalling system N

 +0
Advanced caution signal
The following signal S_2 shows caution. At S_2 the speed must be reduced so much that the train is able to stop at the third signal S_3. Used with short block distances when no precise intermediate speed indication is possible

S_1

 -0

S_2

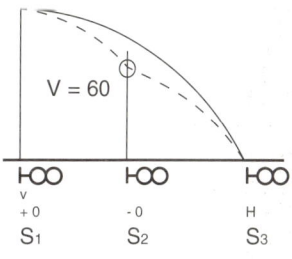

$V = 60$

V   -0   H
+0
S_1 S_2 S_3

 B
Entry into an occupied track
an obstacle is to be expected on the next track section; the previous signal displays caution to reduce speed to 40km/h

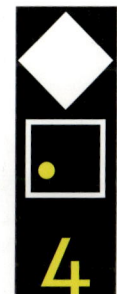 Diamond indicates that the next signal (if not a repeater) is a home signal, or a change of speed is necessary at the next block signal

 Calling-on signal (flashing red) may only be accepted if the driver has noticed a steady red light; speed reduction to 20km/h and move at sight

 'Bn' indicates signal name plate (1 to 4 characters). Provided at home signals

Fig. 1.24D SBB: signal aspects

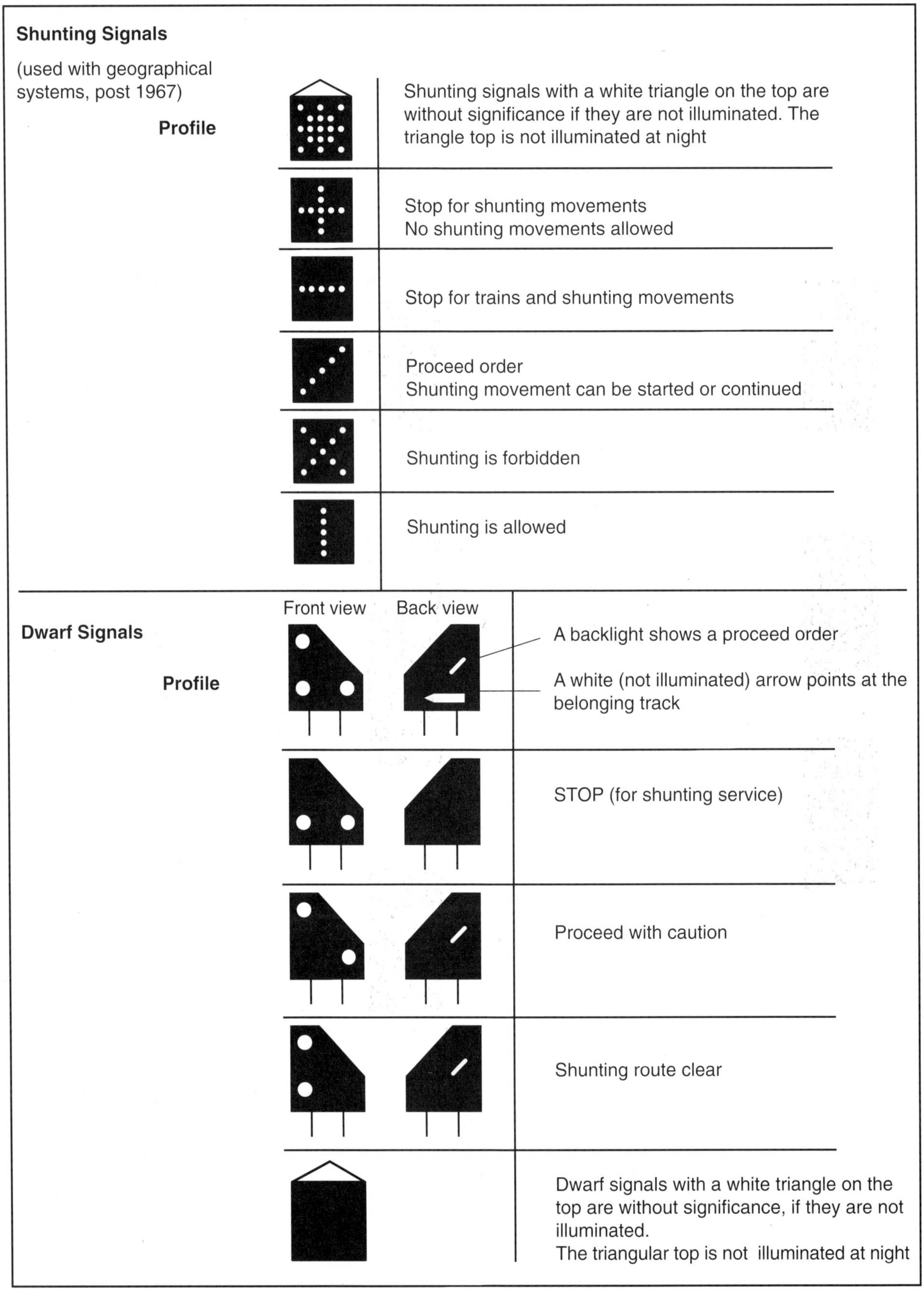

Shunting Signals

(used with geographical systems, post 1967)

Profile

Aspect	Meaning
(triangle top with dots)	Shunting signals with a white triangle on the top are without significance if they are not illuminated. The triangle top is not illuminated at night
(plus/cross of dots)	Stop for shunting movements No shunting movements allowed
(horizontal row of dots)	Stop for trains and shunting movements
(diagonal row of dots)	Proceed order Shunting movement can be started or continued
(X of dots)	Shunting is forbidden
(vertical row of dots)	Shunting is allowed

Dwarf Signals

Profile

Front view / Back view

Aspect	Meaning
(3 white dots front / arrow back)	A backlight shows a proceed order A white (not illuminated) arrow points at the belonging track
(2 dots horizontal)	STOP (for shunting service)
(2 dots diagonal)	Proceed with caution
(2 dots with arrow back)	Shunting route clear
(triangle top)	Dwarf signals with a white triangle on the top are without significance, if they are not illuminated. The triangular top is not illuminated at night

Fig. 1.25A SBB: aspect sequences

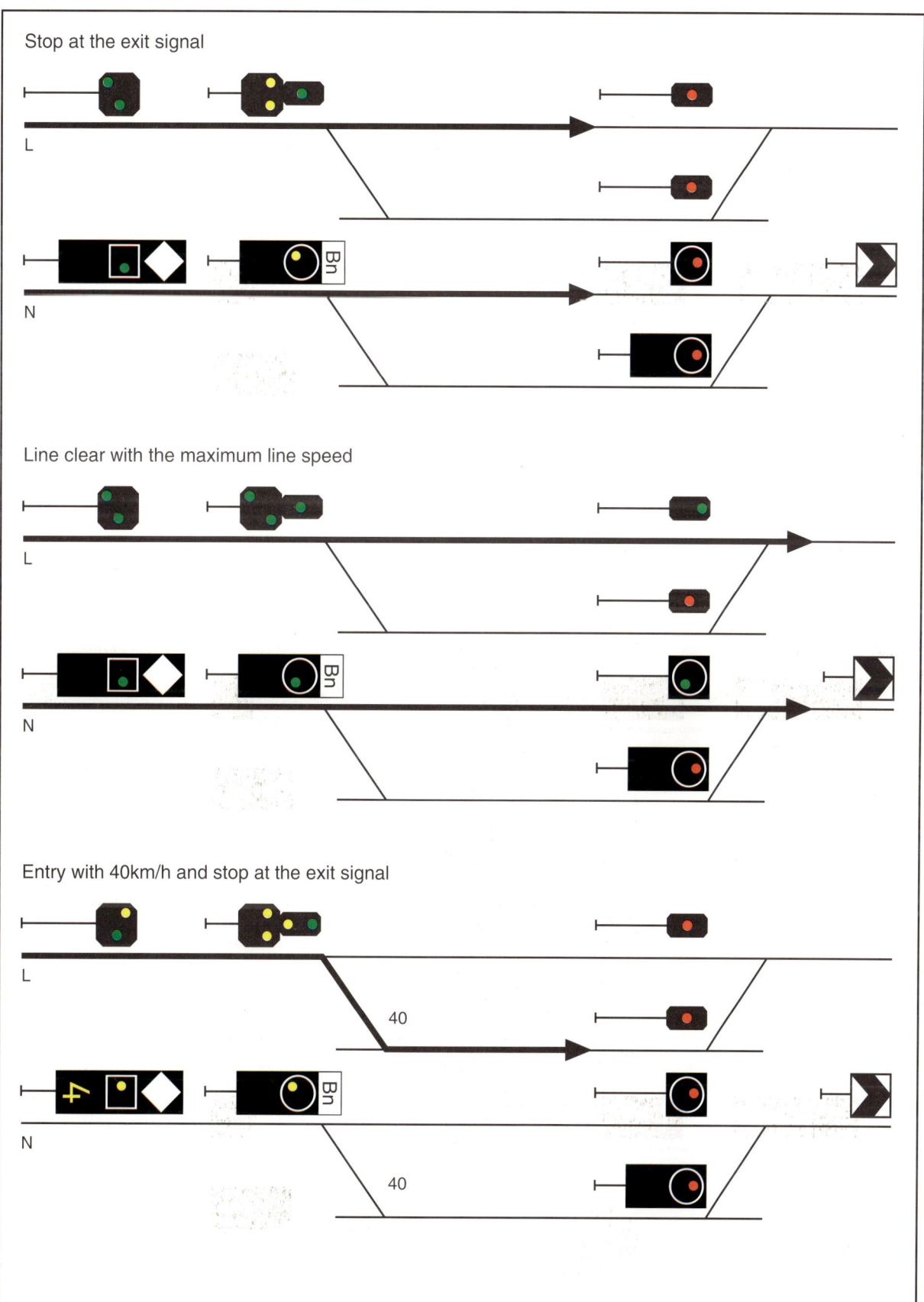

Fig. 1.25B SBB: aspect sequences

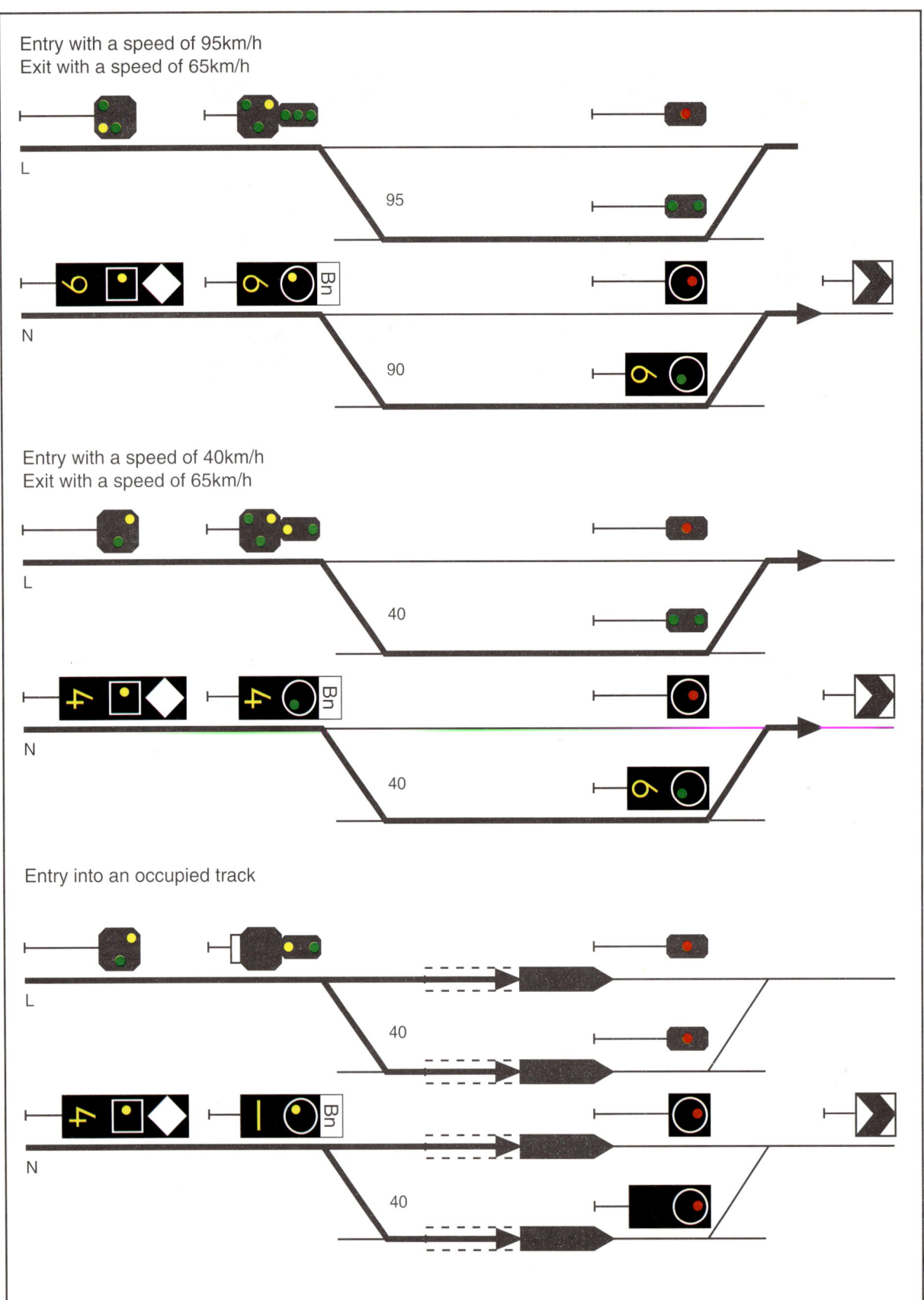

Fig. 1.25C SBB: aspect sequences

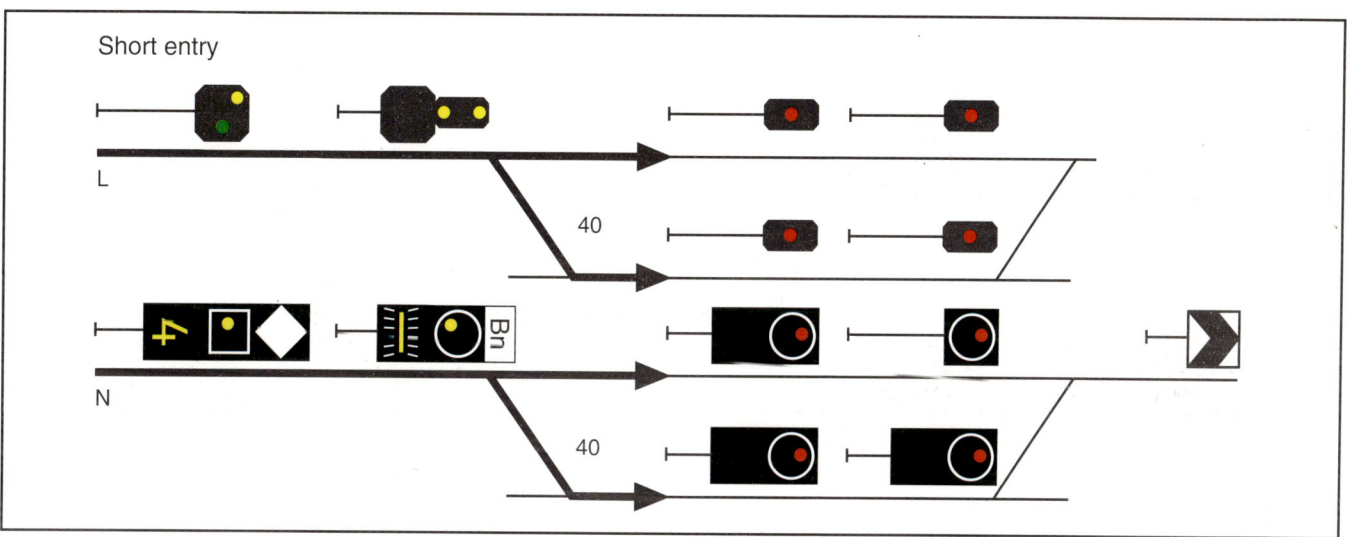

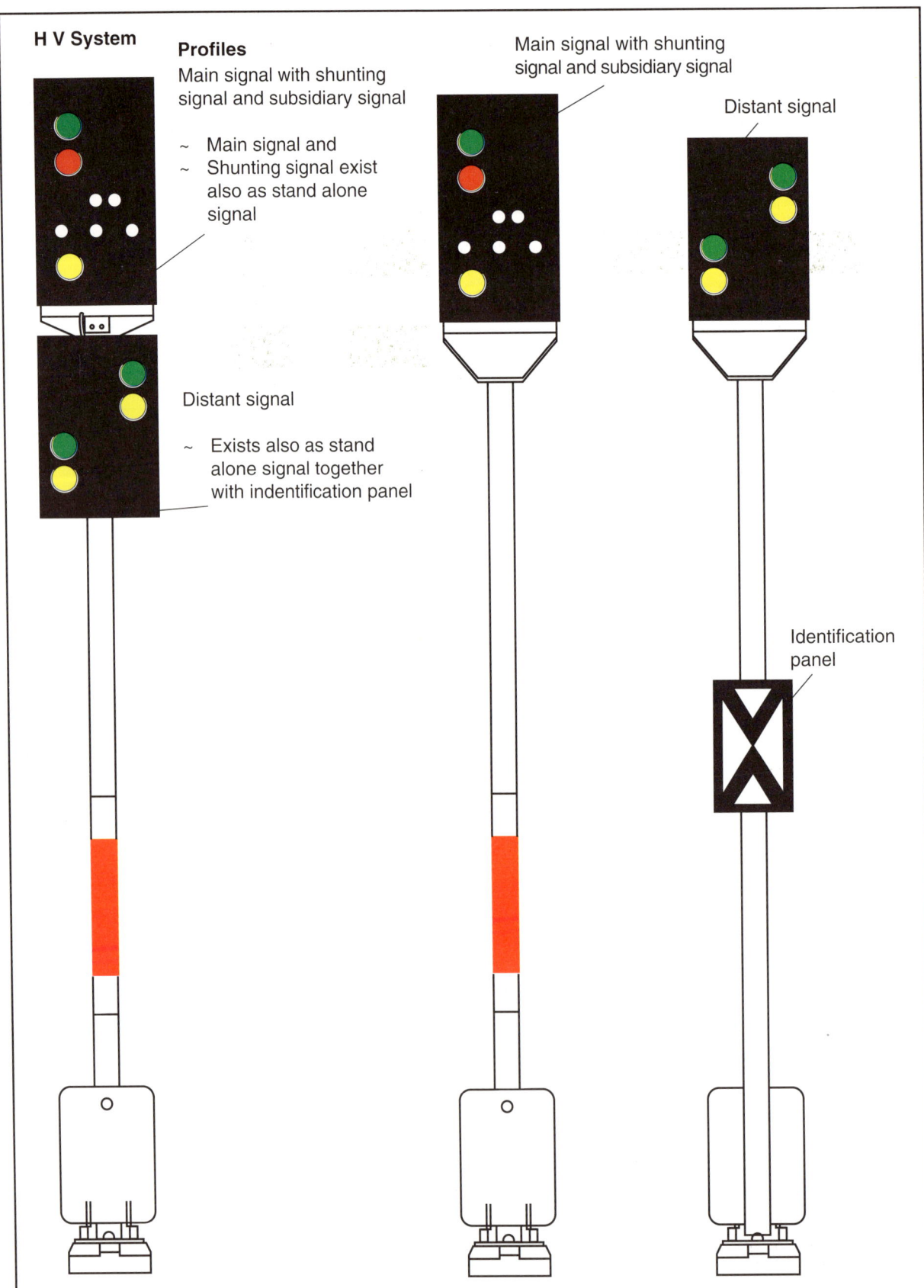

Fig. 1.26A DB AG: signal aspects

Fig. 1.26B DB AG: signal aspects

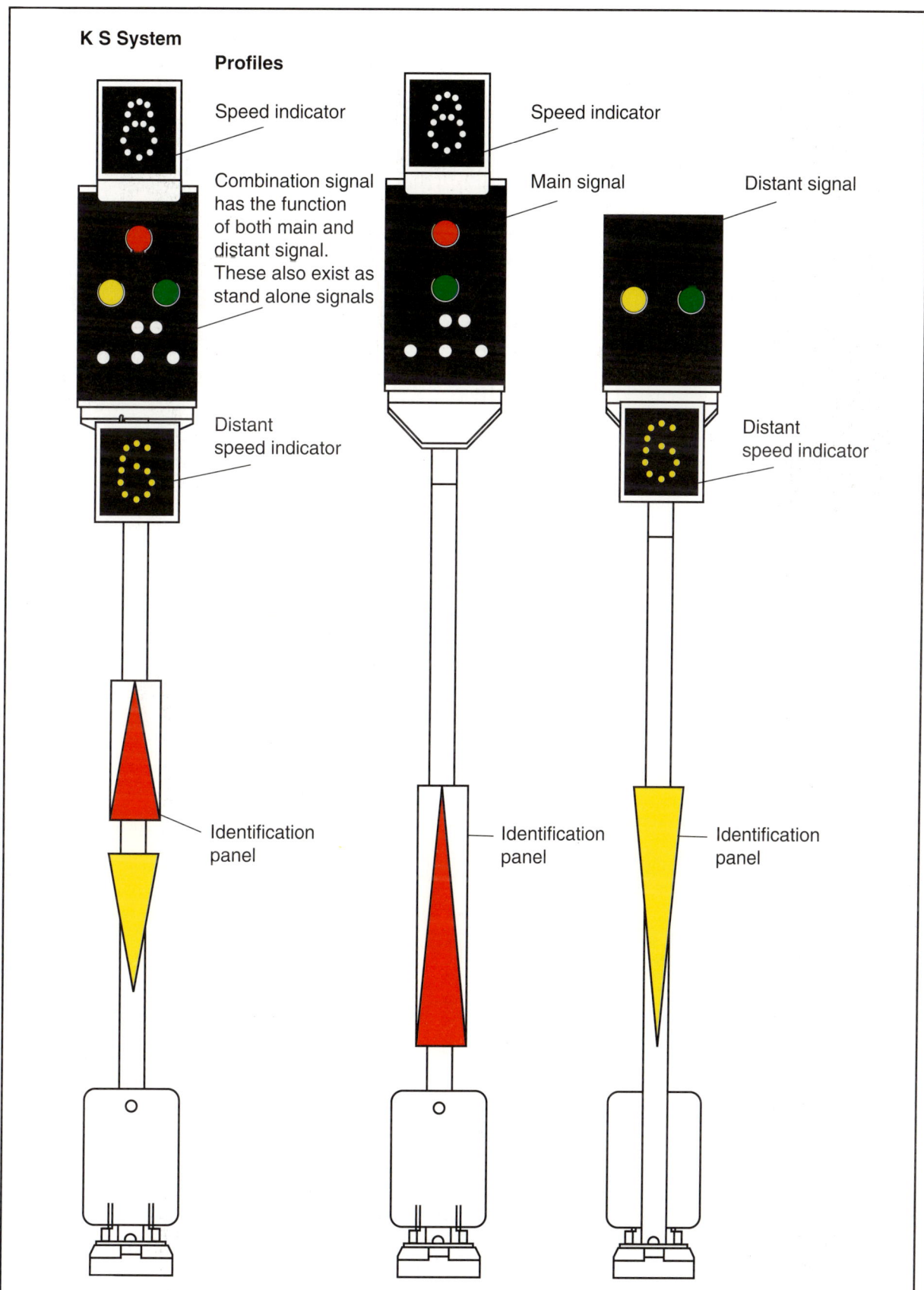

Fig. 1.26C DB AG: signal aspects

H V ~ Signals (applied up to now)
Speeds differing from those given in the timetable are indicated by coded signal aspect and/or by speed indicator signal

Description	Aspect	Meaning
Hauptsignal **Profile**	 green / red / yellow	Main signals indicate whether the track section ahead may be run over. It only serves train traffic
Zughalt Hp 0	 red	The signal prohibits the passing of a train
Fahrt Hp 1	 green	The signal permits the passing at the speed allowed in the timetable unless a speed restriction is indicated in the timetable or by the speed indicator
Langsamfahrt Hp 2	 green / yellow	The signal permits the passing at a speed of 40km/h unless otherwise indicated in the timetable or by the speed indicator (maximum 60km/h)
Vorsignal **Profile**	 green, yellow / green, yellow	Distant signals indicate by the braking distance which signal aspect is to be expected at the corresponding next main signal
Zughalt erwarten Vr 0	 yellow / yellow	The signal calls for start of braking with stop at the next main signal
Fahrt erwarten Vr 1	 green / green	The signal indicates that the corresponding next main signal shows clear (Hp 1)
Langsamfahrt erwarten Vr 2	 green / yellow	The signal indicates that the corresponding next main signal shows speed restriction (Hp 2)

K S ~ Signals (applied at new installations)
Speeds differing from those given in the timetable are indicated by speed signal indicator only

Description	Aspect	Meaning
Mehrap schmitts-signal **Profile**	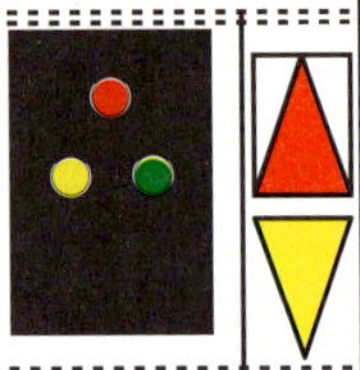 red / yellow, green + red triangle / yellow triangle	Combination signals have the function of both main signal and distant signal. The different functions are indicated by reflecting identification panels at the signal post
Hauptsignal **Profile**	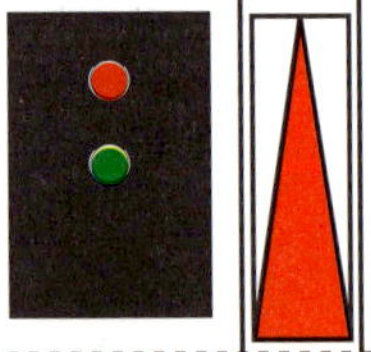 red / green + red triangle	Main signals indicate whether the track section ahead may be run over. The function of a main signal is indicated by a triangular red field with white border
Vorsignal **Profile**	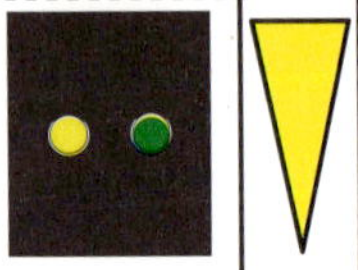 yellow, green + yellow triangle	Distant signals indicate by the braking distance which signal aspect is to be expected at the corresponding next main signal The function of a distant signal is indicated by a triangular yellow field
Halt	 red	The signal prohibits the passing of a train; it is analogue to Hp 0
Fahrt Ks 1	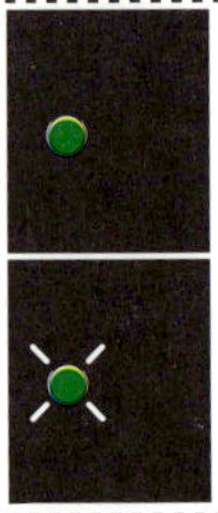 green / green blinking	1. The signal permits the passing at the speed allowed in the timetable unless a speed restriction is indicated by speed indicator 2. The green luminous spot blinks if a distant speed indicator shows in the signal
Halt erwarten Ks 2	 yellow	The signal permits the passing and indicates a stop at the next main signal

Fig. 1.26D DB AG: signal aspects

Supplementary signals, shunting signals and signals controlling level crossings

Supplementary signals			Shunting and level crossing signals		
Description	Aspect	Meaning	Description	Aspect	Meaning
Geschwindig-Keitsanzeiger Zs 3	 (white figure)	The speed indicator indicates by a white figure the maximum speed (divided by 10) permitted here. This indication is normally shown in connection with the main signal	Bahnübergangs-signal **Profile**	 (yellow panel with white dot above yellow dot)	Signals controlling level crossings are positioned at braking distance before level crossings which are equipped with flashing lights or other luminous signs (with or without half barriers)
Geschwindig-keitsvoran-zeiger Zs 3v	 (yellow figure)	The distant speed indicator shows by a yellow figure the maximum speed (divided by 10) permitted after the next speed indicator. The distant speed indicator is normally used in conection with the distant signal	Bahnupergang garf befahren werden Bü 1	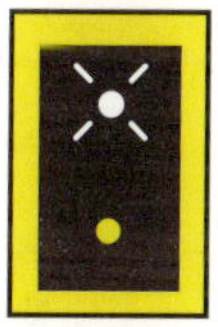 (flashing white X with yellow dot below)	The signal indicates that the technical protection of the level crossing is effective. The white luminous spot is flashing
Ersatzsignal Zs 1	 (3 white lights, A-shape)	The substitution signal~3 white lights forming an A~indicates that a main signal showing stop (Hp 0, Hp 00) or which is faulty may be passed without written instruction	Halt vor dem Bahn-übergang Weiterfahrt nach Sicherung Bü 0	 (yellow panel with yellow dot)	The signal calls for the start of braking with stop before the level crossing. Continuation of train/traffic is possible after the crossing has been protected
Vorsicht-signal Zs7	 (3 yellow lights, V-shape)	The caution signal~3 yellow lights forming a V~shows that a main signal showing stop (Hp 0, Hp 00) or which is faulty may be passed without written instruction and the driver must proceed with caution	Rangier-signal **Profile**	 (two red dots above, white dot below)	Shunting signals serve to block off a track, to order a stop or cancel a prohibition to proceed
Gleiswechsel-anzeiger Zs 6	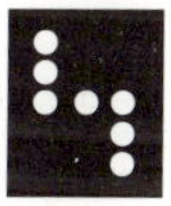 (6 white dots)	The change-to-counter-track signal indicates that in two-way track operations a train is directed by signal to the adjacent continuous main track (countertrack) of the clear line	Halt, Fahrverbot Sh 0	 (two red dots)	The signal prohibits train traffic and shunting
Endesignal Zs 10	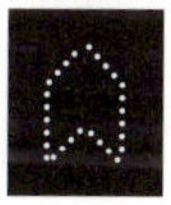 (dotted arch shape)	The end signal indicates the end of speed limit imposed by Zs3 or Hp2 before a point region ends	Fahrverbot aufgehoben Sh 1	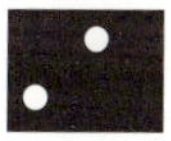 (two white dots diagonal)	The signal permits train traffic and shunting. Also serves as a clear signal for shunting

Fig. 1.27A DB AG: aspect sequences

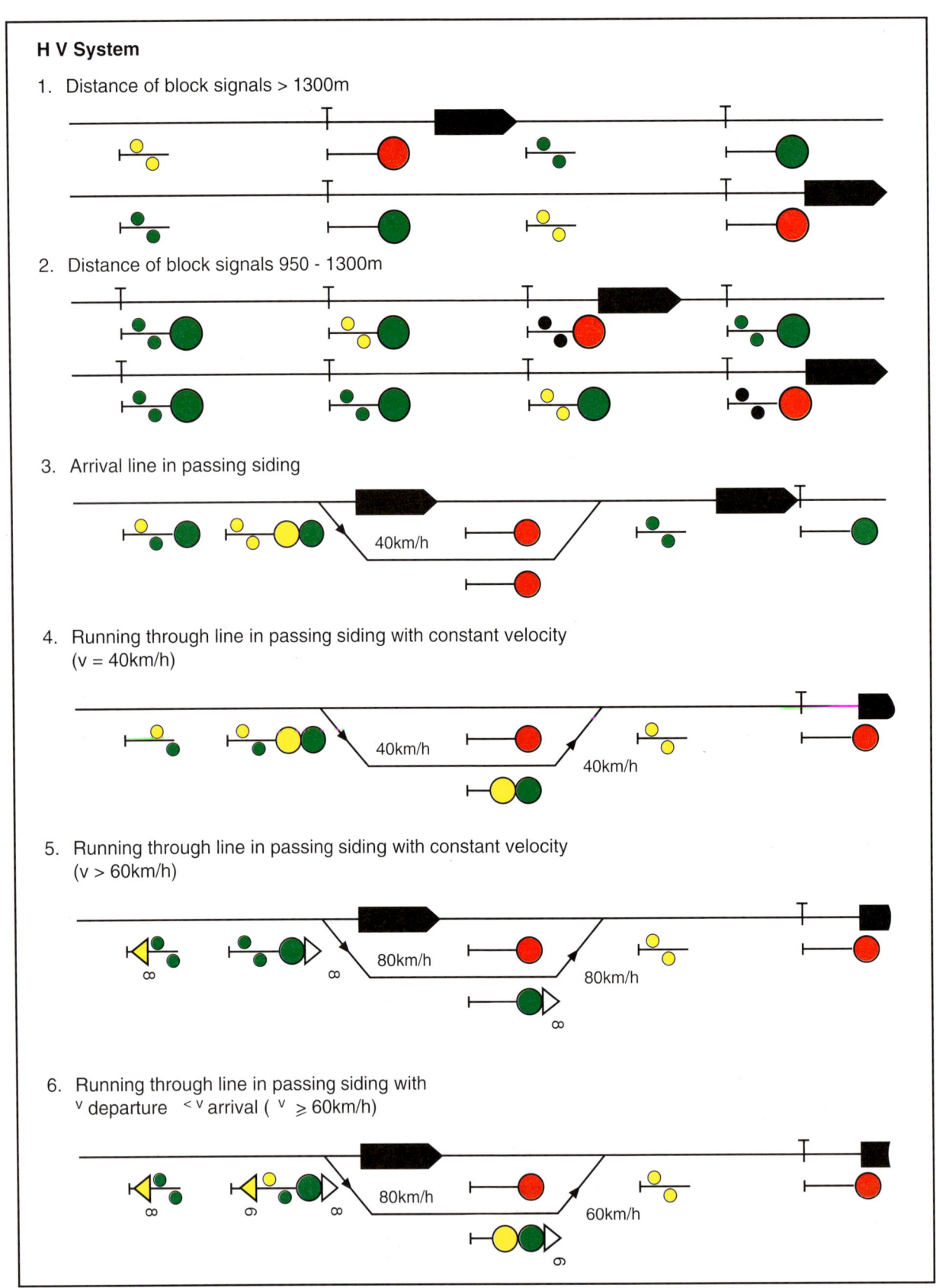

Fig. 1.27B DB AG: aspect sequences

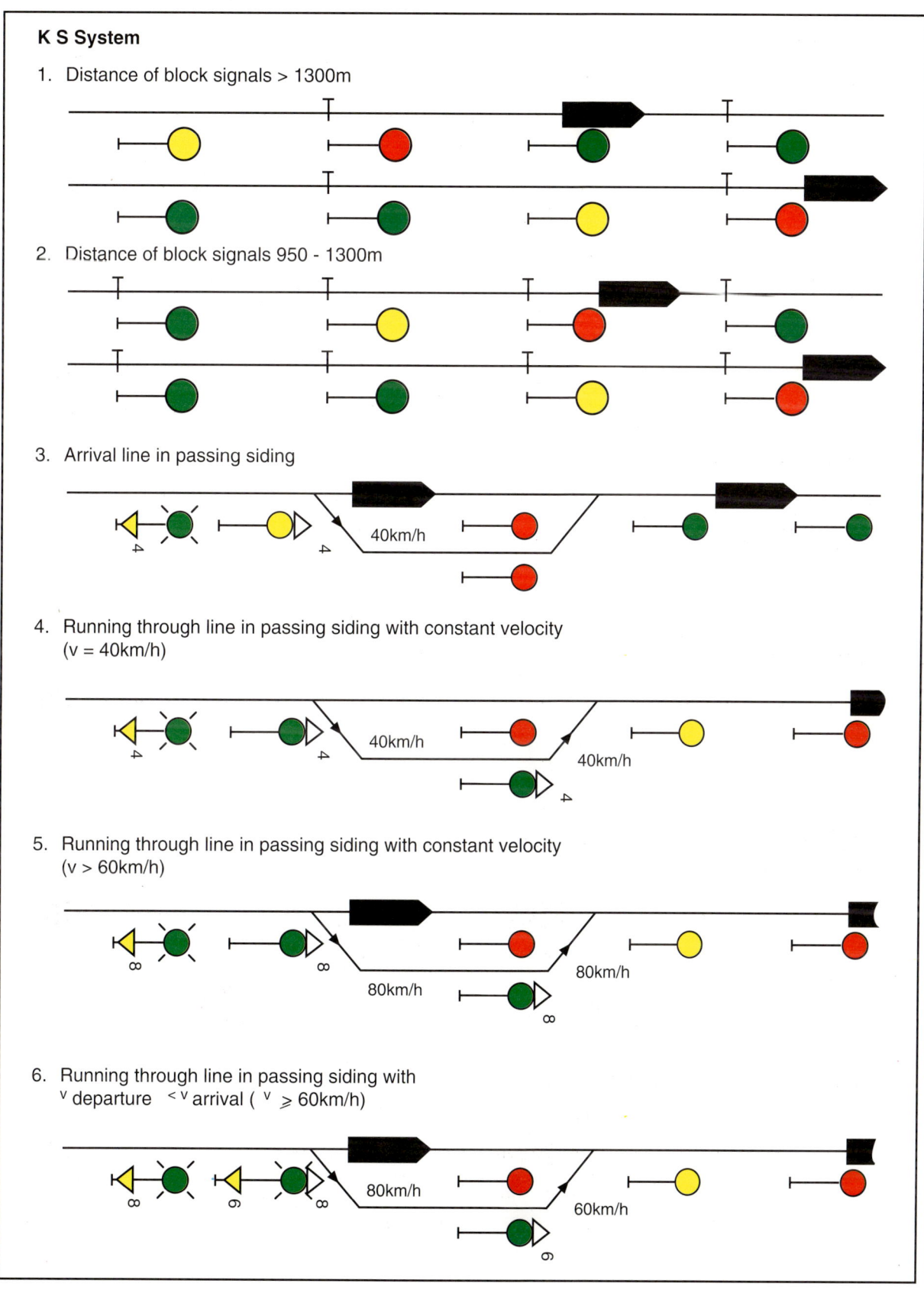

Fig. 1.28A RENFE: signal aspects

Description	Aspect	Meaning
Profiles Warning Signal	 (green over yellow)	Can also exhibit directional indication
Permissive Stop Signal	 (green, red, yellow with P plate)	May be passed at danger after 1min stop
Mandatory Stop Signal with directional indication and white auxilary aspect for shunting	 (green, red, yellow with white below)	For possible aspects, see aspects sequences chart
Ground Mounted Stop Signal	 (green, red, yellow, white)	Can also give shunting aspects. Used at exit from loops

Fig. 1.28B RENFE: signal aspects

Description	Aspect	Meaning
Mandatory Stop Signal with Identity Plate	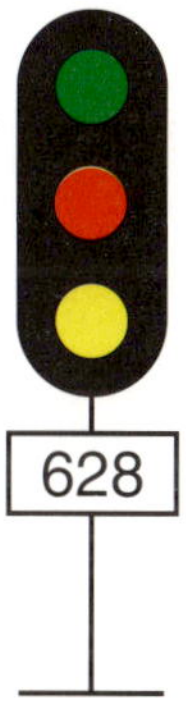 (Green/Red/Yellow signal with plate "628")	Signal located at 62.8km
Running Shunt Signal	 (Red/White over White/White)	Aspects: Danger: Red Straight Route: White over White Diverging Route: White/White Local Shunting: Red over White
Siding Shunt Signal	 (Red over White)	Aspects: Danger: Red Local Shunting: Red over White
Speed Restriction Over Turnout	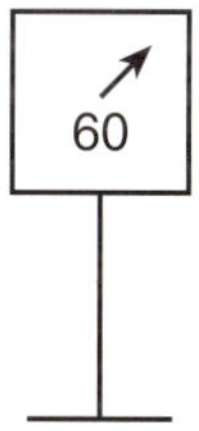 (60 with arrow)	Do not exceed 60km/h when traversing the turnout to the right Note: numerals and arrow are illuminated (White)
Temporary Speed Restriction Indicator	 (90)	Do not exceed 90km/h when traversing the section of line to which this restriction applies
Countdown Marker	 (diagonal stripes)	300m to signal

Fig. 1.29 RENFE: aspect sequences

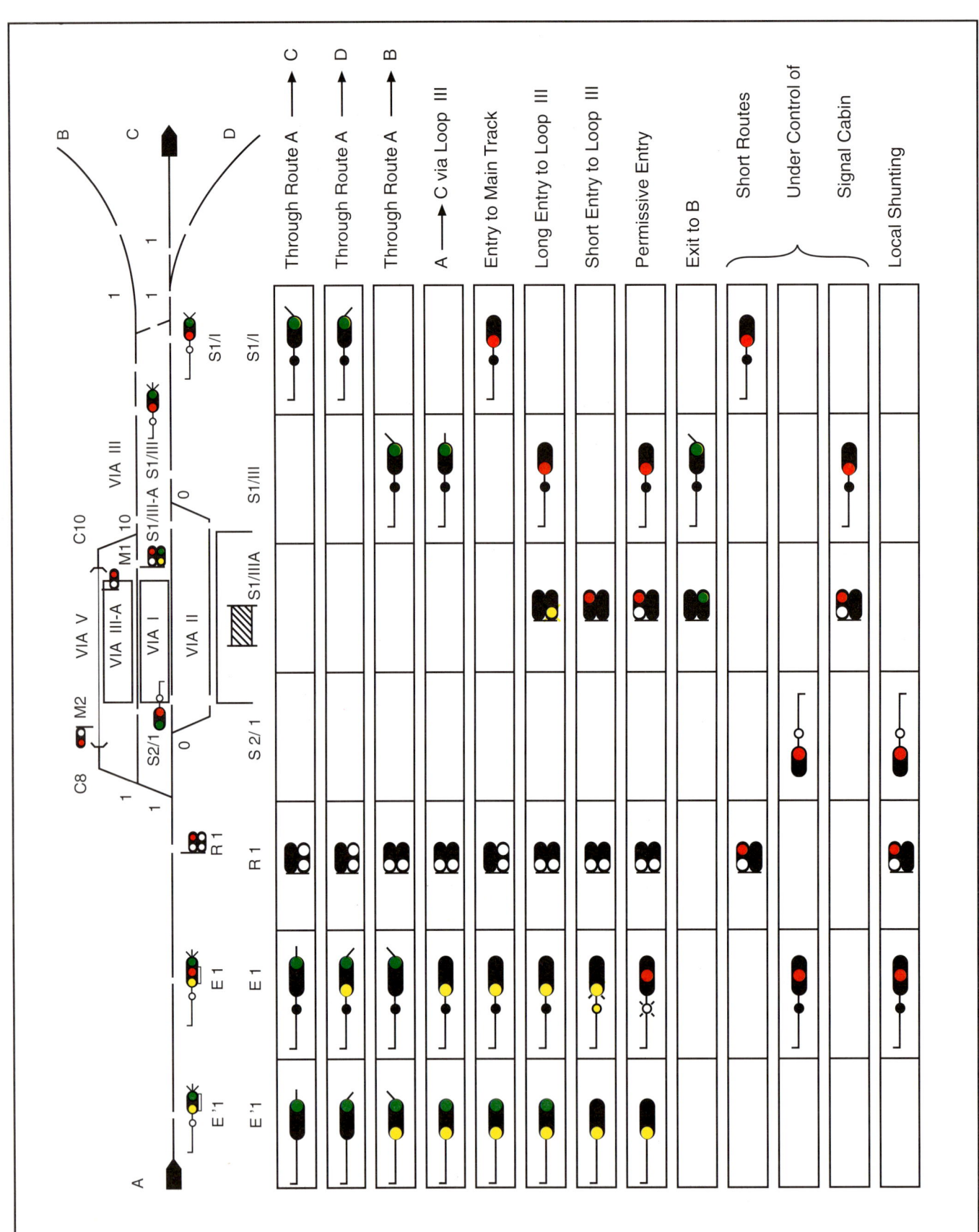

Fig. 1.30A SNCF: signal aspects

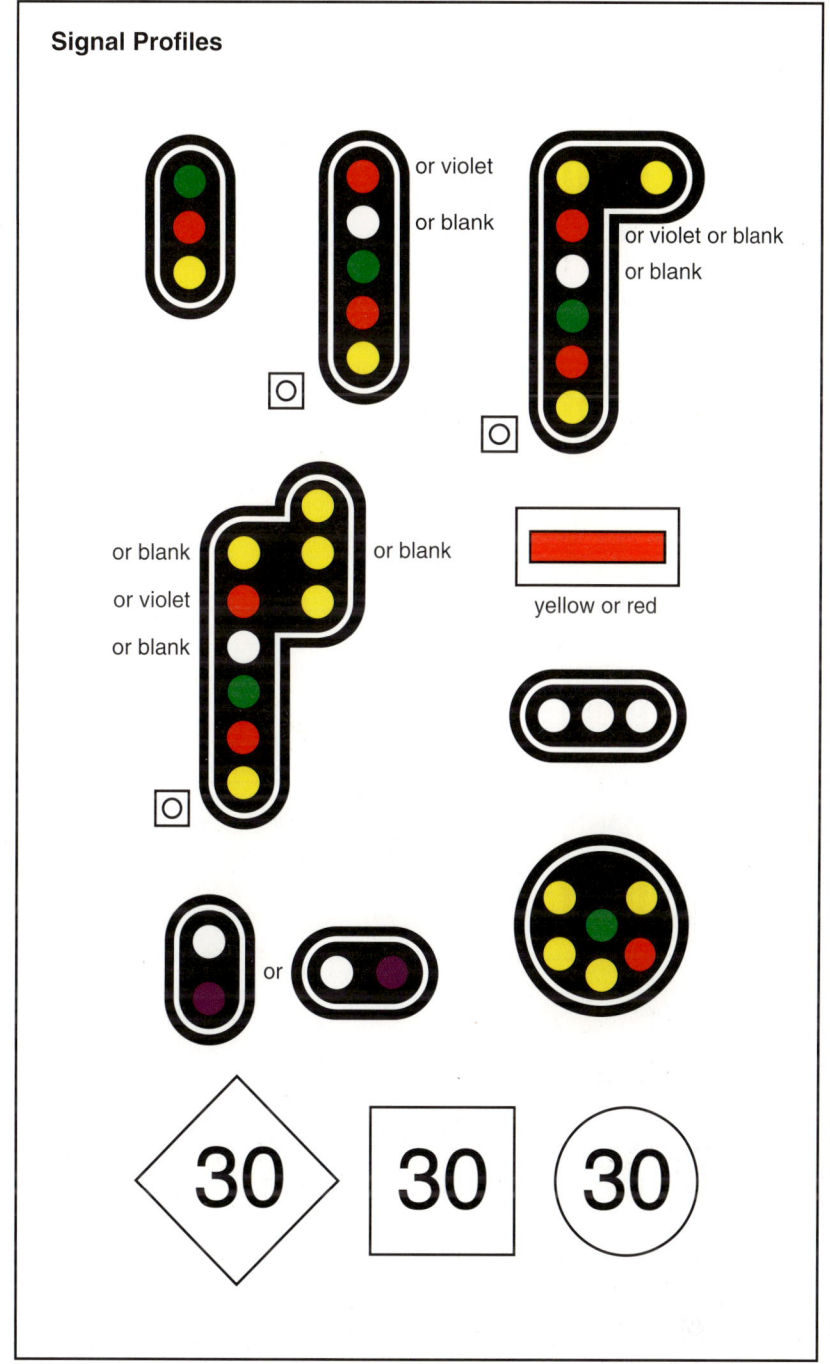

Fig. 1.30B SNCF: signal aspects

Description	Aspect	Meaning
Feu vert 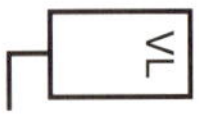	 (steady green)	Steady green: proceed at line speed
Feu vert clignotant 	 (flashing green)	Flashing green: equivalent to steady green for trains limited to 160km/h. Any train running at more than 160km/h must reduce speed before passing the warning (*avertissment*) signal
Feu jaune clignotant 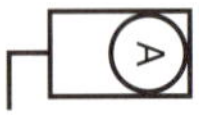	 (flashing yellow)	Flashing yellow: the next signal is at steady yellow and is at reduced distance to the stop signal
Avertissment 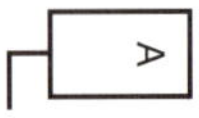	 (steady yellow)	Steady yellow: prepare to stop at the next signal (*Carré*, *sémaphore* or flashing red)
Carré (2) 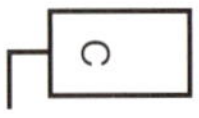	 (double red)	Double red: absolute stop. Plated 'Nf' (*non-franchissable*)
Sémaphore 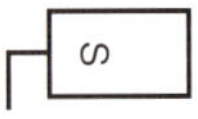	 (steady red)	Steady red: stop and proceed prepared to stop on sight. Plated 'F' (*franchissable*)
Feu rouge clignotant 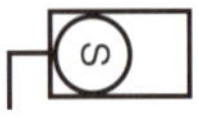	 (flashing red)	Flashing red: may be passed without stopping but speed not to exceed 15km/h, prepared to stop on sight

Fig. 1.30C SNCF: signal aspects

Description	Aspect	Meaning
Ralentissement 30 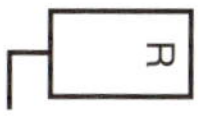	 (two yellow lights, horizontal)	Reduce speed to 30km/h when traversing the junction(s) ahead of next signal
Ralentissement 60 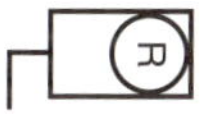	 (two flashing yellow lights, horizontal)	Reduce speed to 60km/h when traversing the junction(s) ahead of *next* signal
Rappel 30 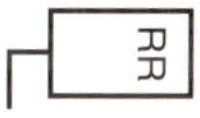	 (two yellow lights, vertical)	Do not exceed 30km/h when traversing the junction(s) ahead of this signal
Rappel 60 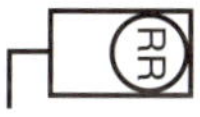	 (two flashing yellow lights, vertical)	Do not exceed 60km/h when traversing the junction(s) ahead of this signal
TIV de rappel 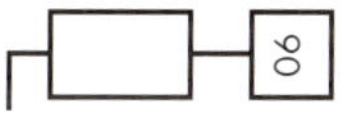	 90 (square sign)	Do not exceed speed indicated in km/h when traversing the junction(s) ahead of this sign. (Sign displayed when diverging route set)
TIV mobile à distance 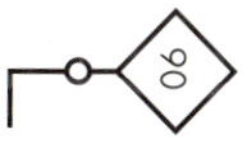	 90 (diamond sign)	Reduce speed to that indicated in km/h when traversing the junction(s) ahead (Sign displayed when diverging route set)
TIV du fixes type à ordinaire distance	  30 (diamond sign) / 90 (square sign)	Reduce speed to that indicated (in km/h) when traversing the track section or junction ahead (fixed sign)

Fig. 1.30D SNCF: signal aspects

Description	Aspect	Meaning
Feu blanc 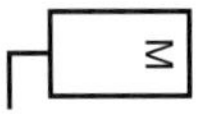	 ⬤ (white)	Steady white: proceed as shunting movement
Feu blanc clignotant 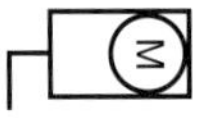	 ✦ (flashing white)	Flashing white: restricted (short) shunt; departure to the main line prohibited
Carré violet 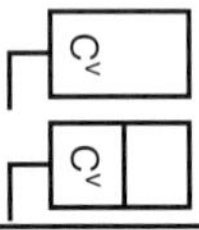 bas	 ⬤ (violet)	Steady violet: absolute stop

Description	Aspect	Meaning
Pancartes Z et R	 Z  / R	The 'Z' board defines the beginning and the 'R' board the end of a speed restriction. (Fixed signs)

Fig. 1.31A SNCF: aspect sequences

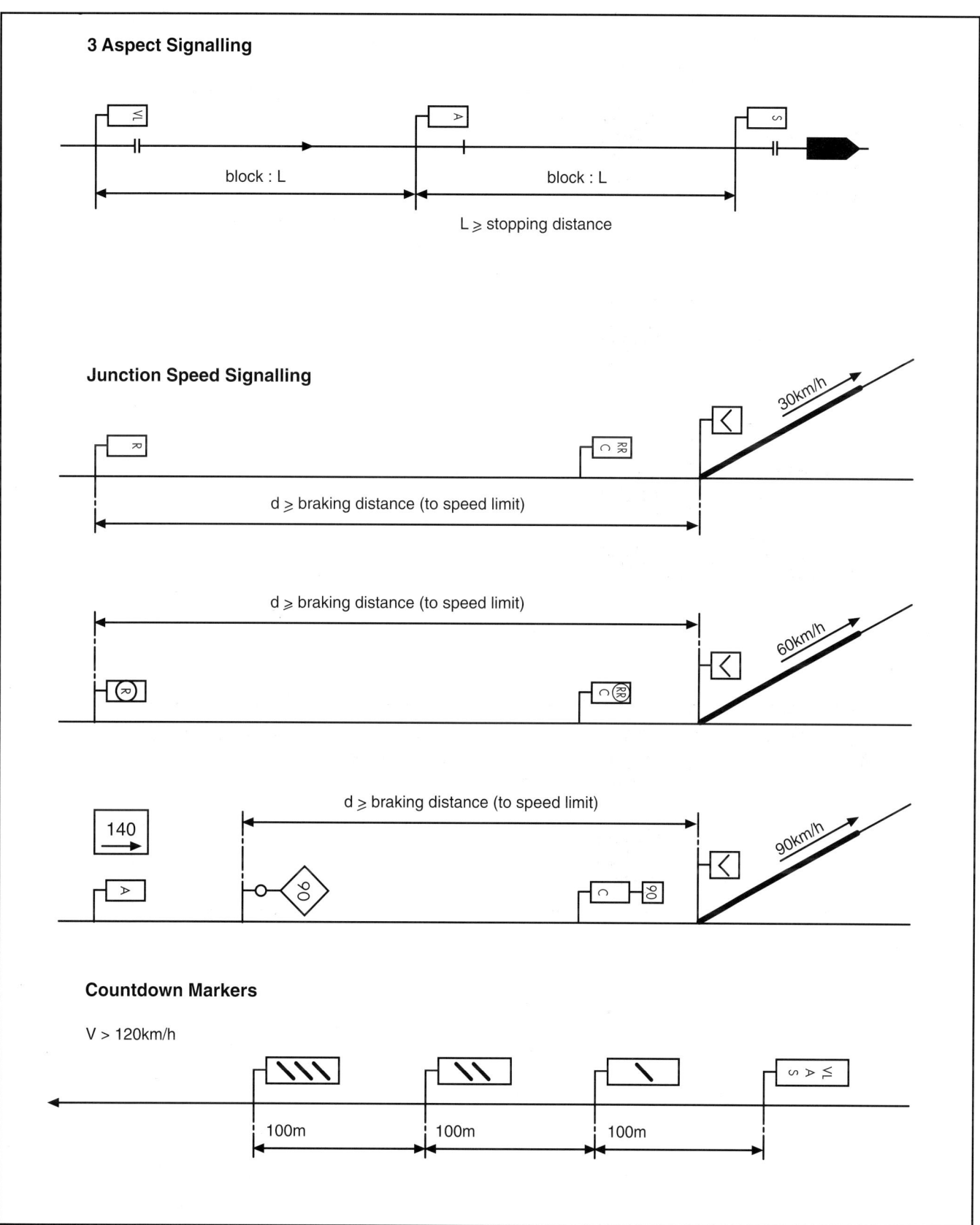

Fig. 1.31B SNCF: aspect sequences

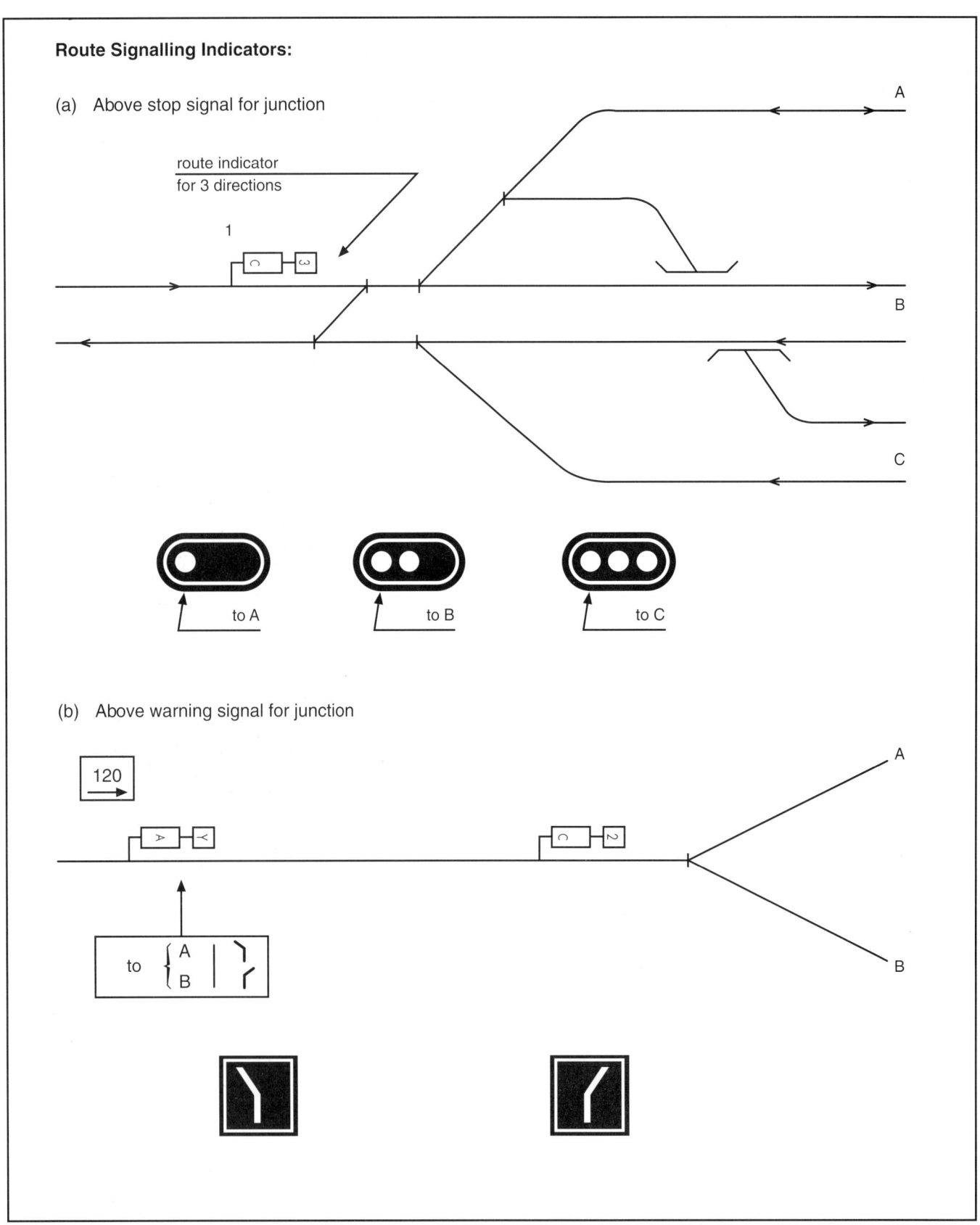

Fig. 1.31C SNCF: aspect sequences

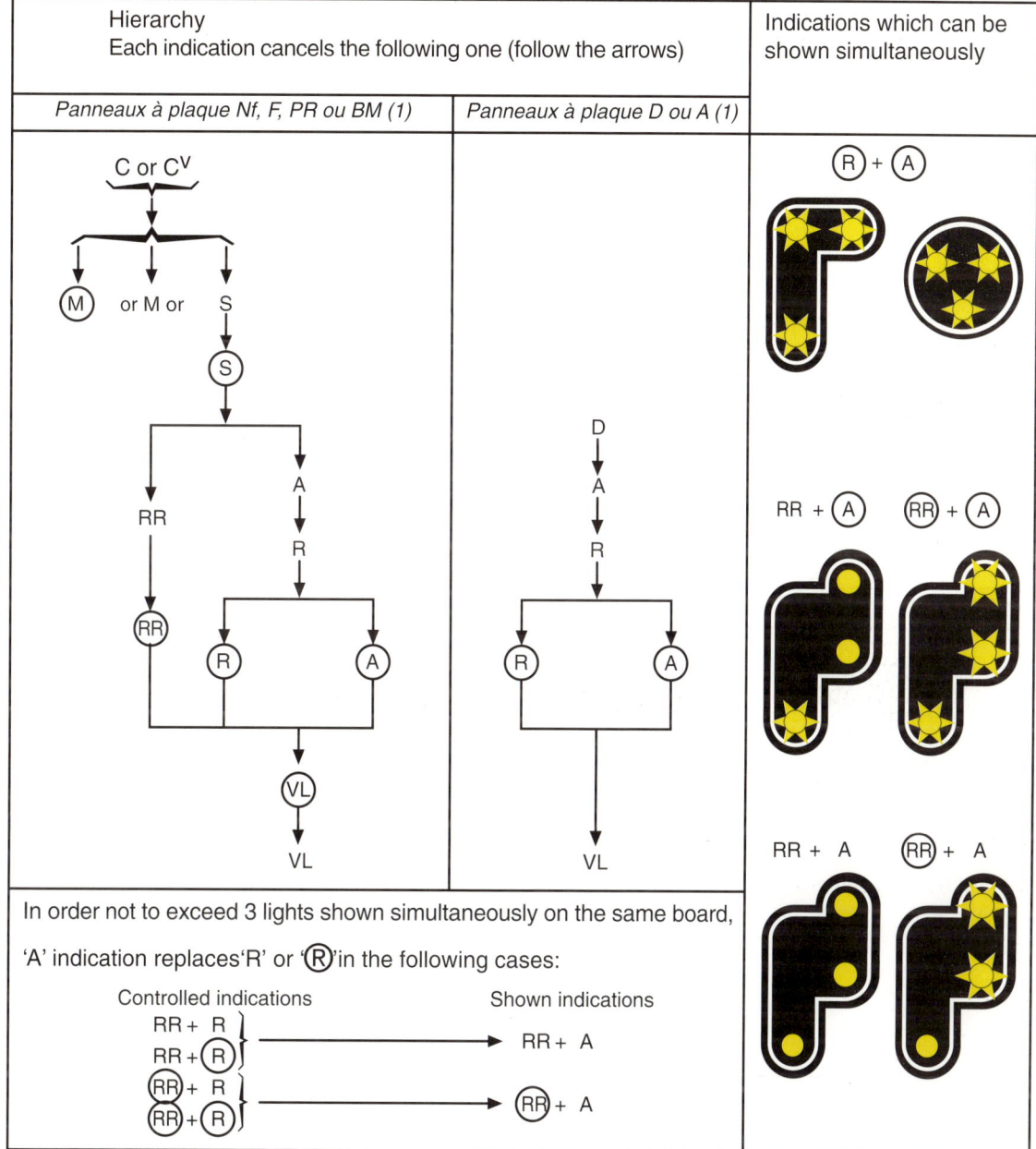

Fig. 1.31D SNCF: aspect sequences

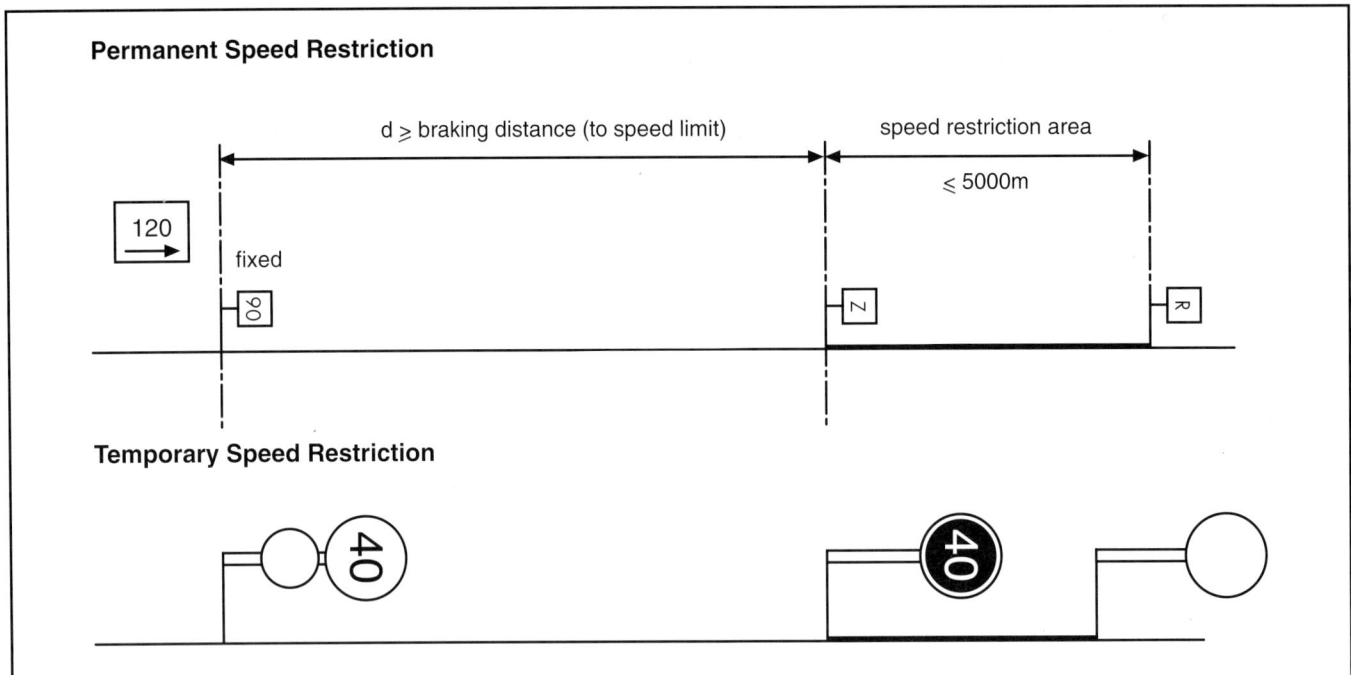

RAILWAY SIGNALLING PRINCIPLES 73

Fig. 1.32A BR: signal aspects

Description	Aspect	Meaning
Main Signal **Profiles**	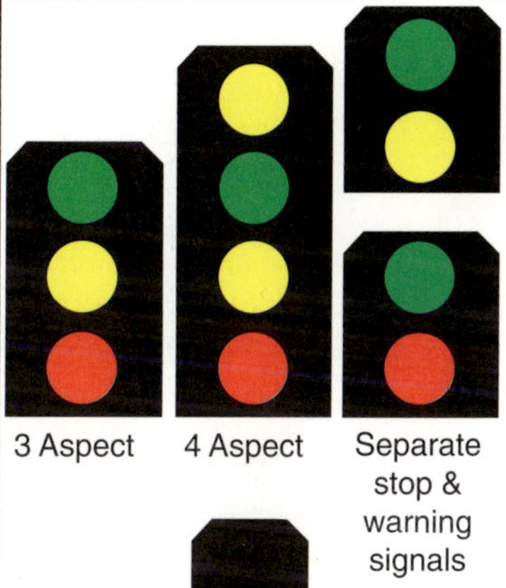 3 Aspect / 4 Aspect / Separate stop & warning signals	Note: when signals are mounted at ground level, the order of the aspects is reversed ie red at the top
Red	 (red)	Danger: Stop
Green	 (green)	Clear: Next signal is exhibiting a proceed aspect
Single Yellow	 (yellow)	Caution: Be prepared to stop at the next signal
Double Yellow	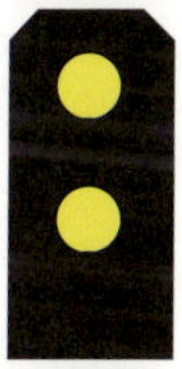 (two yellows)	Preliminary Caution: Be prepared to find the next signal exhibiting one yellow
Flashing Single Yellow	 (flashing yellow)	Preliminary Caution: Be prepared to find the next signal exhibiting one yellow with junction indicator for the highest speed diverging route
Flashing Double Yellow	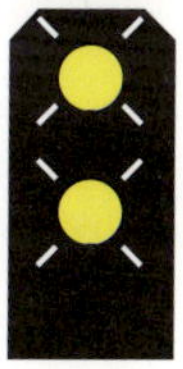 (flashing double yellow)	Next signal is exhibiting flashing single yellow aspect

Fig. 1.32B BR: signal aspects

Description	Aspect	Meaning
Position Light Junction Indicator		A junction indicator displays a line of white lights normally above the signal when a proceed aspect is exhibited. No route indication is given for movements along a straight route. Indications 1, 2 and 3 are for routes progressively to the left of the straight route and indications 4, 5 and 6 for routes progressively to the right.
		Junction indicators may be provided for routes to the left or right of the highest speed route where this is not the straight route. Where there are routes of equal speed and none can be indentified as the straight route, a junction indicator may be provided for each route
Position Light Shunt Signals Associated with a Main Signal		Line ahead may be occupied. Proceed cautiously to next signal or buffer stops. Be prepared to stop short of obstruction (main aspect may be passed at danger)
Independent shunt (ground or elevated) ON Aspect (red/white)		Danger, Stop
ON Aspect (yellow/white)		Danger, Stop (applies to movements in the direction to which the signal can be cleared; other movements may pass this signal without it being cleared)
OFF Aspect		Line ahead may be occupied. Proceed cautiously to the next signal or buffer stops. Be prepared to stop short of obstruction

Fig. 1.32C BR: signal aspects

Description	Aspect	Meaning
Position Light Indicators controlling loading/unloading movements	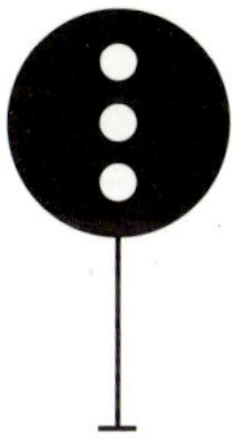 (three white dots vertical)	Move slowly in the **normal** direction for loading or unloading
	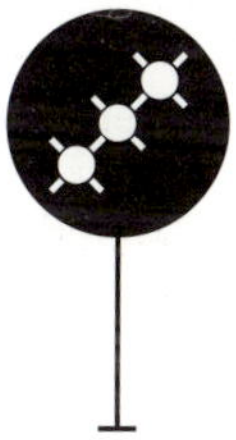 (three white dots diagonal)	Move slowly in the **opposite** direction to that required for loading or unloading
	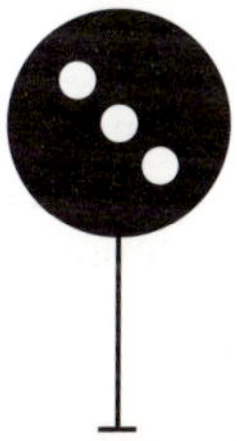 (three white dots)	Prepare to stop
	 (three red dots horizontal)	Stop immediately, irrespective of distance from the indicator
'BANNER' Repeater 'ON' Aspect	(horizontal black bar on white circle)	Be prepared to find the stop signal ahead at danger
'OFF' Aspect	 (diagonal black bar on white circle)	Stop signal ahead is exhibiting a proceed aspect

Fig. 1.33A BR: aspect sequences

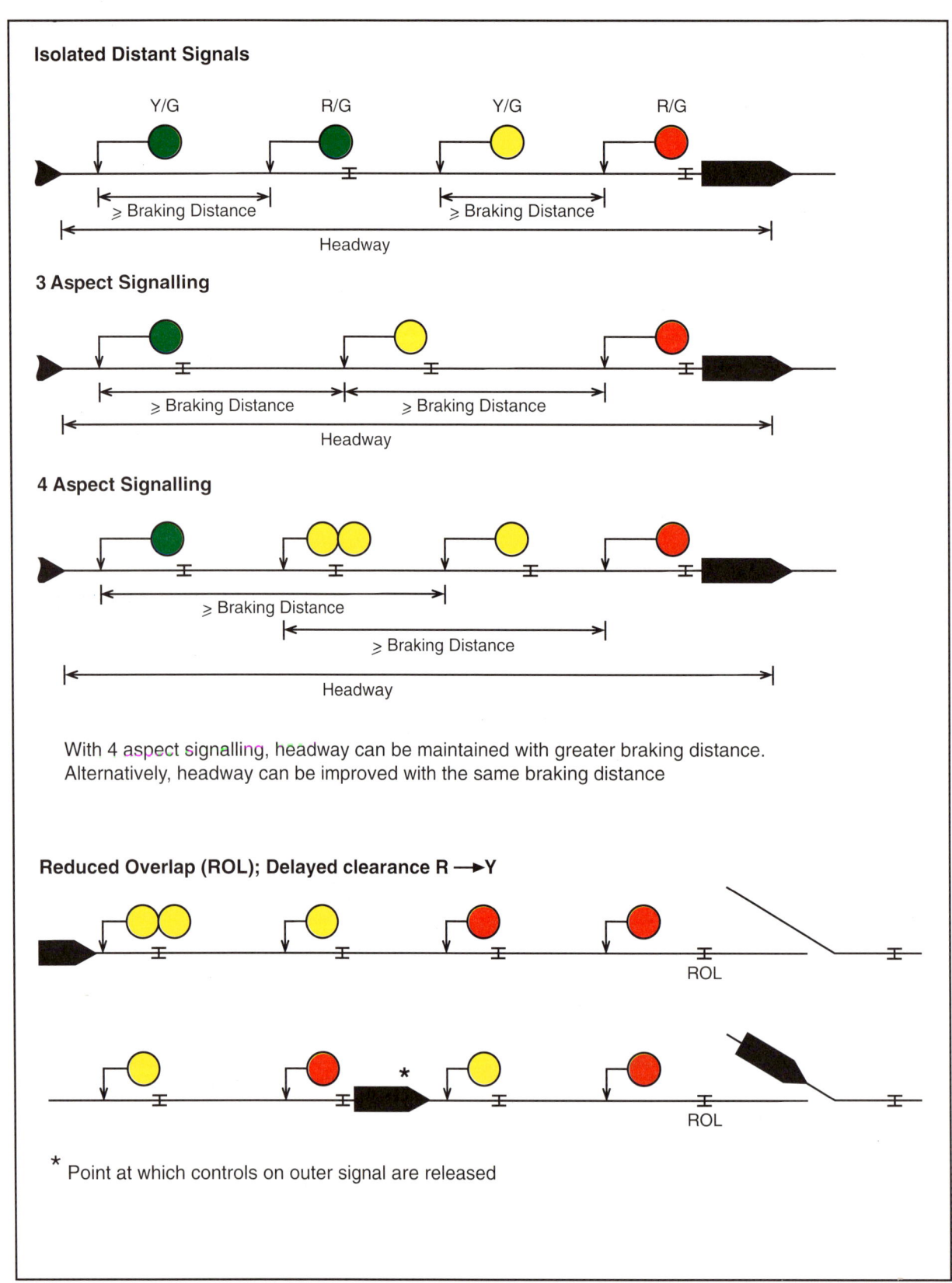

Fig. 1.33B BR: aspect sequences

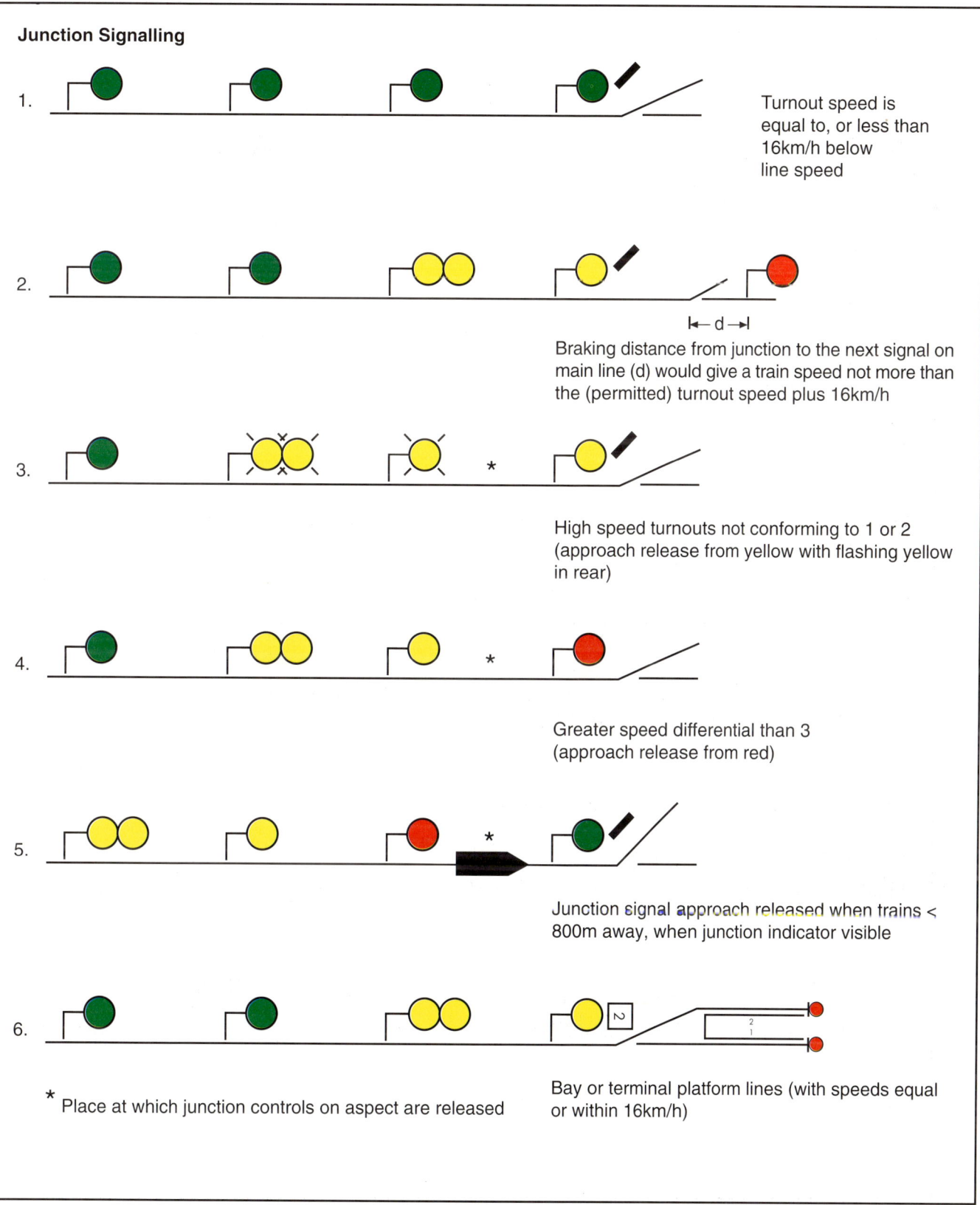

Fig. 1.33C BR: aspect sequences

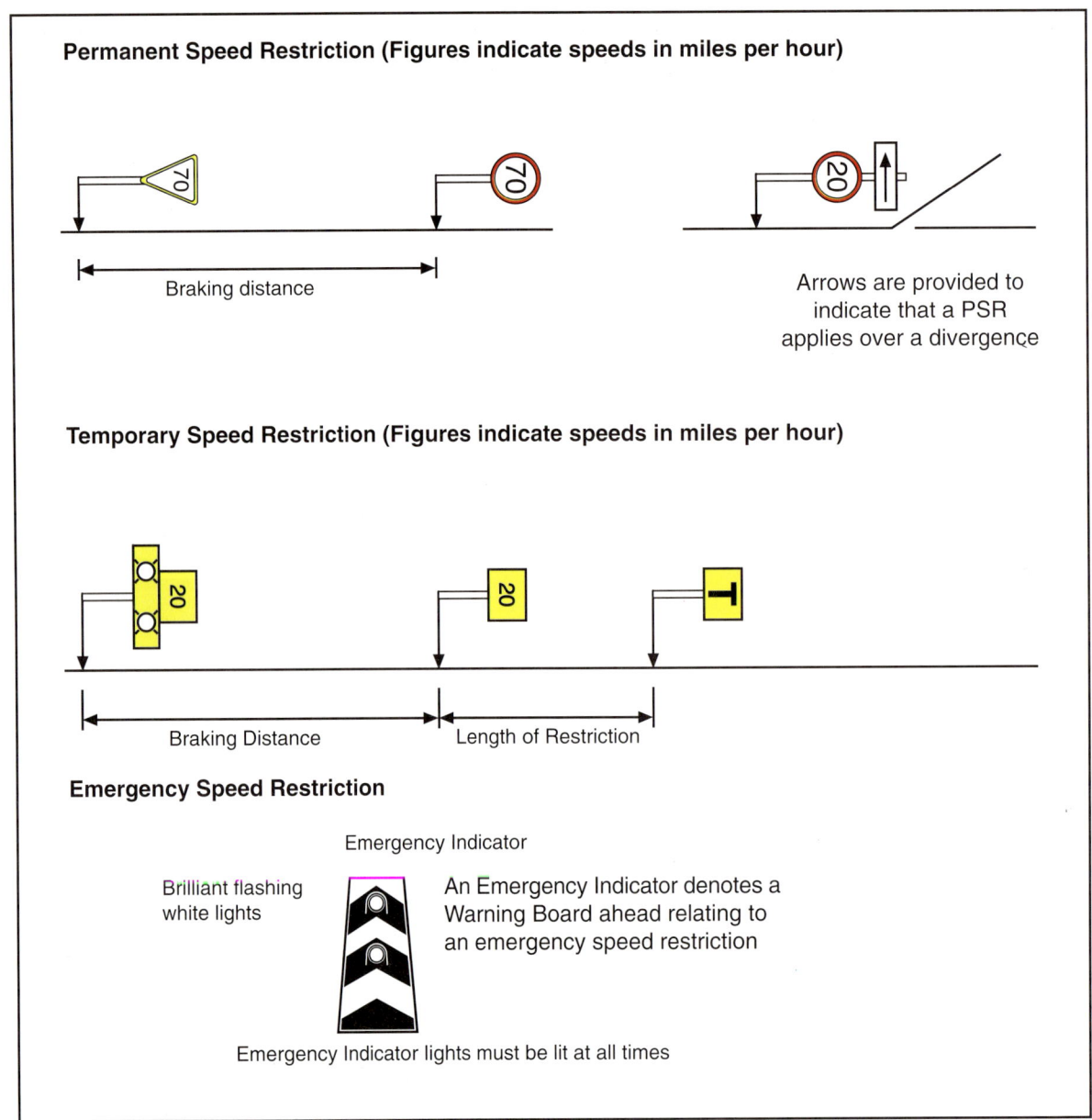

Fig. 1.33D BR: aspect sequences

Automatic Open Level Crossing, Locally Monitored (AOCL)

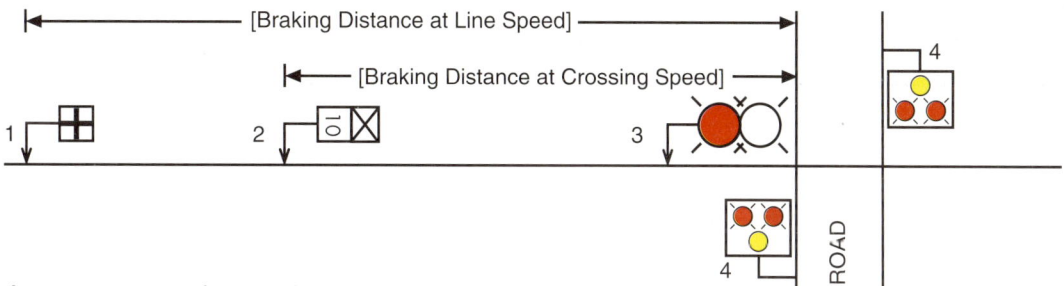

Aspect sequence for one direction only

1 Advanced Warning board (St George's Cross)

2 Special Speed Restriction board (shows crossing speed in miles/h) located at sighting point of crossing

3 Flashing red light changes to flashing white when road signals operating and crossing proved healthy

4 Road traffic signals (detail omitted)

Radio Electronic Token Block (RETB)

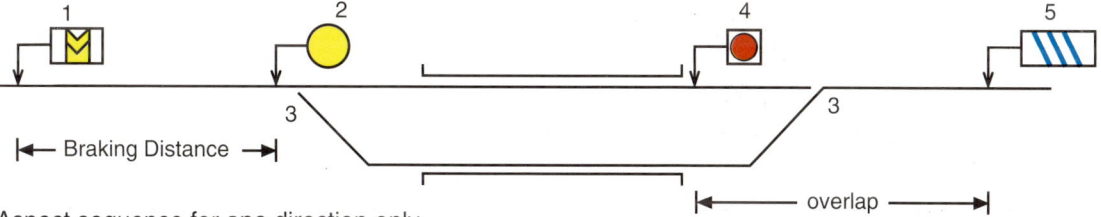

Aspect sequence for one direction only

1 Reflectorized distant board

2 Point indicator (yellow aspect)

3 Train operated (trailable) points

4 Stop & obtain token / permission to proceed board (red target)

5 Loop clear board (blue stripes)

Fig. 1.34A FS: signal aspects

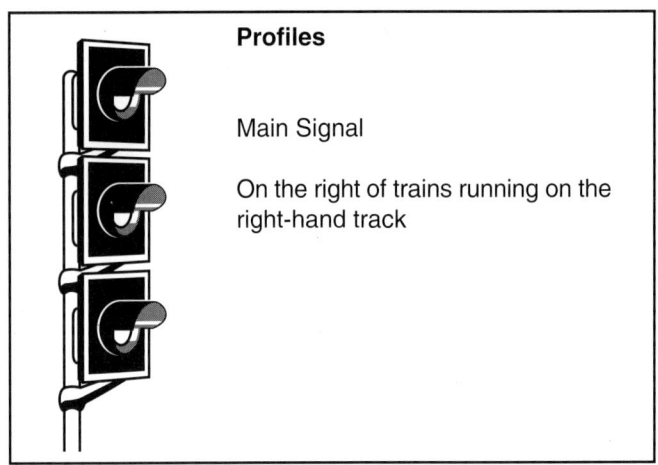

Profiles

Main Signal

On the right of trains running on the right-hand track

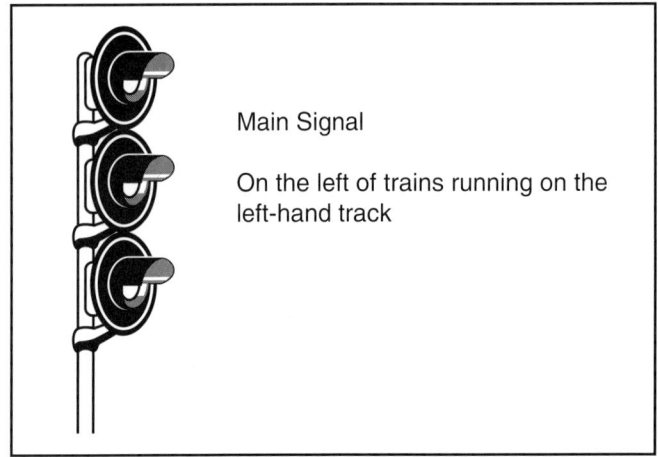

Main Signal

On the left of trains running on the left-hand track

Fig. 1.34B FS: signal aspects

Stop Signals

Signal name	Signal aspect	Signal meaning	Behaviour notes
1.1 Red light	 🔴	Occupied track	The driver has to stop the train without overrunning the signal
1.2 Green light	 🟢	Line clear	When a proceed aspect is displayed, the driver is given permission to continue to run, if the train is running; if, on the contrary, the train is standing, the driver can move forward after he received permission in accordance with Signal Regulations
1.3 Red light over green light	 🔴🟢	Line clear and confirmation of speed reduction at 30, 60 or 100km/h according to the indications of the preceeding warning signal	

Warning Signals

Signal name	Signal aspect	Signal meaning	Behaviour notes
2.1 Yellow light	 🟡	Warning of occupied track	The stop signal is at danger: so the driver must regulate running so as to be able to stop at the next signal Signal Regulations define the behaviour to be followed if the next signal is found at a proceed aspect
2.2 Flashing yellow light	 🟡 (flashing)	Pre-warning of occupied track	The next signal displays a proceed aspect, but it is located at a reduced distance from the succeding signal which is displaying a red aspect or a proceed aspect for a diverging route

Fig. 1.34C FS: signal aspects

Warning Signals (cont.)

Signal name	Signal aspect	Signal meaning	Behaviour notes
2.3 Group of yellow and green lights	 (yellow over green)	Warning of line clear at 30km/h	The driver must not exceed such speed when running past the next signal and throughout the running on the following group of points
2.4 Group of simultaneously flashing yellow and green lights	 (flashing yellow over green)	Warning of line clear at 60km/h	
2.5 Group of alternately flashing yellow and green lights	 (alternately flashing yellow and green)	Warning of line clear at 100km/h	
2.6 Group of two yellow lights	 (two yellow)	Warning of occupied track at an abnormally reduced distance or reception on a blocked or short track	The distance from the stop signal cannot be less than 350m. The reason for a white mast is that this aspect is always combined with a stop signal
2.7 Green light	 (green)	Warning of line clear at maximum permissible speed	The next stop signal is displaying a proceed aspect for a route which does not require reduced speed
2.8 Shunt signal: 'ON' Aspect		Stop	
'OFF' Aspect		Proceed (max. 30km/h)	

Fig. 1.34D FS: signal aspects

Signal name	Signal aspect	Signal meaning	Behaviour notes
3.1 Subsidiary indication of reduced speed	 (a) red over green with single bar	(a) Line clear with confirmation of reduced speed at 60km/h	
	(b) red over green with double bar	(b) Line clear with confirmation of reduced speed at 100km/h	
3.2 Speed indication at departure signals	 ▽ 30km/h ▽60 60km/h	Do not exceed the indicated speed along the exit route	
3.3 Permissive Stop Signal	 Red light with P plate	May be passed at danger after 3min stop	
3.4 Countdown Marker	For warning signals / For stop or combined signal	Signal Ahead	
3.5 High Shunting Signals	Stop / Proceed to the next signal / Proceed, the next signal is at proceed aspect / Proceed track ahead is occupied		

Fig. 1.35A FS: aspect sequences

Combined signals can display all aspects which can be displayed by stop signals and warning signals with the same meaning. There are 15 available combinations. Some significant examples follow.

Signal aspect		Signal meaning	Example
4.1 	🔴	Occupied Track	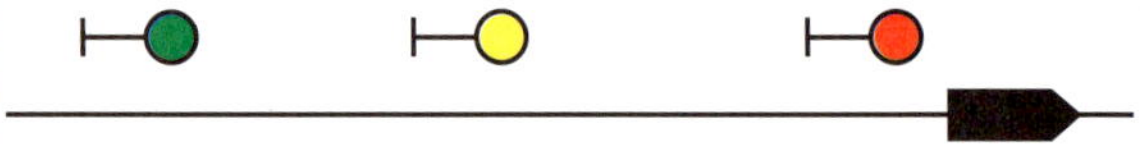
4.2 	🟢	Line clear. Moreover, it warns of a signal which displays a proceed aspect for a main line route	
4.3 	🟡	Line clear. Moreover, it warns of the next signal at danger	
4.4 	🟡(flashing)	Line clear. Moreover, it warns that the next signal is displaying a proceed aspect, but is located at a reduced distance from the succeeding signal, which is displaying a red aspect or a proceed aspect for a diverted route	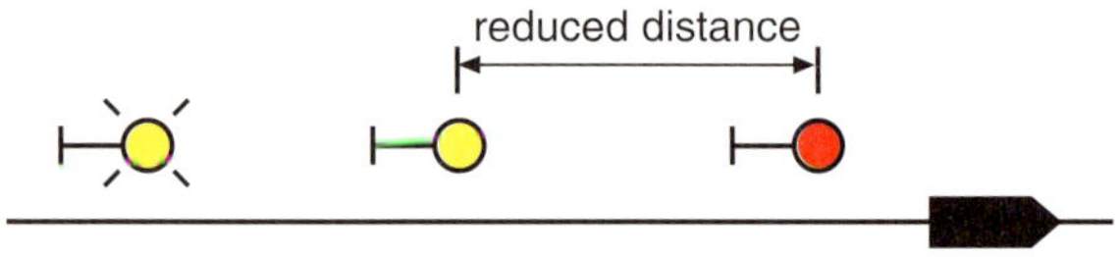 reduced distance
4.5 	🟡🟡	Line clear. Moreover, it warns an occupied track at an abnormally reduced distance or a reception on a blocked or short track	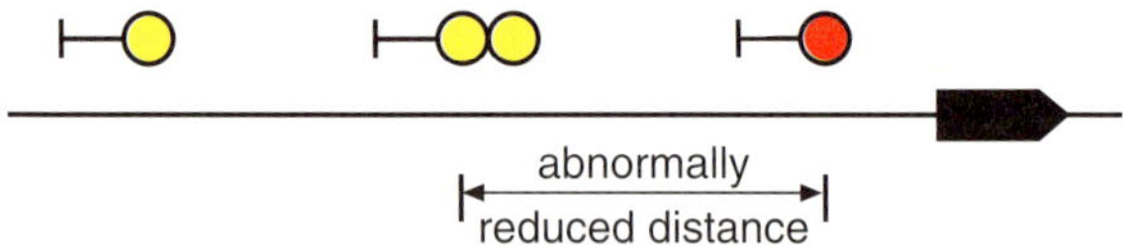 abnormally reduced distance
4.6 	🟡🟢	Line clear. Moreover, it warns that the next signal is at a proceed aspect for a diverted route to be run at a speed not exceeding 30km/h	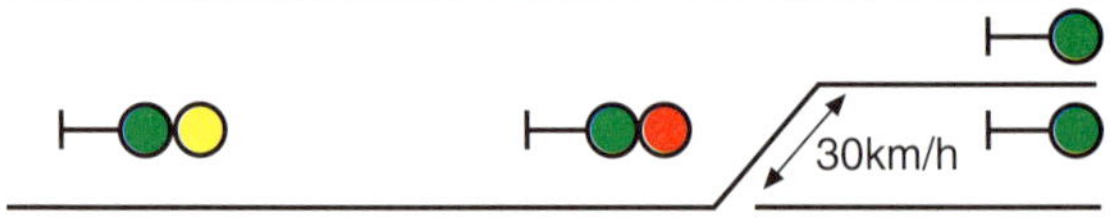 30km/h

Fig. 1.35B FS: aspect sequences

Signal aspect		Signal meaning	Example
4.7 	(flashing yellow over green)	Line clear. Moreover, it warns that the next signal is at a proceed aspect for a diverted route to be run at a speed not exceeding 60km/h	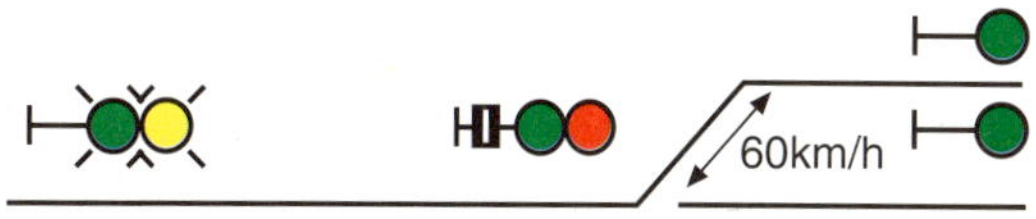 (green+yellow flashing) ... (green+red) ... 60km/h → (green)
4.8	(flashing yellow over green)	Line clear. Moreover, it warns that the next signal is at a proceed aspect for a diverted route to be run at a speed not exceeding 100km/h	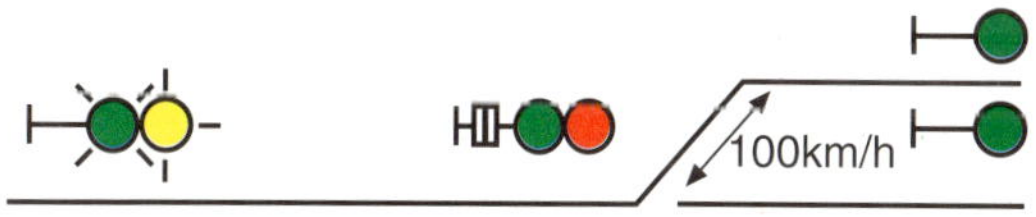 (green+yellow flashing) ... (green+red) ... 100km/h → (green)
4.9	(red over green)	Line clear and confirmation of speed reductions at 30, 60 or 100km/h according to the indications of the preceding warning signal. Moreover, it warns that the next signal is displaying a proceed aspect for a main line route	
4.10 	(red over yellow)	Line clear and confirmation of speed reduction at 30, 60 or 100km/h according to the indications or the preceding warning signal. Moreover, it warns that the next signal is at danger	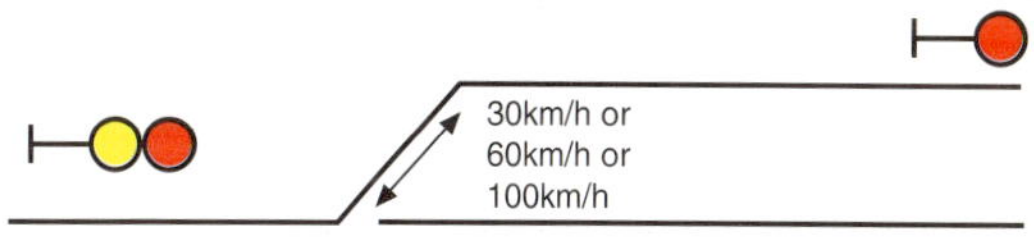 (yellow+red) ... 30km/h or 60km/h or 100km/h → (red)
4.11	(red over flashing yellow)	Line clear and confirmation of speed reduction at 30, 60 or 100km/h according to the indications of the preceding warning signal. Moreover, it warns that the next signal is displaying a proceed aspect for a main line route but it is located at a reduced distance from the succeeding signal, which is displaying a red aspect or a proceed aspect for a diverted route	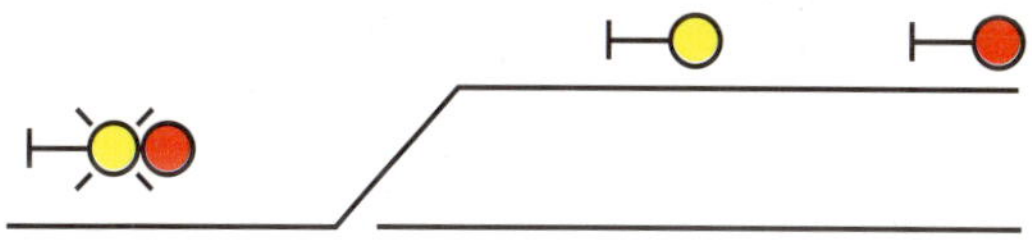 (yellow+red flashing) ... (yellow) ... (red)
4.12	(red over yellow over yellow)	Line clear and confirmation of speed reduction at 30km/h. Moreover, it warns an occupied track at an abnormally reduced distance or a blocked or short track	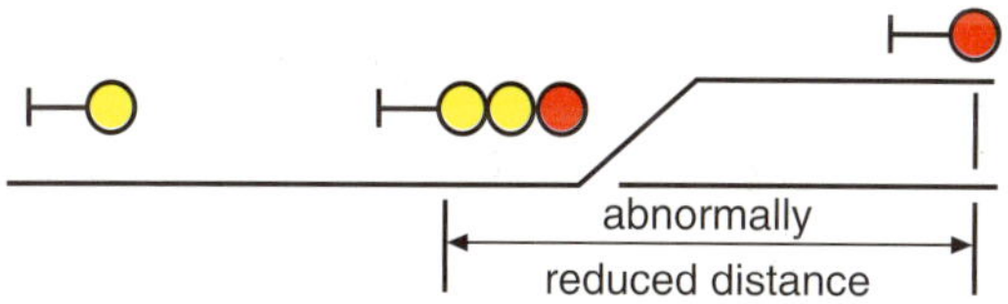 (yellow) ... (yellow+yellow+red) ... abnormally reduced distance → (red)

Fig. 1.35C FS: aspect sequences

Signal aspect		Signal meaning	Example
4.13	(red, yellow, green)	Line clear and confirmation of speed reduction at 30, 60 or 100km/h according to the indications of the preceding warning sgnal. Moreover, it warns that the next signal is at a proceed aspect for a diverted route to be run at a speed not exceeding 30km/h	30km/h
4.14	(red, yellow flashing, green)	Line clear and confirmation of speed reduction at 30, 60 or 100km/h according to the indications of the preceding warning sgnal. Moreover, it warns that the next signal is at a proceed aspect for a diverted route to be run at a speed not exceeding 60km/h	60km/h
4.15	(red, yellow flashing bright, green)	Line clear and confirmation of speed reduction at 30, 60 or 100km/h according to the indications of the preceding warning sgnal. Moreover, it warns that the next signal is at a proceed aspect for a diverted route to be run at a speed not exceeding 100km/h	100km/h

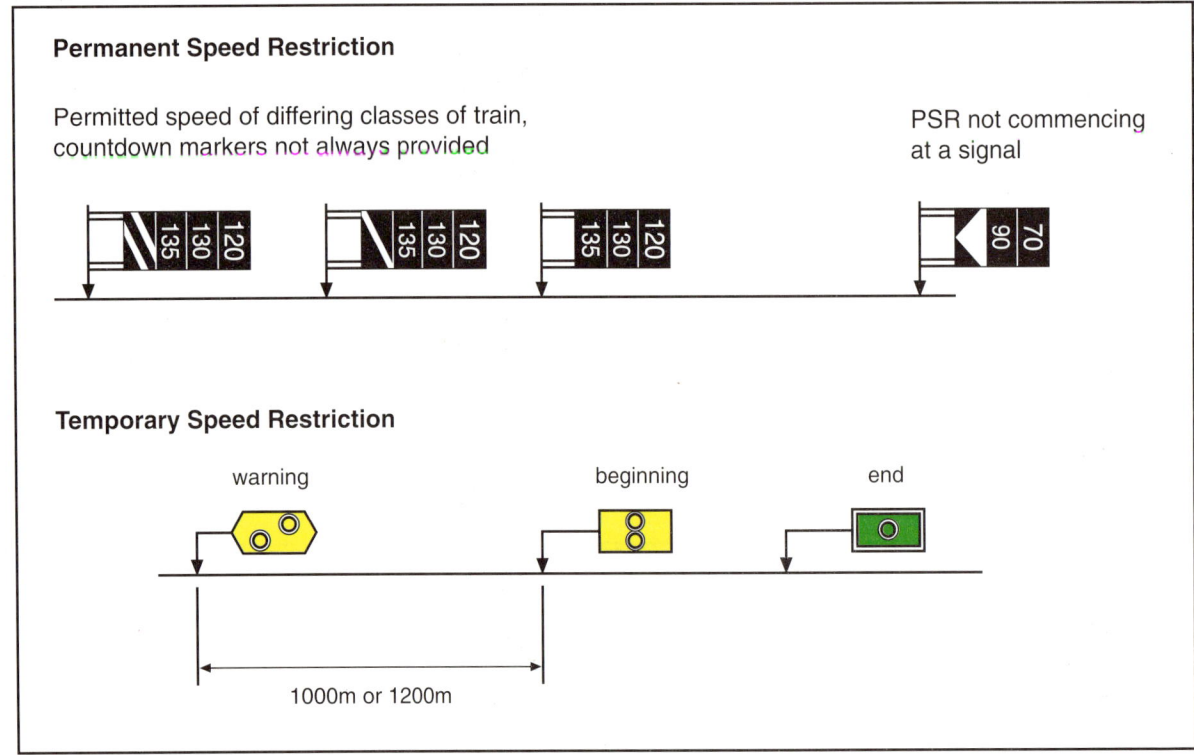

Permanent Speed Restriction

Permitted speed of differing classes of train, countdown markers not always provided

PSR not commencing at a signal

Temporary Speed Restriction

warning — beginning — end

1000m or 1200m

Fig. 1.36A CFL: signal aspects

Signal Profiles:

 Advanced (Warning)
 Repeater of Advanced
 Main (Stop)
 Stop with Shunt
Shunt

Distant signal	Interpretation	Aspect	Main signal	Interpretation	Aspect
(yellow)	Warning, expect to stop at next signal	SFAv 1	(red)	Stop	SFP 1
(green)	Line clear	SFAv 2	(green)	Line clear	SFP 2
(yellow yellow)	Reduce speed to 30km/h	SFAv 3	(yellow over yellow)	Clear, speed limit 30km/h	SFP 3
(yellow yellow + E)	Reduce speed to 60km/h	SFAv 3 + SFAv I	(E over yellow over yellow)	Clear, speed limit 60km/h	SFP 3 + SFI/L
(yellow yellow + Vo)	Reduce speed to 30km/h	SFAv 3 + SFAvVo	(⊔ over yellow over yellow)	Clear, speed limit 30km/h, proceed with caution from the beginning of the platform	SFP 3 + SFVo

Fig. 1.36B CFL: signal aspects

Repeater signal	Interpretation	Aspect
 yellow top-left, white bottom-right	Repetition of SFP 1	SFAvR 1
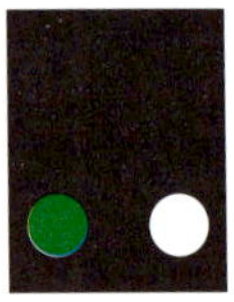 green bottom-left, white bottom-right	Repetition of SFP 2	SFAvR 2
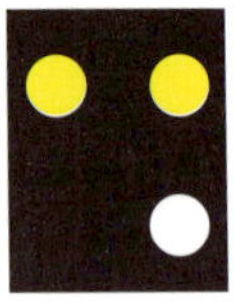 two yellow top, white bottom-right	Repetition of SFP 3, SFP 3 + SFI/L, SFP 3 + SFVo	SFAvR 3

Shunting signal	Interpretation	Aspect
 blue	Stop	SFVb 1
 white	Clear, speed limit 25km/h, proceed with caution	SFVb 2

Fig. 1.36C CFL: signal aspects

Description	Aspect	Meaning
Permanent Speed Restriction: Announcement	 60	Service braking distance from a permanent speed restriction of 60km/h
Execution	 Z	Commencement of permanent speed restriction
Release	 R	End of permanent speed restriction
Temporary Speed Restriction: Announcement	 60 (yellow disc)	Service braking distance from a temporary speed restriction
Execution	 60	Commencement of temporary speed restriction
Release	(blank)	End of temporary speed restriction

RAILWAY SIGNALLING PRINCIPLES

Fig. 1.37 CFL: aspect sequences

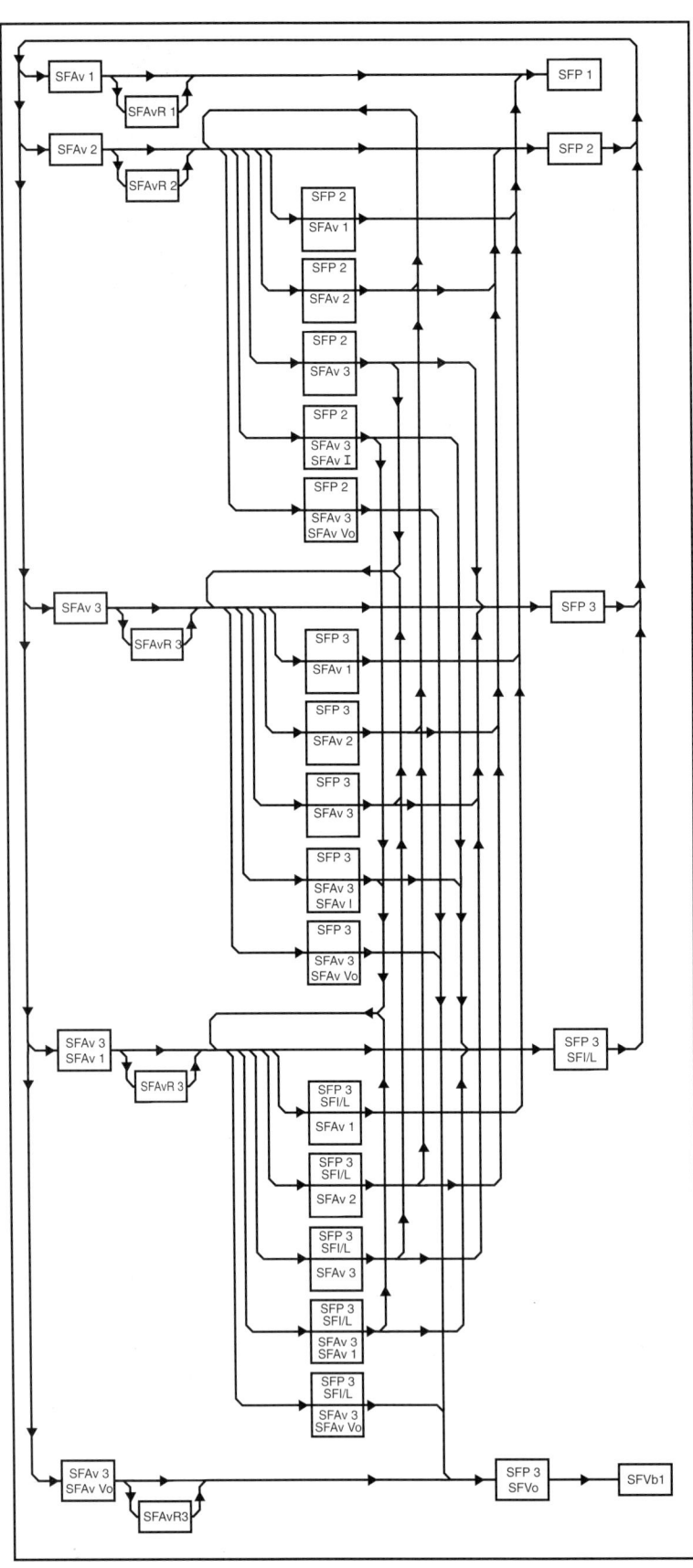

Fig. 1.38A NSB: signal aspects

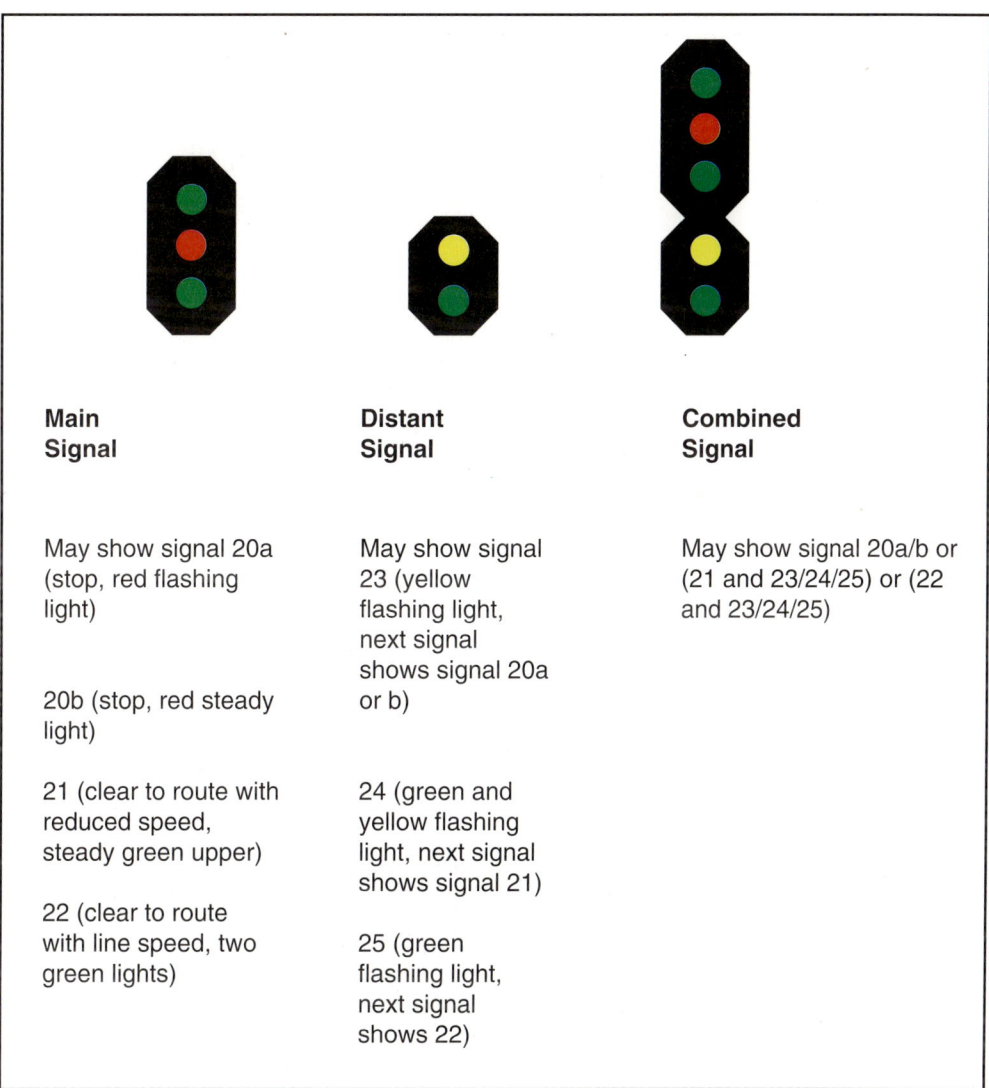

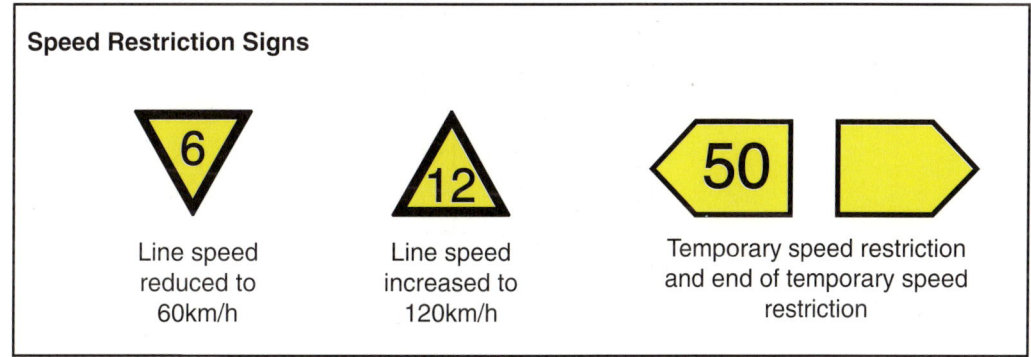

Fig. 1.38B NSB: signal aspects

Description	Aspect	Meaning
Shunting Signals **Tall Signals**		'Tall shunting signals' are mounted 3-4m above ground level and usually in conjunction with starter signals. They can show aspect 41 or 42. They are shown by white lamps on a black background
41	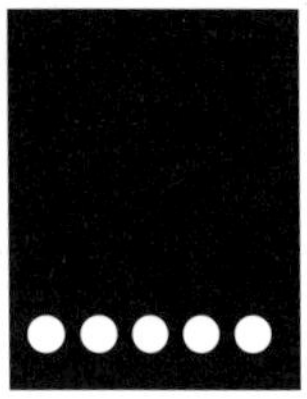 (five white lamps in a horizontal row)	Shunting is not permitted
42	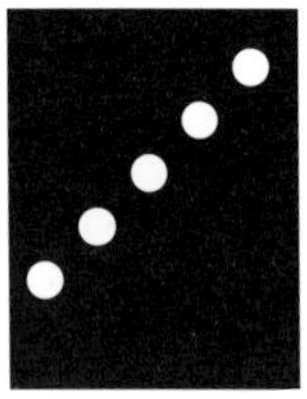 (five white lamps in a diagonal line)	Careful shunting is permitted Points must be operated locally
Dwarf Signals		'Dwarf signals' are usually mounted 60cm above ground level
43	 (two white lamps horizontal)	Shunting is not permitted
44	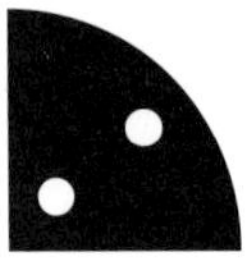 (two white lamps diagonal)	Careful shunting is permitted
45	 (two white lamps vertical)	Shunting is permitted
46	 (two white lamps, one top one right)	Careful shunting is permitted Points must be operated locally

Fig. 1.39 NSB: aspect sequences

Present Block Signalling

X/Y: 'Drive with speed X' / Expect next signal to show 'drive with speed Y'

L: Line speed (50~160km/h)
R: Reduce speed (40~100km/h)
S: Stop
-: No such message given

D1 and D2: Signalling distance. Dependent on signal configuration. In this case both must be greater than the longest stopping distance.

Future Block Signalling

Optical:	160/160	160/160	160/100	100/0	0/-
Cab signal:	200/200	200/160	160/100	100/0	0/-

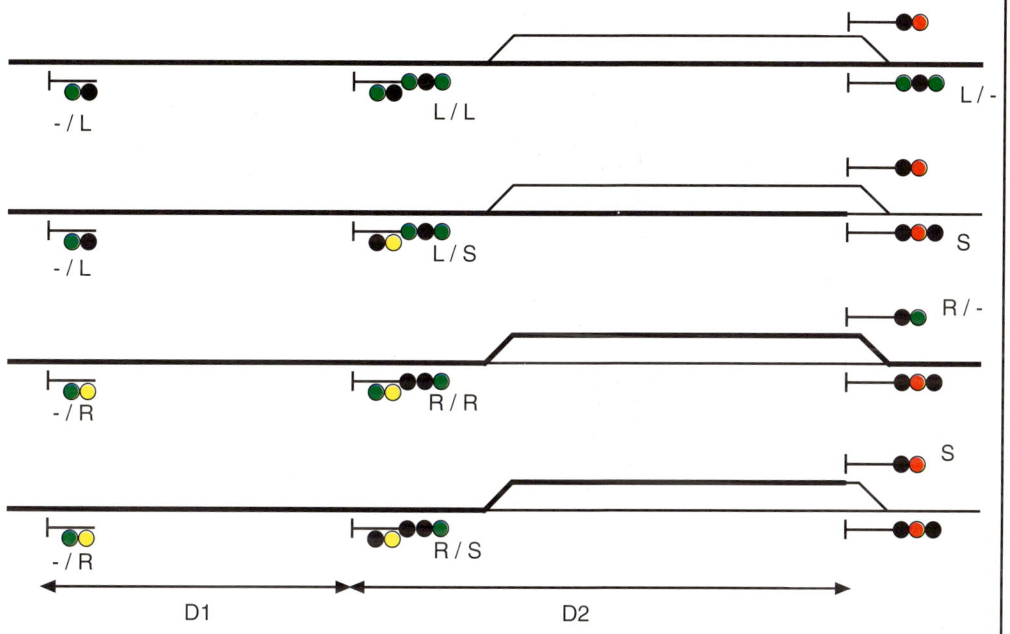

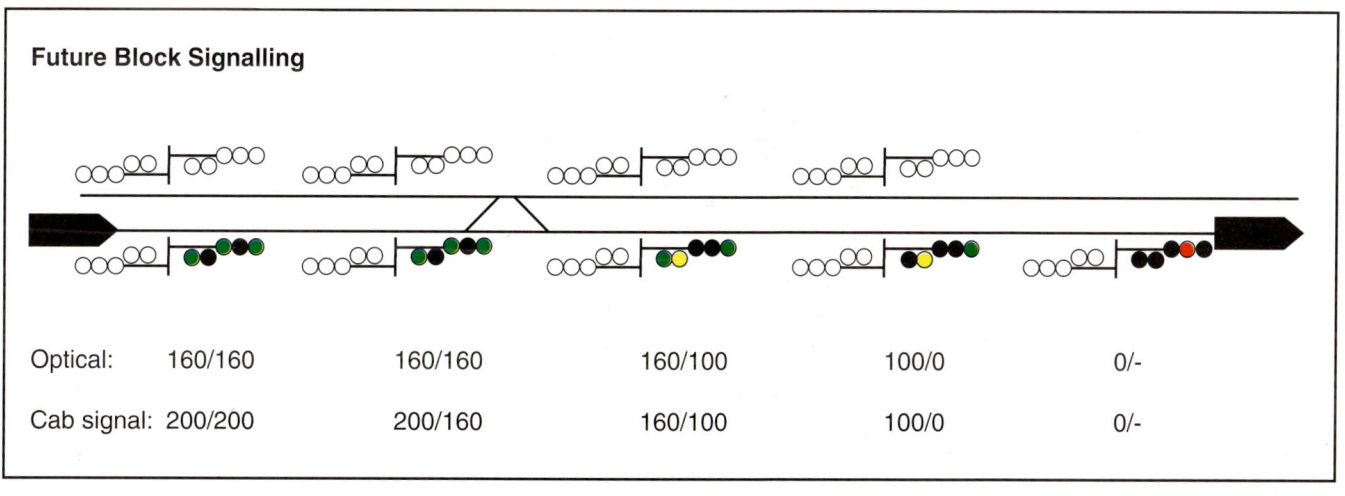

Fig. 1.40A NS: signal aspects

Profiles of 'High' and 'Low' (dwarf) signals

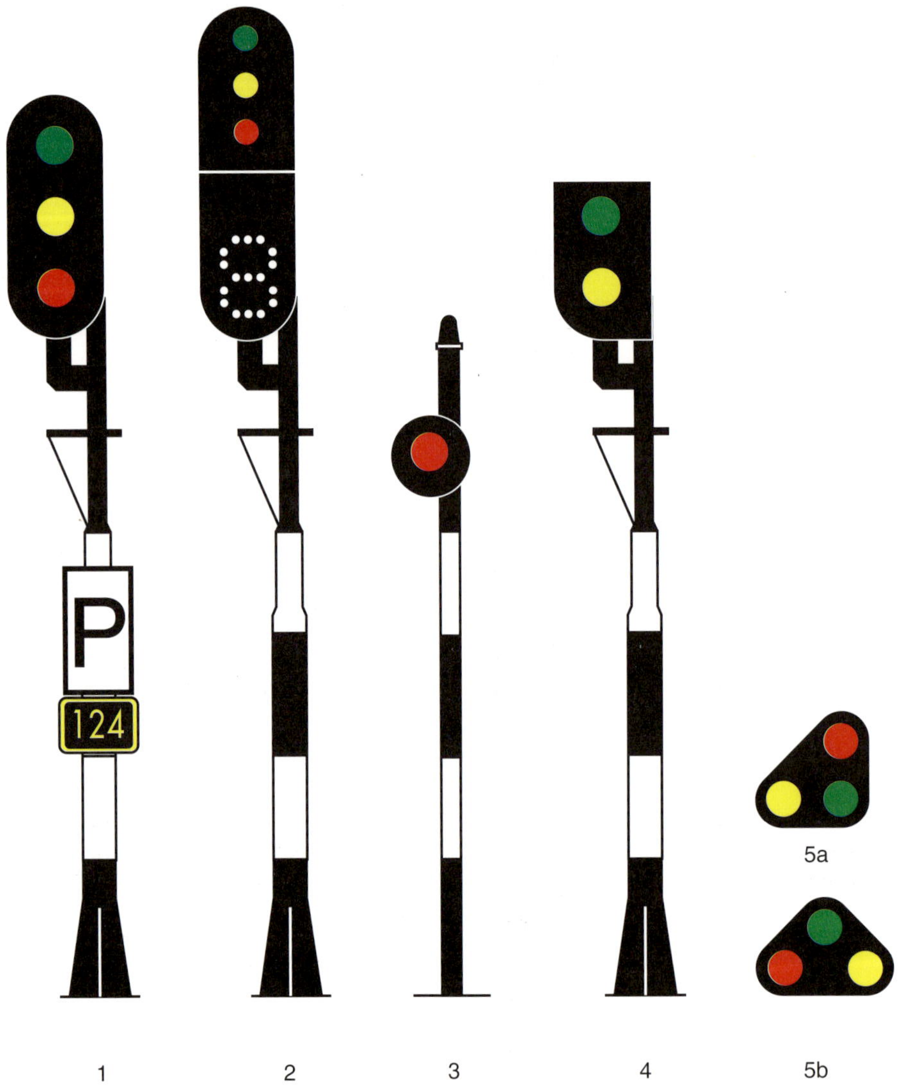

The red/yellow/green searchlight signal (Nº 3) is no longer in widespread use

Signals 1, 2, 3 and 5 are stop signals (*Hoofdsein*)

Signal 4 is a warning signal (*Voorsein*)

Fig. 1.40B NS: signal aspects

Description	Aspect	Meaning
Red	 🔴	Order to stop before the signal
Flashing Yellow	 🟡 (flashing)	Permission to drive on sight at a maximum speed of 30km/h, in order to be able to stop before a signal showing RED or at any point that signal as track may be occupied
Yellow	 🟡	Order to reduce speed to 40km/h, then permission to drive on sight with a maximum speed of 40km/h, in order to be able to stop before a signal showing RED If train speed equals or is lower than 40km/h or if dwarf signal: permission to drive on sight at a maximum speed of 40km/h, in order to be able to stop before a signal showing RED
Yellow + Speed Indicator	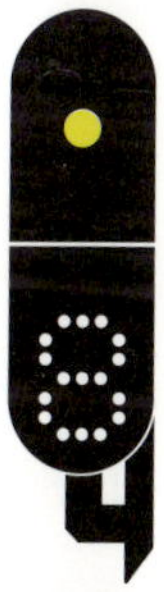 🟡 + indicator	Order to reduce speed to speed indicated by the speed indicator. The lower speed imposed must be reached before the next signal If train speed equals or is lower than speed indicated by the speed indicator: permission to proceed at the speed indicated by the speed indicator
Yellow + Flashing Speed Indicator	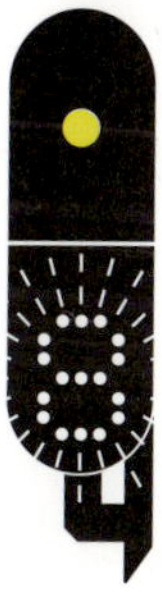 🟡 + flashing indicator	Order to reduce speed to that indicated by the speed indicator. The distance to the next signal is shorter than the braking distance required for a regular deceleration from local maximum speed to the lower speed imposed. An uninterrupted brake application, however, guarantees that the order given by the next signal can be carried out in time If train speed equals or is lower than speed indicated by the speed indicator: permission to proceed at the speed indicated by the speed indicator

Fig. 1.40C NS: signal aspects

Description	Aspect	Meaning
Flashing Green		Permission to proceed at 40km/h
Flashing Green + Speed Indicator	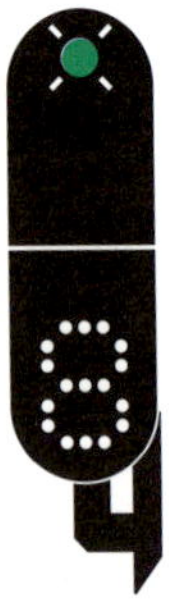	Permission to proceed at the speed indicated by the speed indicator
Green		Permission to proceed at the speed indicated by fixed speed signs along the line
Green		Dwarf signal: permission to proceed at 40km/h Note: Red, Flashing Yellow and Yellow aspects in dwarf signals carry the same meaning as in 'high' signals
Richtingaanwijzer		Route is set to left at junction ahead

Fig. 1.40D NS: signal aspects

Description	Aspect	Meaning
'L' sein	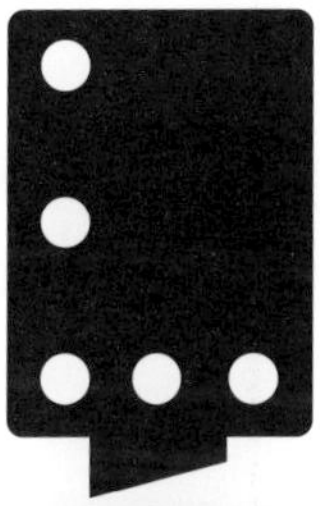	Freight Train Signals (see **1.7.17.2** for an explanation) Signal in advance is displaying an 'H' sign. (Applicable to designated freight trains only)
'H' sein	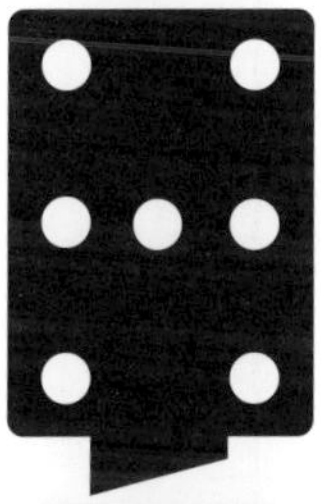	Stop at this signal, although a proceed aspect is shown. (Applicable to designated freight trains only)
Witte 'G'	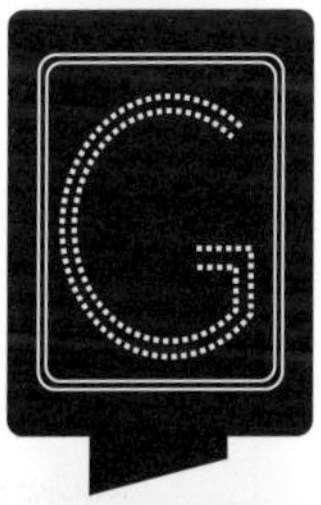	Qualifies proceed aspect in signal. All signals in tunnel section are showing proceed aspect. The tunnel entry speed must not be exceeded. (Appliable to freight and other designated trains only)
Witte 'X'		Steady 'X': Stop at signal, although proceed aspect is shown. (Applicable to freight and other designated trains only) Flashing 'X': Warning signal for steady 'X'
Herhalingssein horizontaal	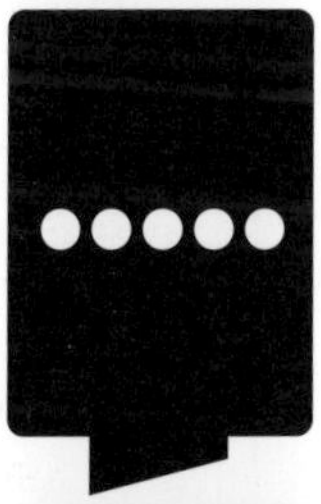	Repeating signal: Associated signal is displaying red or flashing yellow
Herhalingssein diagonaal	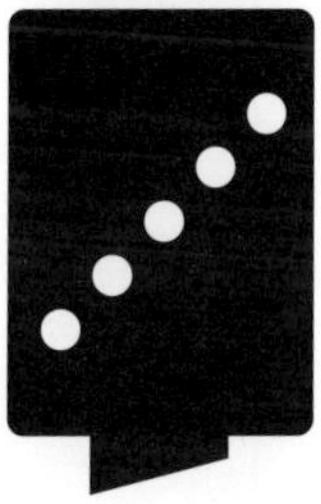	Associated signal is displaying an aspect better than flashing yellow

Fig. 1.40E NS: signal aspects

Description	Aspect	Meaning
Permanent Speed Restrictions		
Snelheidsvermin-deringssbord	 (yellow inverted triangle with 6)	Service braking distance from permanent speed restriction of 60km/h
Snelheidsbord	 (white square with 6)	Commencement of permanent speed restriction of 60km/h
Baanvaksnel-heidsbord	 (green triangle with 11)	Resume line speed of 110km/h
Temporary Speed Restrictions		
'L' bord	 (yellow oval with L)	Service braking distance from temporary speed restriction
'A' bord	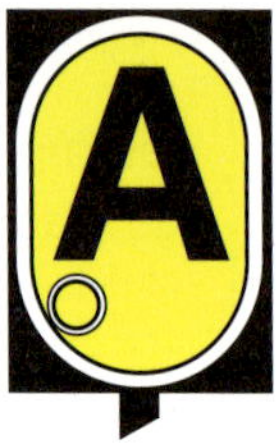 (yellow oval with A)	Commencement of temporary speed restriction
Tijdelijk snelheids-bord	 (octagons with 6, and 6/4)	Speed indicator for temporary speed restriction (km/h x 10). (In differential speed indicator, higher speed applies to passenger trains, lower to freight trains)
'E' bord	 (green oval with E)	End of temporary speed restriction

Fig. 1.40F NS: signal aspects

Description	Aspect	Meaning
Countdown Boards (see 1.4.1.6)		
Reflectorplaatje (Type 'A')	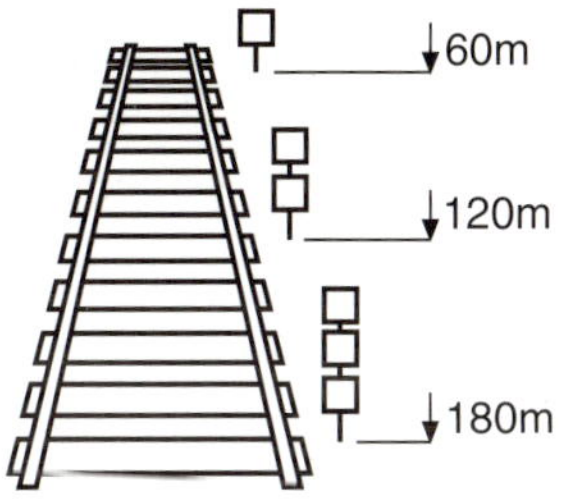 60m / 120m / 180m	Approach to a stop signal
Reflectorplaatje schuin (Type 'B')		Approach to the last permissive signal on a line
Baken (Type 'C')		Approach to a warning signal
Gele baken (Type 'D')		Approach to a signal at braking distance from a junction signal or a signal protecting a movable bridge

RAILWAY SIGNALLING PRINCIPLES

Fig. 1.41 NS: aspect sequences

The aspect to be shown in a signal is determined by:
~ the local maximum speed (V) at the signal
~ the passing speed (VP) at the signal
~ the target speed (VT) at the next signal

G	Green	If VP ≤ VT and VP = V
GFL*	Green Fl + * VP	If VP ≤ VT and 40 < VP < V
GFL	Green Fl	If VP ≤ VT and 40 = VP < V
Y*FL	Yellow + * VT Fl	If VP > VT ≥ 40 and insufficient braking distance to next signal, but sufficient braking distance to the signal, following next signal
Y*	Yellow + * VT	If VP > VT ≥ 40 and sufficient braking distance to next signal
Y	Yellow	If VP > VT < 40
YFL	Yellow Fl	If VP ≤ 30 (route setting for drive on sight)
R	Red	If VP = 0 (No signal clearance)

* Speed indicator can show as a rule 4, 6, 8 or 13, according to the ATC-speed levels 40, 60, 80 and 130km/h

Some examples of aspect sequences:

\# = Regular Braking Distance

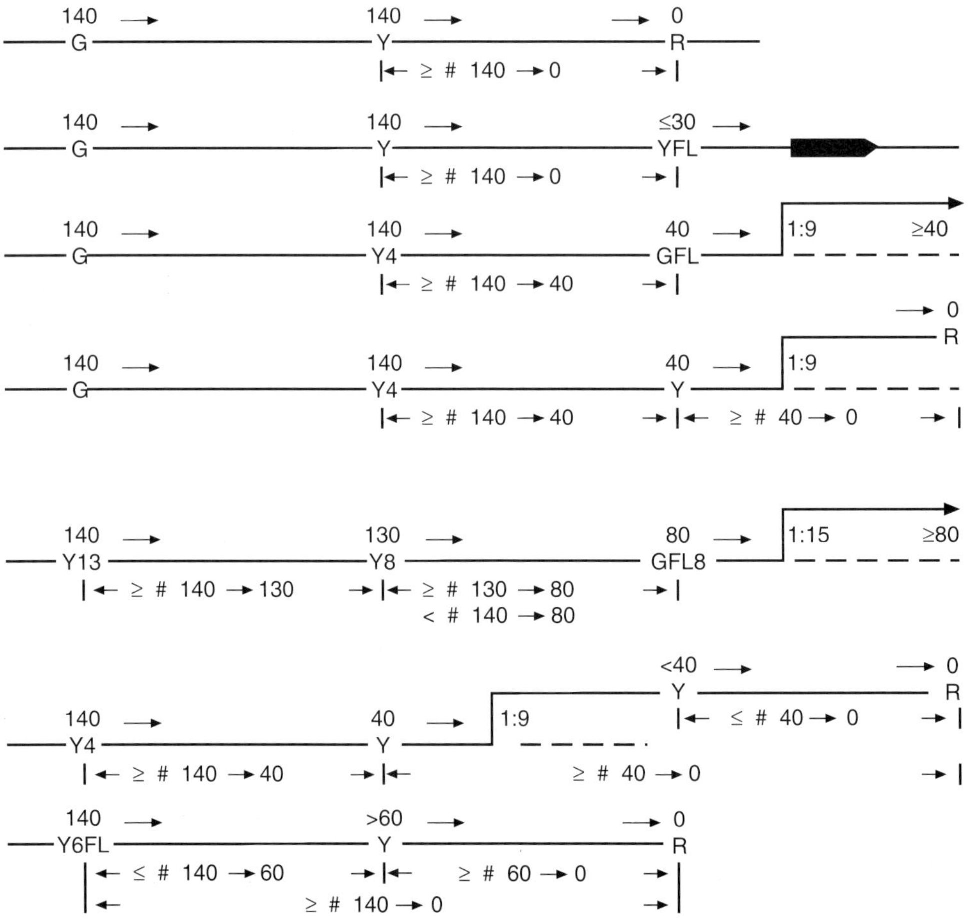

```
   140 →            140 →                  → 0
───G──────────────── Y ───────────────────── R ─────
                    |← ≥ # 140 → 0        →|

   140 →            140 →                 ≤30 →
───G──────────────── Y ───────────────────── YFL ─────▶
                    |← ≥ # 140 → 0        →|

   140 →            140 →                  40 →    1:9    ≥40
───G──────────────── Y4 ──────────────────── GFL ──── ─ ─ ─ ─
                    |← ≥ # 140 → 40       →|
                                                           → 0
                                                           ─ R
   140 →            140 →                  40 →    1:9
───G──────────────── Y4 ──────────────────── Y ──── ─ ─ ─ ─
                    |← ≥ # 140 → 40       →|← ≥ # 40 → 0   →|

   140 →            130 →                  80 →    1:15   ≥80
───Y13─────────────── Y8 ──────────────────── GFL8 ─── ─ ─ ─ ─
   |← ≥ # 140 → 130 →|← ≥ # 130 → 80       →|
                      < # 140 → 80

                                          <40 →            → 0
                                            Y ─────────────── R
   140 →            40 →   1:9            |← ≤ # 40 → 0   →|
───Y4──────────────── Y ───── ─ ─ ─ ─
   |← ≥ # 140 → 40 →|←               ≥ # 40 → 0            →|

   140 →            >60 →                  → 0
───Y6FL─────────────── Y ───────────────────── R
   |← ≤ # 140 → 60 →|← ≥ # 60 → 0         →|
   |←                ≥ # 140 → 0          →|
```

Fig. 1.42A BV: signal aspects

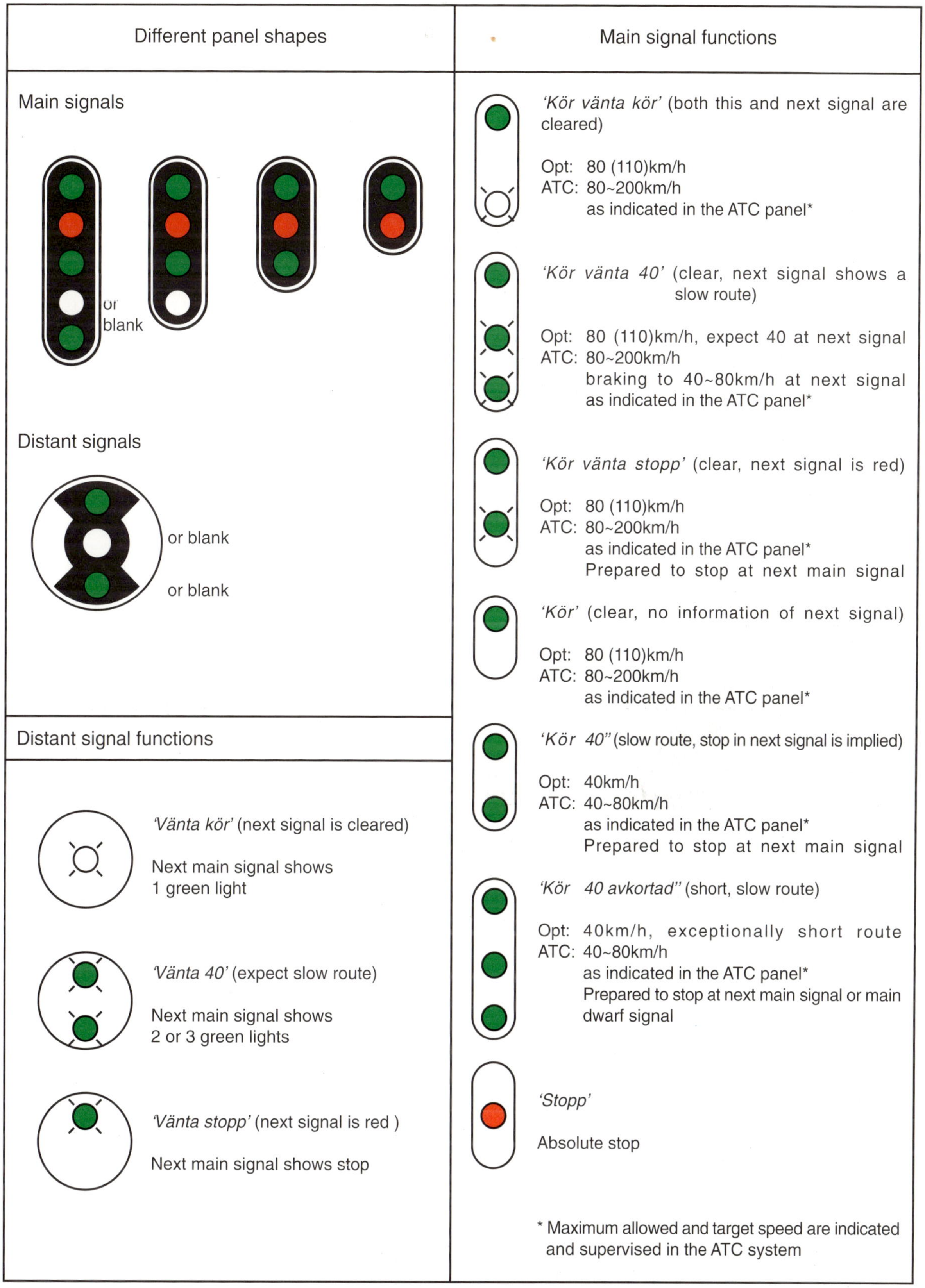

Different panel shapes	Main signal functions
Main signals [four main signal panels shown: green/red/green/white-or-blank/green; green/red/green/white; green/red/green; green/red] **Distant signals** [circular panel with green/white/green — "or blank / or blank"]	**'Kör vänta kör'** (both this and next signal are cleared) Opt: 80 (110)km/h ATC: 80~200km/h as indicated in the ATC panel* **'Kör vänta 40'** (clear, next signal shows a slow route) Opt: 80 (110)km/h, expect 40 at next signal ATC: 80~200km/h braking to 40~80km/h at next signal as indicated in the ATC panel* **'Kör vänta stopp'** (clear, next signal is red) Opt: 80 (110)km/h ATC: 80~200km/h as indicated in the ATC panel* Prepared to stop at next main signal **'Kör'** (clear, no information of next signal) Opt: 80 (110)km/h ATC: 80~200km/h as indicated in the ATC panel* **'Kör 40'** (slow route, stop in next signal is implied) Opt: 40km/h ATC: 40~80km/h as indicated in the ATC panel* Prepared to stop at next main signal **'Kör 40 avkortad'** (short, slow route) Opt: 40km/h, exceptionally short route ATC: 40~80km/h as indicated in the ATC panel* Prepared to stop at next main signal or main dwarf signal **'Stopp'** Absolute stop
Distant signal functions **'Vänta kör'** (next signal is cleared) Next main signal shows 1 green light **'Vänta 40'** (expect slow route) Next main signal shows 2 or 3 green lights **'Vänta stopp'** (next signal is red) Next main signal shows stop	* Maximum allowed and target speed are indicated and supervised in the ATC system

Fig. 1.42B BV: signal aspects

Shunting signals

To control shunting movements on larger junctions position light dwarf signals are used. These are small signals standing on ground between tracks.

Dwarf signal aspects

The same type of signal is used as a subsidiary signal too

Sh 0 = stop for all kinds of vehicles

Sh 1 = shunting route set, route may be occupied by other vehicles

Sh 2 = shunting route set, route free

Sh 3 = shunting allowed, turnouts are locally controlled

A dwarf signal can have one red and one or two green lights added to make it a '**main dwarf signal**'. That kind of signal is used as a starter signal from low speed tracks (40km/h).

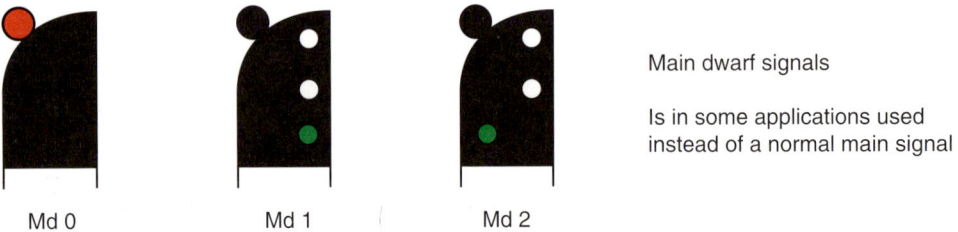

Main dwarf signals

Is in some applications used instead of a normal main signal

The coloured lights have no meaning for shunting movements.

The position of the green light indicates if the signal is at the beginning of a high speed route (Md 1) or of a low speed route (Md 2).

The green light is flashing if the following main signal is red.

Fig. 1.43A BV: aspect sequences

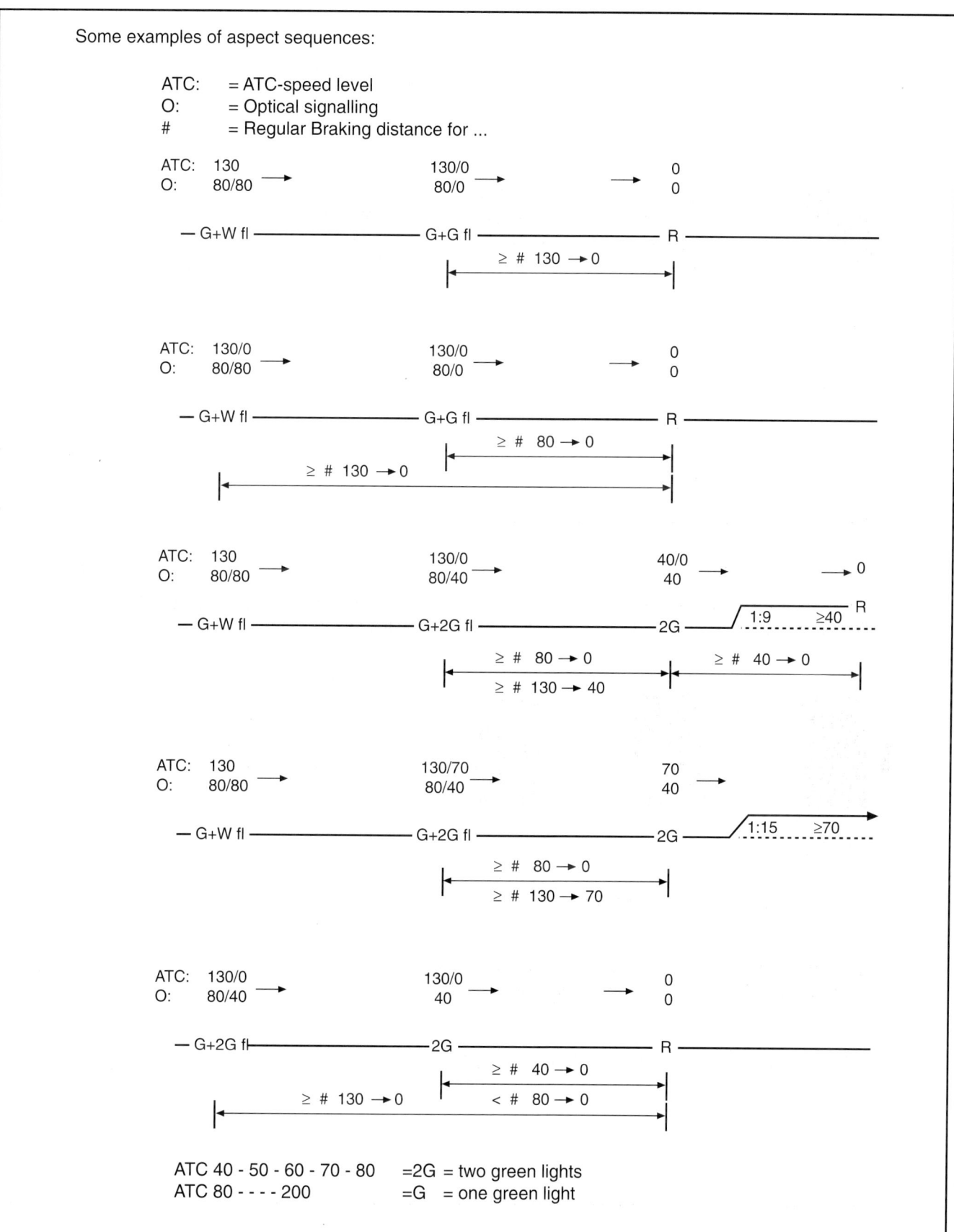

Fig. 1.43B BV: aspect sequences

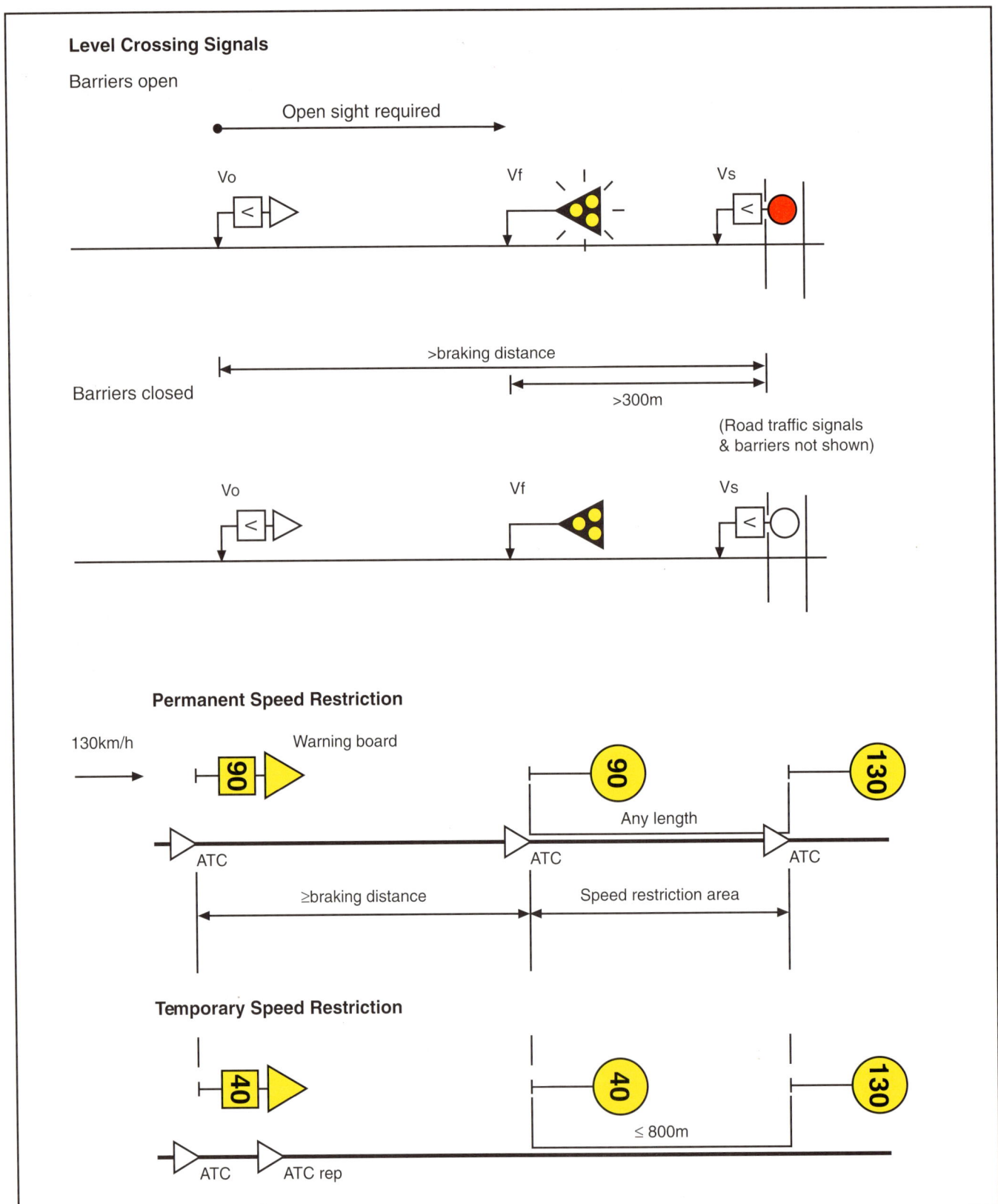

1.6 Bidirectional working

Bidirectional working is being practised increasingly on double track routes, due partly to the advent of mechanized track maintenance and partly to commercial pressures to maintain signalled running at or near line speed during engineering work. The degree of sophistication of the additional facilities can vary from the provision of low speed crossovers with little or no extra signalling, to higher speed, fully signalled crossovers with full block signalling between them, enabling traffic to run at line speed at the same headway on each line. This latter option is generally called banalization.

However, there is a range of options, in increasing order of cost/complexity:

- manually operated crossovers, with single line working by instruction, no signalling being provided for contraflow traffic (BR, FS).
- power operated crossovers, but still with single line working by instruction or substitution signals (ÖBB, FS).
- signalling for contraflow traffic onto and off the opposing track at crossovers. The contraflow headway is still governed by the distance apart of the crossovers, but contraflow traffic can be operated without stopping for instructions and without any staff specially stationed at the crossovers. On BR, this is called SIMBIDS (*SIM*plified *BID*irectional *S*ignalling).
- provision of intermediate signal sections for contraflow traffic (BR, DB AG, FS).
- full banalization (ÖBB, NMBS/SNCB, SBB, RENFE, FS, CFL, NSB, BV).

Unless the terrain is particularly featureless, however, the spacing of the crossovers and intermediate signalling is governed by the locations of stations, level crossings and the like.

In particular, automatic level crossings require controls for both directions of traffic on each line, which greatly adds to their complexity. The alternatives are either to require contraflow trains to stop before crossing the road or to appoint a local control operator – neither of which is conducive to efficient operation.

If the contraflow facility is to be used without prior notice to track maintenance and other personnel, for example to work traffic around a temporary obstruction, then national legislation may require enhanced protection. This may take the form of a lockout device to inhibit contraflow movements whilst a specific activity, for instance patrolling, is undertaken. Alternatively, an automatic staff warning system which, like level crossing protection, has to be configured for both directions, may be used. As well as being provided to facilitate engineering work, bidirectional signalling may be provided for other reasons:

- at stations, to enable platforms to be used in both directions.
- to enable out of gauge traffic, for example, large containers, to pass through a tunnel or other tight spot where loading gauge clearance is only available on one line.

1.7 Principles of interlocking and controls

Although computer driven or solid state interlockings are beginning to appear all over Europe, their principles of operation have generally been modelled on those of relay interlockings which form the vast majority of installations and will do so for some years into the future. The description that follows will therefore relate to relay interlockings, but it will be appreciated that continuing experience in the use of software will enable different approaches to be developed – particularly in the field of intelligence. Two examples are the enforcement of sequential operation of track circuits by a train to guard against the loss of trainshunt (NS) – this requires the system to distinguish between a genuine train and a track circuit failure – and the use of the SSI computer to integrate the readings of track mounted axle counter equipment (NMBS/SNCB, BR).

1.7.1 Type of system
There are two broad families of relay interlocking:

- Free wired (or custom built).
- Geographical.

Geographical systems are more appropriate for large interlockings with extensive flank and countermove requirements, whereas small stations can generally be engineered more economically using a free wired system, since non-relevant features can be excluded. Several railways have therefore developed geographical and free wired systems in parallel. For instance, SNCF have PRS free wired and PRG geographical.

It is also possible to provide safety relays only for the core interlocking part of the system, peripheral activities such as route marking, calling and indication being left to non-vital equipment. SNCF (PRCI), RENFE, BR (ERSE), FS, NS and BV, all have installations which work in this manner.

Another factor that influences circuit design principles is the use of safety relays with metal to metal contacts. Most European railways use such relays, which are designed so that the welding on overload of any contact pair will render the relay unable to make any contacts of its opposite state. Such relays in vital positions in the system then require to be cross proved to ensure that a welded contact is detected. In contrast, BR and NS use non-weldable contacts, typically silver/silver impregnated graphite. Such relays do not require to be cross proved, although this feature is sometimes provided for other reasons.

1.7.2 Method of operation

Hitherto, the favoured method of operating relay interlockings has been via a series of buttons or switches mounted geographically on a panel. With the advent of computer graphics, it is now possible to effect control using a video screen with a light pen, mouse, trackerball or keyboard. Such equipment – videopult (ÖBB), EBP (NMBS/SNCB), IECC (BR) – is normally associated with electronic interlockings, but relay systems can be interfaced with it.

The flexibility associated with software driven systems enables a regime where operator intervention is on a 'by exception' basis, with routine operations being undertaken by a timetable-based, automatic route setting program. On those railways where preselection of routes is not permitted, the supervisory system can also be used to perform some of the preliminary route selection tasks. With route setting panels, there are a number of methods of operation.

To set a route:

- Push button at entrance and button at exit simultaneously (ÖBB, SBB, DB AG, RENFE, CFL).
- Push button at entrance, followed by button at exit (RENFE, BR, NS, BV), preceded, on NS by one of several common buttons for respectively a *normal*, *automatic* or *drive on sight* route.
- Push button at exit, followed by button at entrance (SNCF).
- Push separate button for each route (NMBS/SNCB).
- Push button at entrance, separate button for each route (FS, BR – older installations).
- Turn switch at entrance followed by pushing button at exit (BR – older installations).

To cancel a route:

- Push button at entrance, together with a group cancel button (artificial cancel button – RENFE).
- Push button at exit, together with a group cancel button (ÖBB, SBB, DB AG, CFL).
- Pull button at entrance (FS, BR).
- Push emergency button (NMBS/SNCB).
- Push common button, followed by pushing entrance button (NS).

Equivalent operations can also be made by a dispatcher via a keyboard (SBB, DB AG, RENFE, FS, NS, BV).

Extensive use is made by some railways of group buttons which, when operated with the geographically located buttons, perform specific functions, for example set shunt routes, cancel routes, cancel preselection, operate subsidiary signals and independently operate sets of points.

For critical operations, involving bypassing or releasing locking, use of the group button is required to be documented and a counter is provided which is incremented by each group button operation (ÖBB, NMBS/SNCB, SBB, DB AG, RENFE, FS, CFL).

To operate points individually, use can be made of group buttons, to set either (+) or (−) (ÖBB, NMBS/SNCB, SBB, DB AG, RENFE, CFL) or a three position key, operated to set points Normal or Reverse and with a centre position which allows route setting (DB AG, BR, FS, NS). Specific operation of an individual point button/switch renders that set of points unable to be called to the opposite position by the route setting logic.

Indication lamps (or light emitting diodes – leds) are provided to confirm the various steps in the route setting and locking process and to indicate the passage of a train through the route. Indications are also provided to repeat the signals controlled by the interlocking, the positions of points and the status of level crossings, ground frames and other functions. The latter include lockouts to protect staff working on or adjacent to trains, or staff working on the track in areas of limited clearances, for example tunnels or viaducts, or where bidirectional running is in operation.

1.7.3 Route calling

The operator having made his choice, the interlocking system first checks the availability of the chosen route. This is not an essential step, but failure to do so might result in half set routes, premature level crossing barrier operation and operator confusion if he fails to detect a route which incompletely sets.

The availability check typically requires:

- points in line of route in correct position or free to move;
- appropriate flank protection available by points, track circuits clear, or signals proved at danger (not on NS);
- no opposing route called;
- local control not selected;
- ground frames proved Normal;
- staff protection systems not in use;
- no route already selected from the entrance signal.

Some railways require the individual setting of sub-routes as part of the calling of a main route (SBB, RENFE); in other cases, calling the main route automatically presets any intervening shunt signals (RENFE, BR, FS, BV).

The availability check may be undertaken by non-vital relays or by commercial electronics and software (NMBS/SNCB–EBP, BR–IECC, FS). However, if this is so the route locking circuit level must then re-check all the conditions before the signal is allowed to show a proceed aspect.

The subject of preselection of routes has been hotly debated; some railways allow it (SBB, BV); some do not (RENFE, BR, FS, NS). The case against preselection is that a momentary track circuit false clearance, causing route locking to be prematurely released, could enable an already preselected route to set. This can be countered on modern installations by enforcing a time delay of say five seconds after track circuit clearance or by giving the interlockings intelligence, for instance by proving the track circuit in advance occupied before the rear track circuit can clear during a signalled movement.

1.7.4 Route setting

Having been proved available, the route is then set, in some cases without any further action on the part of the operator (ÖBB, DB AG, RENFE, BR, CFL, NS), or by the operator continuing to hold down the route button (FS) or by a further button operation, which may be used to select the class of route and also lock it (NMBS/SNCB).

Points then move to their required position, level crossing operating sequences are initiated, automatically (NMBS/SMCB, RENFE, FS, CFL, NS – if the train has passed the strike-in point for the crossing), or require separate actions by the operator (BR).

This activity of converting a provisional route to a real route is registered on the control panel, typically by a row of white lights (SBB, DB AG, BR, CFL) or green lights (NS) and by the release of the route calling circuits to initiate routes elsewhere in the interlocking.

Some railways include a check of appropriate track circuits clear, signal ahead illuminated and other conditions required for signal clearance to be satisfied before the route will set (DB AG, RENFE); others do not (NMBS/SNCB, BR, FS, NS). The latter option allows a route to be set although conditions may not exist at that instant for the signal to clear. Under these conditions, the route will not become approach locked and may be freely cancelled by the operator.

1.7.5 Route locking

Some railways lock the route immediately it has been set (ÖBB, SBB, DB AG, RENFE, SNCF, BR – part, FS, CFL, NS, BV); others wait until the signal controls have been proved and the signal is ready to clear before applying route locking (NMBS/SNCB, BR – part). However, in all cases, proof of locking the route is included in the final check before the signal is actually allowed to show a proceed aspect.

ÖBB do not impose route locking in respect of shunt routes – this facilitates shunting operations when the movement stops before reaching the exit signal. Other railways require the train to reach the exit signal or, in special locally defined circumstances, to occupy intermediate track circuits for a time sufficiently long to have come to a stand.

1.7.6 Approach locking

Having successfully called, set and locked a route, the usual method of release is by the passage of a train. However, in situations where a train does not traverse the route and a conflicting route is required to be set, the original route must first be cancelled. There are different ways of achieving this, but all involve a

Railway	Method of Route Release
ÖBB	Registered Command (special button). No time delay
NMBS/SNCB	Special Release Key – to be used by Inspector
SBB	Manual Release Key – no route setting then possible for 2min (shunt routes are exempt)
DB AG	Registered Operation, no time delay
RENFE	Release immediate if no train approaching. Time delay if train approaching
SNCF	Operation of special button. Release immediate if no train approaching; 3min delay if train within 500m of *Avertissment* signal
BR	2min delay, sometimes extended to 3 or even 4min; 30s for shunt routes (In some installations, delay waived if no train within sighting point of distant signal – this is termed comprehensive approach locking)
FS	Immediate release if no train approaching. Emergency lever plus 20s delay if no train has passed the warning signal; 5min delay if interlocking operational remotely
CFL	Registered Operation
NSB	90s delay
NS	At least 2min delay; sometimes depends on whether a train is approaching
BV	1min delay on ordinary lines; 2min on high speed lines. 30s for shunt routes

Fig. 1.44 Method of route release

procedure to ensure that a route is not released in the face of an approaching train (see **Fig. 1.44**).

Where the imposition of a time delay is dependent upon the approach of a train – sufficiently close to have seen the warning signal displaying a proceed aspect before the route was cancelled – the selection circuits must be designed to avoid locking the route if, for instance a train is occupying the track circuits but is travelling away from the signal concerned, or is approaching but will be diverted onto another line before reaching the signal. In the vicinity of large stations and yards, these approach locking circuits can become very complex to design (and even more complex to test) so that their provision must be balanced against the increased operating flexibility which they give.

1.7.7 Aspect controls – stop signals

Before a signal can show a proceed aspect many conditions need to be satisfied. This section deals with the controls required for running signals. Those for shunting and other types of signal are different and are described in subsequent sections.

1.7.7.1 Line of route controls

Line of route controls fall into two categories:

(a) Continuous checking that the conditions in the interlocking required for the route to set are still extant:

- points locked in required position;
- ground frames locked normal;
- non-automatic level crossing barriers locked in the closed position;
- movable bridges and other structures locked in the closed position;
- opposing routes and route locking proved normal.

(b) Continuous checking that the elements of the signalling system *actually are* in the required condition:

- points are detected in the required position and locked;
- ground frames are detected locked;
- non-automatic level crossing barriers are actually detected in the lowered position;
- movable bridges are in the correct position and locked;
- track circuits or other train detection systems in the line of route are indicating clear.

Amongst the European railways there is virtually no variation in these requirements.

1.7.7.2 Signal ahead controls
Some railways prove the signal ahead is illuminated (SBB, DB AG, RENFE, BR, FS, CFL) whilst others apply no such lamp proving (ÖBB, NMBS/SNCB, NS) – but on ÖBB and BV the failure of a proceed aspect lamp will illuminate the stop aspect.

Of those that do prove the signal ahead, SBB and DB AG prove it illuminated (a stop or proceed aspect), before allowing the signal in rear to show a proceed aspect. RENFE operate a comprehensive system of lamp proving, causing the signal in rear to assume a more restrictive aspect as do FS who, like RENFE and SNCF, use single filament lamps. BR, using double filament lamps, switch in the auxiliary filament on failure of the main one, but hold the signal in rear at red only if both filaments have failed. SBB prove that the signal ahead is illuminated (a stop or proceed aspect).

ÖBB and CFL indicate each signal by a white/red/white reflectorized bar on the signal post; this marker is considered sufficient to alert drivers to the presence of a dark signal. Similarly, NMBS/SNCB provide a white or yellow reflectorized board on each signal post.

It will be appreciated that all railways which use a track to train warning system will have the location of all signals monitored in the cab.

1.7.7.3 Foul track and flank controls
Foul tracks, or selection of points to avoid foul tracks, are always included in the controls, except for flashing yellow routes on NS. However, flank protection is subject to a wide variation in practice.

Flank protection, by setting points, proving signals at danger and/or intervening tracks clear, is always applied on ÖBB, SBB, DB AG, RENFE, SNCF, CFL and BV, and is applied where simple and effective by BR. FS and NMBS/SNCB apply it whenever possible, so long as otherwise non-conflicting routes are not inhibited. Flank protection is not used on NS, although common numbering of crossovers, where appropriate, does provide some security.

1.7.7.4 Overlap controls
There is a wide variation in the extent of controls provided in the overlap. These are shown in **Fig. 1.45**.

It can be seen that whilst approximately 200m is the target for DB AG, BR, CFL, NSB and BV, an overlap length of 100m is sufficient for SBB, SNCF and FS, whilst ÖBB, NMBS/SNCB and RENFE are content with 50m. Whether to lock points in the overlap also calls forth different practices, the railways being approximately equally divided in this respect. Of those who do provide locking, ÖBB, DB AG, BR and CFL extend it to flank points in the overlap, whilst BR and BV (at least) do not lock facing points if there is an alternative overlap available. NS do not provide overlaps, but exceptionally, where a signal is protecting a movable bridge or important junction, it is sited 200m in rear of the fouling point.

The very variable treatment accorded overlaps and their locking reflects the difficulty in assessing the risk of a signal being passed at danger – and the consequent hazard thereby created. At best an overlap can only mitigate the effect of a misjudgement or poor rail adhesion but it is significant that with the exception of NS, those administrations that have most experience in automatic train protection systems still retain overlaps.

Railway	Length of Overlap	Locking Provided	Flank	Comments
ÖBB	0m up to 40km/h 50m over 40km/h	Yes	Yes	
NMBS/SNCB	0m up to 40km/h 50m over 40km/h 100m for signals protecting junctions	No		Distance between signal and fouling point
SBB	40m up to 45km/h 100m over 45km/h (up to 165km/h)	Yes	No	On metre gauge: 30m up to 45km/h 60m over 45km/h (up to 95km/h)
DB AG	50m up to 40km/h 100m up to 60km/h 200m over 60km/h	Yes	Yes	With 30km/h, a shorter than 50m overlap allowed
RENFE	50m	Yes	Yes	
SNCF	0m	No		100m margin for trailing points
BR	46m up to 24km/h then increasing incrementally with speed to 183m over 96km/h	Yes	Yes	Facing points leading to an acceptable alternative overlap are free
FS	50m for platform starting signals 100m for other signals			Overlaps not generally used where length of block section much greater than length of train
CFL	0m up to 60km/h (<2.5% falling, or rising gradient) 50m up to 60km/h (>2.5% falling gradient) 100–200m above 60km/h	Yes	Yes	
NSB	0–400m	No		No overlap for block signals
NS	0m (200m for signals protecting junctions, movable bridges etc)	No		
BV	200m where possible 100m permissible up to 80km/h	Yes	No	Facing points are free. Generally no overlap for block signals

Fig. 1.45 Overlap controls

1.7.7.5 Preceding shunts

On most railways, a shunt signal facing a train proceeding under the authority of a running signal must itself be proved to be showing a proceed aspect. Whether such shunts carry the controls applicable to their part of the route, or whether it is all carried directly by the main signal depends on the practice adopted by the railway concerned and more precisely on the design of the interlocking system.

On NMBS/SNCB however, where the shunting signals display white lights for both on and off aspects, they do not show a proceed aspect when main routes are set over them.

1.7.7.6 Approach control of aspect

Notwithstanding the establishment of interlocking and controls for the signal to show a proceed aspect, there may be a requirement to delay the clearance of the signal until either the train is close to the signal and its speed has been reduced, or for a specific time.

(a) Delayed clearance until the train approaches the signal due to a speed restriction in the route ahead is necessary only in route signalling, since in speed signalling the necessary information is displayed to the driver by the signal aspect. On BR, signal aspects may be approach controlled due to:

- the presence of a speed restriction in the route over a diverging junction which is more than 16km/h less than the speed over the straight route from the signal;
- a shorter than normal overlap;
- the presence of a train in the section ahead (a calling-on route);
- a running signal having a shunt route set from it.

(b) Delayed clearance to enable an automatic level crossing sequence to operate, the level crossing being close beyond the signal. This control only operates if the train is approaching the signal at the time that the route is set, if there is insufficient time for the level crossing sequence to operate fully before its arrival at the crossing (NMBS/SNCB, BR, FS, NS, BV).

1.7.8 Aspect controls – shunting signals

The above principles apply to main or running signal routes. However, they must also apply in large measure to shunting routes, particularly where a mixture of main and shunt routes may be in use simultaneously at an interlocking. This is to avoid shunting routes conflicting with main routes or a derailment during a shunting movement obstructing other running lines.

Therefore, point setting and opposing locking apply to shunting routes. There are differences, however, in providing other controls (see **Fig. 1.46 overleaf**). Route setting may be suspended to allow shunting on the responsibility of the shunter (ÖBB, RENFE, FS, NS), points being set individually or moved by hand. Such local control inhibits all route setting in the area and on FS points may be configured to be trailable for shunting only. NS extinguishes the normal signal aspects, illuminating a white light in each signal as an indication that the main aspects have not failed and that local shunting is in operation.

On ÖBB, NMBS/SNCB and BR, where repeated forward and backwards shunting movements occur over a crossover, but clear of the running line, opposing shunting signals may sometimes be allowed to show proceed aspects at the same time. This feature is called Opposing Locking Omitted.

The BR practice is shown in **Fig. 1.47**. Signals X and Y may both show proceed at the same time for routes between the yard and the shunting neck. This is considered a better arrangement than signal X showing a yellow/white ON aspect and not providing a signal at Y.

Opposing shunting signals can also be simultaneously cleared at certain places on RENFE, but only with hand control of the points.

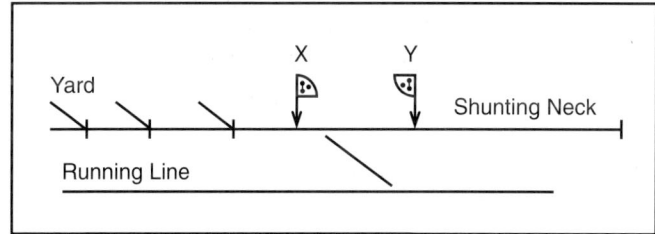

Fig. 1.47 Opposing locking omitted

Fig. 1.46 Shunting signal controls

Railway	Line of Route Tracks	Max. Speed	Flank Protection	Route Locking	Overlap	Trailable Points
ÖBB	No	40km/h	*1	No	No	Yes
NMBS/SNCB	Switching area only	40km/h	Yes	Yes	No	
SBB	Switching area only	40km/h	Yes	Yes	No	Yes
DB AG	No	25km/h	No	Yes	No	Yes
RENFE		25km/h	Yes	Yes	Yes	No
SNCF				Yes		
BR	Switching area only		Yes	Yes	Trailing Points <46m locked	No
FS	No	30km/h	No	Yes	No	Yes
CFL	No	25km/h	No	Yes	No	*2
NSB		40km/h				
NS*3	No	30km/h	No	Yes	No	
BV		30km/h	Yes	Yes	No	No

*1 For ÖBB, flank protection is provided in relay interlockings, but only exceptionally in electronic interlockings.
*2 On CFL, points are trailable, but this feature is not used in normal movements.
*3 NS do not use shunting signals. All shunting in signalled areas is done by normal route setting, making use of the Drive on Sight aspect if necessary (occupied track).

1.7.9 Aspect controls – warning signals

On all railways, where a combined main and warning signal is indicating stop, the warning signal aspect is extinguished, even where a separate backboard is provided. When the main signal shows a proceed aspect, those railways that use separate backboards for warning signals (ÖBB, SBB (System L), DB AG (System HV), CFL, NSB) allow it to show its own aspect in addition to that of the former. Where the warning is integrated with that of the main signal, the warning aspect effectively takes over the signal and shows caution or clear as appropriate. Isolated warning signals are controlled by the aspect displayed by the main signals in advance; some railways replace the signal to its most restrictive aspect as soon as it is passed by a train, others do not.

Most railways use combinations of yellow and green for warning signal aspects (BV use white and green). NSB and BV distinguish warning aspects by flashing them. Most other railways flash aspects in warning signals; they carry a variety of meanings and are discussed fully in **1.5**.

1.7.10 Aspect controls – other signals

1.7.10.1 Repeating signals
Repeating signals are provided to indicate to the driver the aspect of a signal of which he has reduced sighting, due perhaps, to curvature of the line, an overbridge or a tunnel. They may be provided instead of, or in addition to, countdown markers. They are controlled directly by the aspect of the signal which they repeat.

On DB AG and CFL the normal warning signal aspects are used, the addition of a small white light denoting that the signal is at less than braking distance from its associated stop signal.

1.7.10.2 Protection signals
Sometimes a depot or marshalling yard is provided with a full signal only at the yard exit. Protection signals are provided for each siding to enable the yard supervisor to indicate to drivers from which siding a move is to be made towards the yard exit signal. Points may be trailable and controls to prevent conflicting and opposite routes are imposed. ÖBB, NMBS/SNCB, DB AG and, in some cases, BR, use such signals.

ÖBB and DB AG also use protection signals to indicate that a train should stop short of the next main signal, for example in the middle of a platform.

1.7.10.3 Substitution signals
When, due to failure or other cause it is not possible to signal trains with the normal proceed aspects, a substitution signal may be used. This signal carries minimal controls – to enable its use in as wide a spectrum of failure conditions as possible – and authorizes movement at slow speed only; drivers to be able to stop on sight of an obstruction. (On DB AG (System HV), this restriction applies at the *Vorsicht* signal – 3 yellow lights, but not at the *Ersatz* signal – 3 white lights.)

SBB enable a main aspect to be displayed after:
- the route has been set; and
- a special emergency button (appropriate to the failure) has been pressed, together with the route entrance button (track failures), or exit button (block failures) or barrier track button (barrier failures).

BR, CFL and NS do not use substitution signals.

1.7.10.4 Platform Duties signals
Platform Duties signals comprise, on various railways, Brake Test (ÖBB, FS, DB AG), Close Doors (BR) and Right Away (most). They are operated by the supervisor in charge of the platform and are free of signalling controls except for the Right Away indicator which requires the platform starting signal to be showing a proceed aspect. In many instances the aspect of the latter signal is itself repeated on the platform – as in the OFF indicator of DB AG and BR.

1.7.11 Aspect replacement
Replacement of the aspect of a stop signal to danger on passage of a train usually occurs when the front of the train is a few metres past the signal post. For example:

- ÖBB : minimum 25m
- NMBS/SNCB : between 6 and 12m
- RENFE : between 0 and 6m
- DB AG : 50m
- BR : between 5 and 20m
- FS : between 20 and 38m
- CFL : 50m
- NS : between 9 and 15m

It is of course, desirable to locate a track circuit joint as near as possible beyond the post to register the train's presence in the route; this changes the status of the route from being approach locked to being route locked.

Where there is need for the joint to be so close to the signal that the minimum distance cannot be achieved, the proceed aspect may be maintained until the second track circuit in the route becomes occupied. Similarly, where trains are assisted by a locomotive in rear, or are propelled by a locomotive being driven from the rear, the proceed aspect may be maintained until the whole of the train has cleared the approach track circuit.

1.7.12 Route release
The release of a route that needs to be cancelled without the passage of a train was dealt with in **1.7.6**. This section considers how the route is released after the passage of a train. It may be sufficient to carry all the interlocking with conflicting routes solely on the route lock relay, no other form of route holding being provided. This route lock relay must not then be permitted to return to Normal, permitting the setting

of conflicting or opposing routes, until the whole train has traversed the whole of the route (NMBS/SNCB, and RENFE simple layouts), or at least until the front of the train has reached the track circuit controlling the last set of points in the route. It may, in addition, be required to operate a treadle or rail circuit located near the exit of the route.

At busy complicated layouts with slow moving trains, for example at large station throats, it is generally necessary to release the route section by section after clearance by the rear of the train. This is termed sectional release route locking. It is the method generally adopted by all administrations except NMBS/SNCB.

1.7.12.1 Release of approach locking

The route has been approach locked, either as soon as the signal cleared, or when the train commenced to occupy track circuits nearer than the sighting point of the outermost signal, the aspect of which would be affected by the signal returning to danger.

With reference to **Fig. 1.48**, release of approach locking occurs when the train has entered the route (B to C) and signal B has been proved to have returned to danger. Some railways require the first track circuit beyond the signal occupied and cleared, plus a treadle operated (NMBS/SNCB, SNCF) or the second track circuit occupied (BR). Other combinations are possible – for example berth track occupied and cleared, first track occupied; first and second tracks occupied (NS). The objective in choosing particular combinations is generally to avoid a false release of approach locking – for instance by a momentary power failure.

Where sectional release route locking is not used, the release of approach locking is stored and does not become effective until the train has reached the end of the route, or in the absence of any direct opposing routes, has reached the last set of points in the route.

1.7.12.2 Train operated route release

Where automatic route setting systems are in use, it is also necessary automatically to release the route. This may be achieved merely by relying on the approach locking release conditions being satisfied. However, it is sometimes considered that an additional check should replace the signalman's manual cancellation of the route and this can be achieved by a separate control which may either:

- check that the approach tracks to the signal are clear; or
- require an additional track circuit beyond the signal to be occupied.

In either case the object is to ensure that the train has entered the route before it is released, and that a false release has not occurred.

1.7.12.3 Sectional release route locking

On the basis that the conditions for releasing the approach locking were (first track occupied and cleared, second track occupied), when this release becomes effective, that is, where the train is shown in **Fig. 1.48**, the controls on the points in the route will be as follows:

- points in 1st track (W)
  – locking released
- points in 2nd track (V, X)
  – locked by the presence of the train on that track
- points in track beyond 2nd track (Y)
  – locked by route holding
- points in the overlap (Z)
  – locked by overlap locking
- routes from signal S
  – locked by route holding
- routes from signal T
  – locked by route holding and overlap locking

Clearance of the train from the 2nd track will release the direct locking on points V and X and also the opposing locking on routes from signal S. Points Y will have become direct track locked when the front of the train occupied the 3rd track.

As the train proceeds through the route, clearance of each track circuit releases locking in rear of the train. The actual logic which performs the release may be merely rear track clear (NS), or rear track clear for n seconds, or rear track clear plus advance track occupied (NS – electronic interlockings only) or rear and advance tracks occupied simultaneously, then rear track clear (DB AG and CFL). The more complex requirements afford protection against a momentary track false clearance due, for instance, to a poor train shunt. For new works BR use a combination (rear track clear for 15 seconds or advance track occupied) for the same reason.

Flank locking is released at the same time as the associated points in the route – flank points P would be released at the same time as route points Y.

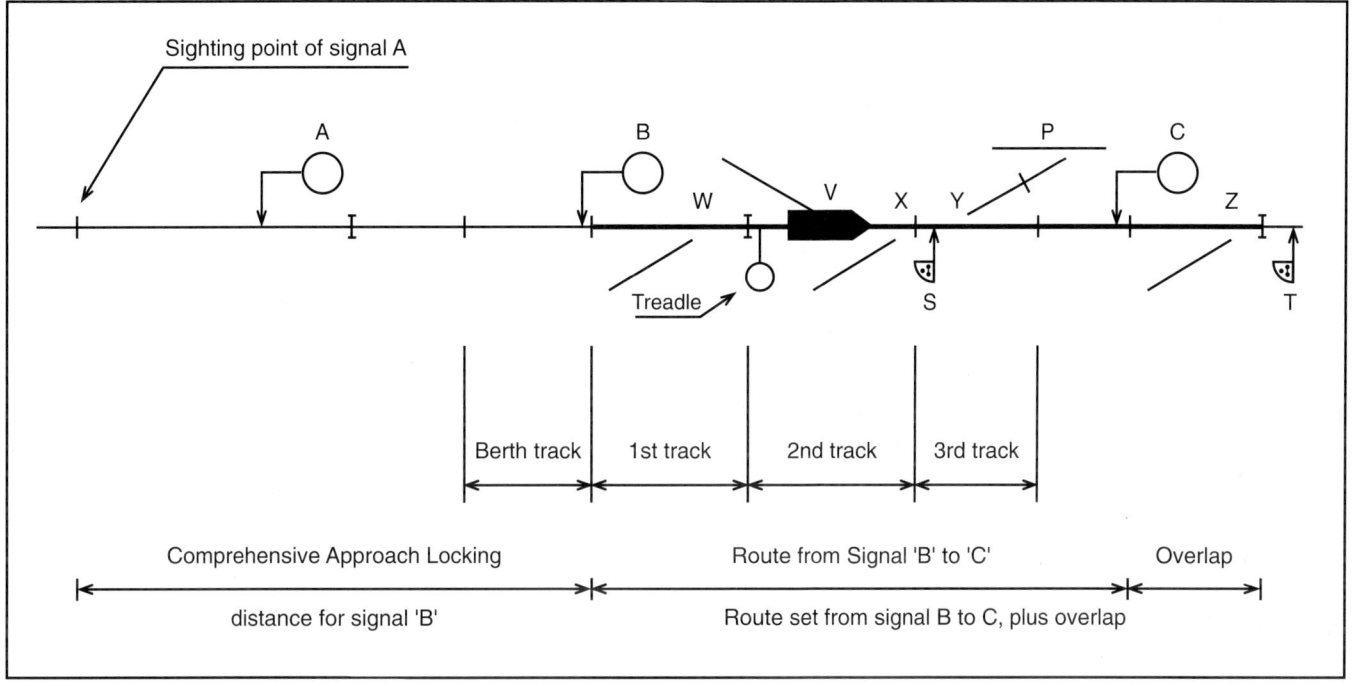

Fig. 1.48 Sectional release of route locking

When a train reverses direction, the remainder of the route locking of the original route is released upon proof that a countermove route (that is, a route in the opposite direction), has been set and the train has commenced to move in the opposite direction. A train from B to C coming to a stand behind shunt signal S to reverse would typically allow release of the route locking towards C when:

- the route has been set from S
- the track circuit in advance of S has been occupied and
- the track circuit in rear of S has cleared

1.7.12.4 Release of overlap locking
On those railways which use locked overlaps, the locking can be released by:

- the train coming to a stand at the signal ahead (C) and the expiry of a time delay after the berth track (to C) has been occupied;
- the train proceeding past C, when the normal rules for sectional release route locking will apply.

However, reliance must not be placed on the locking applicable to the route ahead to secure the overlap, because the train may overrun the signal. The overlap locking must, therefore, be freestanding.

1.7.12.5 Failure of route to release
Some railways provide a facility for the signalman to release manually a route or part of a route which remains locked after the passage of a train. This is strictly controlled and monitored (mechanical or electronic counter incremented by release). NMBS/SNCB also use a microcomputer to register such releases, but on BR, however, the signalman has no such facility and any release must be given by the signal technician.

1.7.13 Point operation
Apart from being set locked and detected by the route setting process, points are able to be set individually by key/button operation from the control panel and sometimes locally for shunting purposes. For a set of points to be moved for route setting, the following conditions must apply:

- track circuits, including foul/flank conditions, which directly lock the points must be clear. (NMBS/SNCB store the 'track clear' condition when the point locking is released by the cancellation of the previous route. This stored condition is then used to control individual point operation.);
- all routes which lock the points in their present position must be normal;

- all route holding which locks the points in their present position must be free;
- point to point locking which locks the points in the present position must be free;
- the signalman's individual point key/button control must not be locking the points in their present position.

Individual point operation by the signalman – for instance, to permit unsignalled moves, must effectively lock the points – even if they are operated from a remote interlocking many kilometres distant. It follows that remote control systems must be designed with sufficient integrity or redundancy to enable this to be reliably achieved.

1.7.14 Level crossing controls

Chapter 12 deals with the generality of level crossings; this section deals only with the principles of interlocking of railway signals with level crossing equipment.

1.7.14.1 Controlled level crossings

Barriers with manual control and extending the full width of the road are located only on the busiest level crossings; the barriers are proved lowered with road traffic signals, where provided, operating before the protecting rail signals are allowed to show a proceed aspect. Release of the barriers subsequently, on passage of the train or cancellation of the route, follows the same principles as for points (see **1.7.12**). For the normal passage of trains, this can be performed automatically. On BR when the protecting signal is passed at danger by a train, an overrun control, either a treadle or a track circuit, causes the road traffic signals to operate but the barriers remain raised to avoid trapping road traffic.

On ÖBB, DB AG and CFL, controlled barriers are called as part of the route setting procedure; BR generally require the signalman or crossing keeper to observe the lowering sequence, to avoid trapping vehicles or pedestrians.

Controlled barriers can be monitored remotely, either by direct vision – up to 200m – or at any distance by closed circuit television. After observing on his monitor that the barriers have correctly lowered and that nothing/nobody is trapped between them, signalmen on BR and FS are required to perform the deliberate act of operating a Crossing Clear button before the protecting signals can display a proceed aspect.

ÖBB do not close level crossings situated in overlaps and this is the present practice on BR. However the increasing use of relatively light multiple unit trains and the advent of heavier road vehicles, including fuel tankers, is forcing a reconsideration of this policy.

1.7.14.2 Automatic level crossings

Automatic level crossings may have full width barriers, half barriers or no barriers at all. However, all level crossings in this category have standard road traffic signals, flashing red (wig-wags) or an equivalent. Where half barriers or no barriers are provided, the principle of keeping the road closed for the minimum time applies; this generally precludes the use of interlocked railway signals. Where rail signals are located close to the crossing, it is necessary to inhibit operation of the latter unless the route is set. Should a train be closely approaching the signal when it is ready to clear, it may be necessary to delay giving a proceed aspect until the level crossing operating sequence is sufficiently far advanced to avoid the train arriving at the crossing prematurely.

On DB AG and BR, certain automatic level crossings are arranged for local monitoring by the train driver. These can be either half barrier crossings (ABCL) or open level crossings (AOCL). In each case the level crossing controls work normally as far as the road user is concerned, but the train driver is assured that the crossing is operating correctly by observing a rail signal at the crossing, which on BR normally shows a flashing red light, changing to flashing white. He must then check that the crossing is clear before proceeding over the crossing. The maximum speed allowed for trains over locally monitored crossings is 90km/h and in many cases it is much lower, depending on intervisibility and level of road and rail traffic.

Since such crossings are often located in rural areas with an uncertain power supply, they are arranged to time-out if activated without the passage of a train after three or four minutes. The driver's white light is returned to red and 30 seconds later the road traffic lights are extinguished. The next train approaching the crossing may then not receive the white light and will, after stopping short of the crossing and ascertaining that it is clear, proceed across the road.

BV have level crossings where the barriers and road traffic signals are monitored by a protecting main signal – if one is near enough – or by a red/white driver's light (Vs) at the crossing. This is preceded by a distant signal at not less than 300m (Vf) but is intended to be seen from a fixed board, Vo (= orientation) which is positioned at braking distance from the crossing. When the crossing is not proved closed to the road, Vs shows a red aspect and Vf a triangular (v shaped) aspect of three flashing yellow lights. Proof that the crossing equipment is correctly functioning enables Vs to change to a white aspect and the triangle of yellow lights in Vf to change from flashing to steady. The arrangements are shown in **Fig. 1.43**.

At line speeds of above 140km/h, all level crossings on BV are protected by a special ATC function (see Chapter 12).

1.7.14.3 Limitations on the use of level crossings

Most railways have limitations on the use of level crossings, either due to speed of rail traffic or density of road traffic. For instance automatic level crossings are not used on BR where the line speed is over 160km/h; as has been mentioned a lower limit of 90km/h applies to locally monitored crossings. Many railways do not permit level crossings at all on lines of over 160km/h; on SNCF, BR and FS the limit is 200km/h.

Automatic level crossings must have no more than two running lines (BR); on NSB, level crossings are not allowed at all except on single lines.

1.7.15 Bidirectional and single line controls

For single line sections with no intermediate signals for flighting purposes, the only controls required are to inhibit entrance of a train after permission has been given for one to enter in the opposite direction until it has either been cancelled or has passed through the section, complete. Where intermediate signals are provided the controls for following trains must be superimposed on the opposing locking for the single line. The controls become even more complex when a train is required to reverse at some point in the single line section and return to its entrance – under signal control – particularly when cab signalling/ATP is involved, as in the case of the Channel Tunnel.

Other complications involve single or bidirectional control of level crossing equipment – including the need to provide protection to ensure that track circuit maloperation does not fool the control logic into believing that an approaching train is one which has already passed over the crossing in the opposite direction and not yet cleared the controlling track circuits. On BR (most) and on FS (all) automatic crossings are provided with bidirectional controls to avoid having to provide a local operator during single line working for engineering maintenance.

Many station tracks are equipped for both ways working, some with intermediate signals, or calling-on facilities to admit a second train from the opposite direction. The first train must then 'time out' by track circuit operation of the appropriate platform tracks. In the absence of mid-platform signals this must be combined with an instruction to drivers that, once having brought their train to a stand, they must not move until properly authorized. Similar rules must apply to the drivers of locomotives remaining at the buffer stops of terminal stations after their incoming train has departed since the signalman will be able to clear a calling-on signal for a second train to enter the platform.

1.7.16 Remote control of interlockings

With the increasing centralization of signalling control, most relay interlockings have become remote, control being exercised by the signalman via some form of data transmission system. Although there are examples of a fail-safe data transmission system being used, enabling the interlocking to be at the control centre (for example vital 'Reed' and SSI on BR), most installations consist of local interlockings and a non-vital data transmission system.

Such a system is entirely adequate for transmission of controls from the signal control panel and of indications from the interlocking back to it, since the integrity of the fail-safe signalling system lies within the interlocking and onwards to the signals, points, track circuits etc, at the trackside. However, when, for any reason, a route cannot be set and the signalman has to instruct a driver to pass a signal at danger, the remote control and indication system at once becomes part of the vital chain of command.

The signalman then has to set points individually from his panel, and in the absence of a route they will not be otherwise locked until the train eventually occupies the deadlocking track. He also has to observe and trust the indications relayed to him by the remote indication system in respect of point detection and track circuit occupancy.

It is also necessary, from time to time, for the signalman to hold a signal at danger, for example to

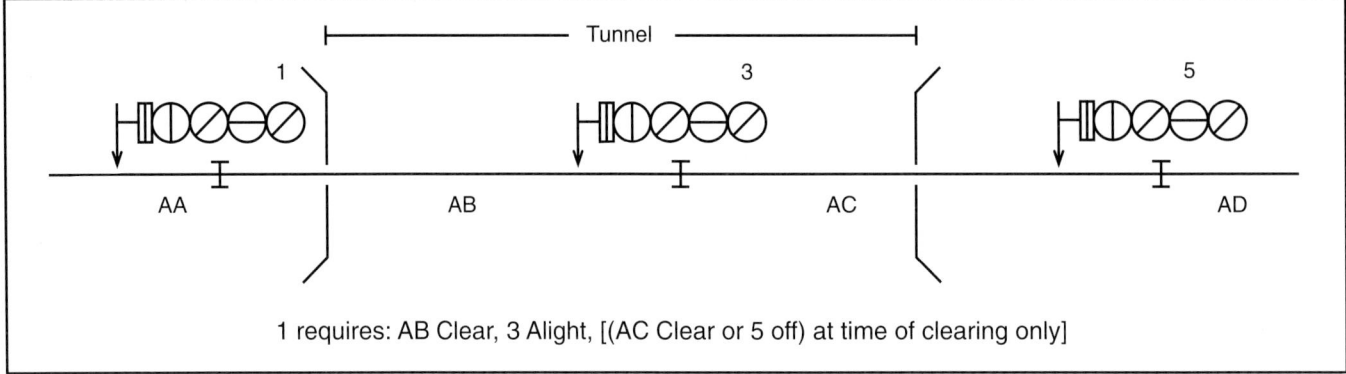

Fig. 1.49 Tunnel controls (BR)

protect staff working on the track or whilst a road vehicle with an exceptional load traverses an automatic level crossing. The remote control system must be robust enough not to cause false operation of the signal under these circumstances. For example, NMBS/SNCB use systems which, although constructed of non-vital components, are designed so as to exhibit 'virtually fail-safe' characteristics, whilst on BR, certain critical functions are duplicated by channels in different addresses.

In the absence of such robustness the operators lose confidence in the remote control system and resort to operating methods which have a restrictive effect on train operation and tend to bring the whole signalling system into disrepute.

1.7.17 Special controls

1.7.17.1 Tunnel controls

On S-bhan lines many signals are in tunnel and it is necessary to legislate for trains being detained thereat, so as to maintain the required frequency of service. However, on main lines it is generally considered unadvisable for trains to be detained in tunnels for any significant period. Until recently therefore, the practice has been to avoid placing stop signals in tunnels wherever possible, at the expense of creating a long block section and consequent lengthening of the ruling headway.

A solution adopted by BR for several long tunnels, principally located on outer suburban routes from London, has been to install automatic stop signals in the tunnel. However, controls are provided to prevent clearance of the signal at the entrance to the tunnel, when the tunnel signal is held at danger by a train in advance, unless the first stop signal beyond the tunnel is showing a proceed aspect for the latter. This is illustrated in **Fig. 1.49**. Under these circumstances, the train in front will be able to leave the tunnel, pass the signal in advance and thus enable the second train to obtain a proceed aspect on the tunnel signal. With successive trains running at line speed under clear signals, there will be no effect on headway, assuming that the signals are evenly spaced. With four aspect signalling, two consecutive tunnel signals can be accommodated without affecting headway.

1.7.17.2 Freight train controls

NS have developed special controls for freight trains, to avoid stopping them on rising gradients and also to prohibit entry to tunnels under rivers – with a consequent rising gradient near the exit – unless the line is clear throughout. Since the signalling system cannot by itself distinguish between freight and passenger trains, the controls are in the form of indicators associated with the appropriate stop signals.

For a rising gradient on which a stop signal is located, for example on the approach to a bridge, the signal in rear is equipped with an indicator – of the same type as the speed indicator – which will show H when that signal is showing a proceed aspect with the gradient signal at danger. The signal in rear of the signal showing the H will itself display an indication L. The L and H indicators apply to freight trains only and mean warning and stop respectively (see **Fig. 1.50**).

For the tunnel situation, as well as freight trains requiring a clear path, a following passenger train must not be allowed to enter the tunnel until the freight train has passed clear. The signalman is, in this instance, required to tell the interlocking whether he is setting the route for a freight or passenger train. For a freight train, the entrance signal will only show a proceed aspect when all intermediate signals in the tunnel have already cleared. The tunnel entrance signal will show a G indication. The freight train is only allowed to enter the tunnel at the tunnel entry speed of less than 40km/h. This speed is such that a train can travel down the falling gradient without braking and that the speed at the bottom will not exceed the maximum line speed. This is to prevent emergency braking being imposed by the ATP.

For a following passenger train, the tunnel entrance signal is only allowed to clear if all track circuits through the tunnel are clear. The indicator will show an X.

For a passenger train following another passenger train, the normal aspect sequence will apply, and the X will again be shown. The signal in rear of the tunnel entrance signal will show a flashing X when the latter is showing proceed aspect, unless the entrance signal is showing G when G will also be

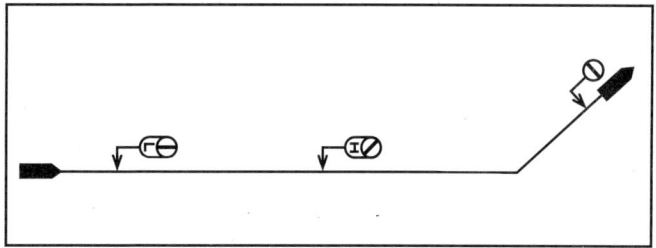

Fig. 1.50 Freight train controls – rising gradient (NS)

shown. The flashing X and X indications, if seen by a freight train, mean warning and stop respectively – because the signalman has made a wrong choice! (See **Fig. 1.51**.)

On NMBS/SNCB, similar controls are generated automatically by the EBP system without any action on the part of the signalman or the need for special indications on the signals concerned.

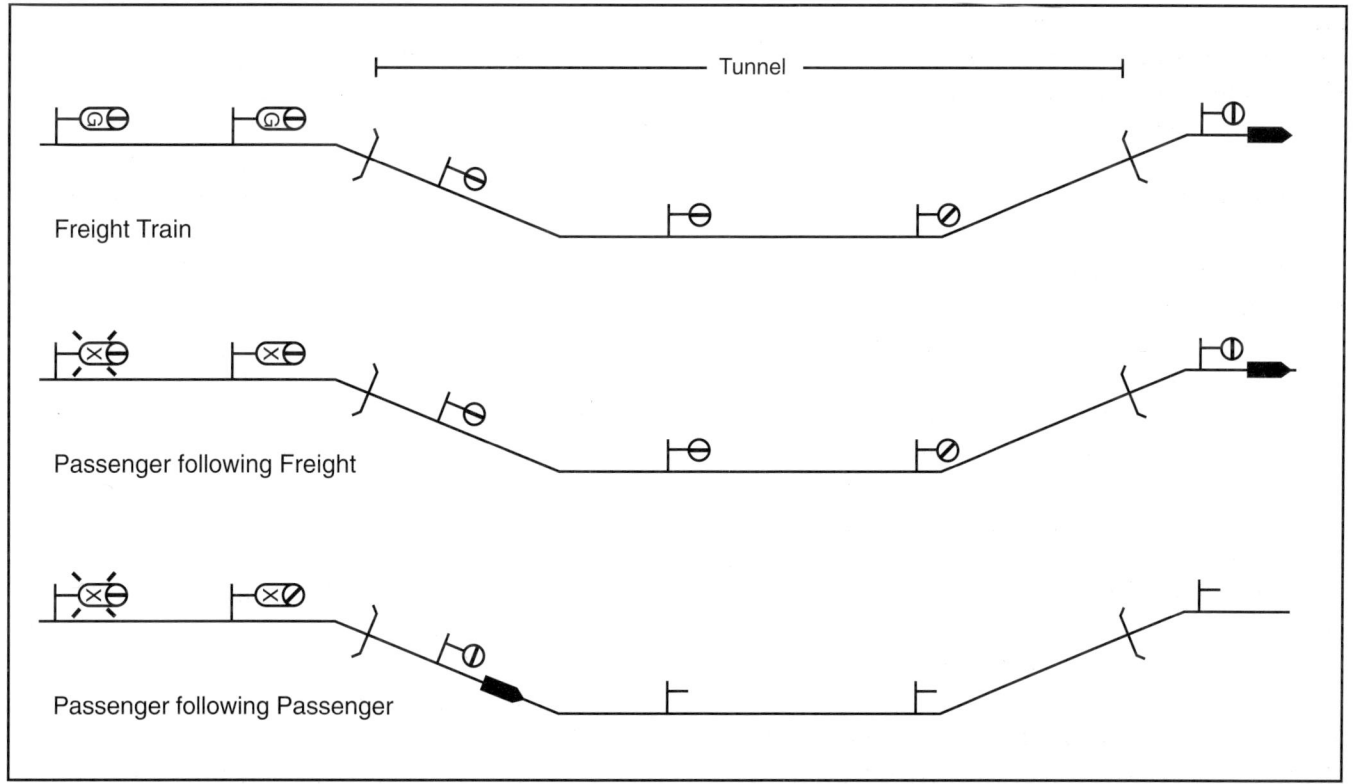

Fig. 1.51 Freight train tunnel controls (NS)

1.7.18 Controls for staff protection systems

The safety of staff working on the line, together with the advent of higher line speeds – particularly over junctions – has led to the development of systems which are linked to the signalling. These systems can take track occupancy and signal aspects as their primary source of data or may themselves control routes and aspects. They may be of the following types:

1.7.18.1 Trackside warning systems

Track occupancy and signal aspect information is processed and arranged to operate either lineside klaxons or personalized bleepers. The object is to give a minimum warning time (25 seconds on BR) sufficient for staff to cease work and move themselves and their tools clear of the line.

Similar warnings may be given by trackside lights which may normally show a steady white aspect, commencing to flash at the approach of a train (NS) or be extinguished (BR).

Comprehensive staff warning systems for complex station or junction layouts rapidly become extremely complicated – more so than the signalling system itself! There is also the problem of a reduced or zero warning being given for unsignalled moves, and when drivers pass signals at danger.

ÖBB operate a system whereby staff working on the track are warned of the setting of a route or request to clear a section signal. They then have to stand clear of the line and return a control called a 'quit signal' to the interlocking before the signal concerned is allowed to show a proceed aspect. The warning and return control messages are transmitted via the lineside telecomms circuits (see Chapter 13).

1.7.18.2 Patrolman's lockout system

Controls may be provided to enable a track patrolman to lockout certain routes, with the agreement or permission of the signalman, to enable track inspection to take place. The control may inhibit contraflow routes on bidirectionally signalled lines or all routes on a particular track. Applications may include lines on viaducts or tunnels where there is insufficient room to stand clear, or on bidirectional lines to avoid being approached from behind.

BV arrange for signals to be manually locked in the stop position to protect a worksite. BR provide a switch, located on the signal post of automatic signals only, for similar reasons. At some locations on FS, special routes may be set and locked for occupation by engineering vehicles. Confirmation that these routes have been set is given by a flashing white letter C on the post of appropriate main signals. Cancellation of the routes by the signalman requires the operation of a device by the maintenance staff.

1.7.18.3 Staff working on trains

Staff working on, or under trains, for example whilst stationary at a platform, may be afforded protection by the release of a key or other token electrically by the signalman. This then inhibits routes into or from the platform, thus ensuring that a second train cannot make contact with the first, neither can the latter be signalled away. This is the modern form of the red flag or lamp attached to the outside of the train whilst work is in progress.

2 Train Detection Systems
F. DE VILDER

Contents

2.1	General	122
2.2	Track circuits	122
	2.2.1 Survey of the European situation	122
	2.2.2 Direct current track circuits	127
	2.2.3 Alternating current track circuits	128
	2.2.4 Track circuits with high voltage impulses (HVI–TCs)	132
	2.2.5 Short jointless track circuits	132
	2.2.6 Audio frequency jointless track circuits	133
2.3	Axle counters	136
	2.3.1 Survey of the European situation	136
	2.3.2 Axle counters in their traditional countries	136
2.4	Treadles	138
	2.4.1 General	138
	2.4.2 Survey of the European situation	138
2.5	Vehicle sensors	141
	2.5.1 General	141
	2.5.2 Survey of the European situation	141
2.6	Miscellaneous	142
	2.6.1 Magnetic relay	142
	2.6.2 Tail detectors	142

Illustrations

Fig. 2.1	Types of track circuits used throughout Europe	124
Fig. 2.2	Basic diagram: DC single track circuit	127
Fig. 2.3	Ballast resistance and drop shunt values	128
Fig. 2.4	AC track circuit principle	129
Fig. 2.5	Impedance bond – BR Southern Network	129
Fig. 2.6	Characteristics of the AC track circuits throughout Europe	130
Fig. 2.7	HVI–TC data	132
Fig. 2.8	Standard AF–TC with S-bond joints	134
Fig. 2.9	AF–TCs – BR	135
Fig. 2.10	Axle counter AZL70 type	137
Fig. 2.11	Simplified outline of electronic axle counters	137
Fig. 2.12	Survey of the degree of application of axle counters in Europe	137
Fig. 2.13	Mechanical treadle: two-way working track Forfex type	138
Fig. 2.14	Characteristics of treadles throughout Europe	139
Fig. 2.15	Characteristics of vehicle sensors throughout Europe	141

2.1 General

It is necessary to detect the presence of trains on the track for several reasons. Signalling personnel speak about track free detection, which has dual status:

- status A means track free; and
- status B means track occupied or failure in the technical system of track free detection.

Maintaining the headway of trains, locking the points and warning of trains' arrival at level crossings are the most commonly used applications. All these functions have to be ensured in a fail-safe manner at all times.

Subsequently, other applications of track free detection were developed, and here too it was important to know whether or not trains were present at a particular location: for example, to trigger the train describer; to activate the passenger information displays on the platforms; or to have an indication of the train length. These applications, however, are less safety related than those mentioned above.

When we approach the question of train detection in a philosophical manner, we will discover two different needs, whether we want to detect the presence of a train on a specified track, or whether we want to detect the passing by of a train at a specified point. The technological means to implement the first need are track circuits and axle counters, whilst treadles and induction loops achieve the second.

2.2 Track circuits

2.2.1 Survey of the European situation

Track circuits were invented a century ago in the United States and afterwards used throughout the world. The first track circuits used by the European railway companies were of the direct current (DC) type. This was because at that time, the only available reliable power supply was provided by dry and wet batteries. Indeed, the local mains supply was neither secure nor guaranteed. The electricity supply companies had problems with their own power plants and high voltage interconnection lines were not then in existence.

Once the local mains supplies became more widespread (approximately in the 1920s), the alternating current (AC) track circuit was developed. It offered the advantage of a very cheap emitter – the transformer. The first track circuits of this type were used as an electrically insulated rail.

Subsequently true track circuits were installed, not only of the 50Hz types but also of other low frequency types.

All these track circuits were delimitated by insulated joints. So when later, in the 1930s, electric traction became popular, the insulated joints created the first conflict between the weak currents of the signalling equipment and the heavy currents of the electric traction. To circumvent the insulated joints and to filter the weak currents out of the heavy currents, impedance bonds were invented.

As the years passed and semiconductor technology became available, new, more complicated, types of track circuits were invented, such as audio frequency, coded, non-coded and even pulsed track circuits. All these enhancements aimed at an improved electrical contact between track and wheel, together with increased protection against return currents. The latter became more and more complex as new types of rolling stock appeared on the market, such as choppers in the 1970s. Moreover, chopper locomotives self-generated low frequency currents.

To correct this problem, filters were introduced on board locomotives. Because the safety of traffic is directly involved, a detector guards the on board filters and cuts the traction current whenever a filter fails.

During the 1980s locomotives were equipped with a three phase drive traction control. These locomotives also generated large currents, sufficient to disturb the track circuits. Even coaches started causing problems when electromagnetic rail brakes were introduced. In the near future, eddy current brakes will be installed on high speed train sets. This will result in a never-ending conflict between the weak currents of the signalling equipment and the heavy currents of the traction and braking.

On high speed lines insulated joints cause maintenance problems and also interfere with passengers' comfort. So the newly available jointless track circuits found a vast application area.

For some special situations, general solutions were applied. For example, because of the disturbances produced by three phase overhead electric power lines, especially in station areas, RENFE replaced 50Hz AC track circuits by low frequency pulses track circuits, between 1985 and 1992. A second example are immunized track circuits being used at interchange stations with different catenary voltages.

For other special situations, however, specific solutions were invented, such as in the northern parts of Scandinavia where earth magnetic effects caused problems on long DC track circuits. NSB and BV avoid these problems by using centre fed track circuits.

Another remarkable item is the track circuit interrupter, as used by BR. At the outlet from sidings or loop lines where catch points exist to effect the throw off of vehicles that *overrun* the outlet signal, a track circuit interrupter is provided. This is a mechanical device fitted to the web of the rail that is ruptured by the passing of a wheel flange. The rupture causes permanent occupation of the relevant track circuit in the vicinity where runaway vehicles may be derailed, by the device forming part of the continuity of that track circuit.

Also from the conceptual point of view, there are unusual application notes for track circuits. Some railway administrations do not rely on them to control a level crossing, for example SBB add 80kHz track circuits or inductive treadles, as a second element in vital applications, whilst BR use treadles as additional train detection devices in association with certain types of level crossing equipment to ensure the correct operation in relation to the approach and passing of trains. Furthermore, ÖBB use an inductive loop or a treadle as second element to control a level crossing.

Some administrations use track circuits for automatic train protection (ATP) and cab signalling purposes. Examples of this are: SNCF with a coded UM71 track circuit, which they call TVM 300 or TVM 430; NS with a coded 75Hz track circuit; and FS track circuits with one or two coded carrier frequencies (50Hz and 178Hz).

Overall, the track circuit has evolved from a rather simple type of equipment to a highly sophisticated one. Consequently, the price has spiralled, and railway administrations tend only to modify their track circuits where and when it is absolutely necessary. In addition to financial considerations, the geographic dispersion of the different catenary tensions causes further difficulties and it is easy to see why there are so many different types of track circuit in use on European railways.

Although there is great diversity in the different administrations, there are some common features. One is the use of single rail versus double rail insulation for electrified lines. Electrified lines use the running rails as a traction current return and for the implementation of safety bonding (high voltage systems). It is easier to use just one rail per track for these purposes. Unfortunately this produces an unbalanced circuitry which maximizes interference to the equipment. It also imposes length restrictions. If both running rails are used for traction return, the disadvantages noted above are avoided and there is the possibility of implementing jointless track circuits, with the elimination of insulated rail joints.

However, double rail traction current is more difficult through complex points and crossings requiring more connection bonds and in such situations a single rail traction return is often used.

Another important consideration is in the application to the detection of broken rails. On the one hand, UIC, question A174 says that it is not compulsory to detect broken rails. Conversely there are the different railway administrations, each with their own history and philosophy:

- *ÖBB, SBB, DB AG, NS*: A welcome fringe benefit;
- *NMBS/SNCB*: We wish to detect broken rails;
- *RENFE, SNCF*: A safety factor much appreciated. We insist that a track circuit identifies broken rails under all circumstances;
- *BR*: Not a mandatory requirement;
- *FS*: We try to detect broken rails. This can be achieved only when the broken rail causes an electric break of the track circuit;
- *CFL, NSB*: AC–TC gives the thrown in benefit of broken rail detection;
- *BV*: The absence of broken rail detection has not caused any great problem. We had, during more than 60 years, only a few minor accidents and we think a functioning system is of more value. The most dangerous situation is not a clean break but a rail that has lost part of its head and that can't be detected by a track circuit.

In future, DB AG will detect broken rails on their high speed lines. In order to do this, DB AG is to develop a separate 'broken rail detection equipment' which is to be used on lines equipped with track circuits as well as on lines with axle counters.

Fig. 2.1 illustrates the results of a survey of all the different types of track circuits in use throughout Europe. At first glance it looks like a patchwork, but gradually five main sets can be discerned:

1. direct current track circuits
2. alternating current track circuits
3. high voltage impulses track circuits
4. short jointless track circuits
5. audio frequency jointless track circuits

Fig. 2.1 Types of track circuits used throughout Europe

Railway	Types used	Description section	Number of items installed on the network	Suited for electrified lines	Insulation of the track	Shape of signal in the track
ÖBB	AC–TC with electronic track relay	2.2.3	6000	15kV 16⅔Hz	single and double rail	sinusoïdal 100Hz or 50Hz or 106⅔Hz
	AC–TC with motor track relay	2.2.3	10 000	15kV 16⅔Hz	single and double rail	sinusoïdal 100Hz
NMBS/ SNCB	AC–TC 50Hz	2.2.3	12 700	3kV DC	single and double rail	sinusoïdal 50Hz
	Short jointless TC	2.2.5	2750	3kV DC	jointless	pulsed sinus base freq. 15kHz repeat freq. 85kHz
	TC with high voltage pulses	2.2.4	1500	25kV DC / 3kV DC	double rail	impulses 100V 1 or 2ms repeat freq. 3 or 4Hz
	DC–TC	–	160[1]	diesel	single and double rail	DC
SBB	DC–TC without converter	–	10 000	15kV 16⅔Hz	single rail	DC
	DC–TC with converter	2.2.2	≈1000			
	AC–TC 125Hz	2.2.3	≈1000	15kV 16⅔Hz	single and double rail	sinus 125 Hz
	Short jointless TC Jointless voice frequency TC (FTGS)	2.2.5 2.2.6	≈5000 ≈100	15kV 16⅔Hz 15kV 16⅔Hz	jointless jointless	sinus 80kHz audio freq. 4kHz to 17kHz coded FSK
	AC–TC	2.2.3	≈2000	15kV 16⅔Hz	single rail	sinus 42 or 100Hz
DB AG	AC–TC	2.2.3	3000	15kV 16⅔Hz	single and double rail	sinus 42 or 50 or 100Hz
	Jointless voice frequency TC (FTGS)	2.2.6	1500	15kV 16⅔Hz	jointless	audio freq. 4kHz to 17kHz coded FSK
	DC–TC	–	none	–	–	DC

TRAIN DETECTION SYSTEMS

Railway	Types used	Description section	Number of items installed on the network	Suited for electrified lines	Insulation of the track	Shape of signal in the track
RENFE	DC–TC	–	1	diesel	double rail	DC
	Jointless voice frequency TC	–	116	3kV DC	jointless	Vpp = 20V Sin. 15–19kHz Repeat 380ms
	Phase relay AC–TC	2.2.3	11 200	3kV DC	single and double rail	sinus 50Hz
	TC with high voltage pulses	2.2.4	1976	3kV DC	double rail	100V–3ms repeat freq. 3Hz
	TC with low frequency pulses	–	1	3kV DC	jointless	7V—0.3ms repeat freq. 800Hz
	Jointless voice frequency TC (FTGS)	2.2.6	560 1040	3kV DC 25kV 50Hz	jointless	audio freq. 4kHz to 17kHz coded FSK
SNCF	DC–TC	2.2.2	3000	25kV 50Hz	single and double rail	DC
	AC–TC	2.2.3	6000[1]	25kV 50Hz or 1.5kV DC	single and double rail	50Hz or 83Hz
	Short jointless TCs	2.2.5	11 500	25kV 50Hz or 1.5kV DC	jointless	sinus 8700Hz
	Audio jointless TC (UM71)	2.2.6	20 000	25kV 50Hz or 1.5kV DC	jointless	audio freq. 1.7kHz to 2.6kHz coded FSK
	High voltage impulse TCs	2.2.4	38 000	25kV 50Hz or 1.5kV DC	single and double rail	impulses 100V 3ms repeat freq. 3Hz

Railway	Types used	Description section	Number of items installed on the network	Suited for electrified lines	Insulation of the track	Shape of signal in the track
BR	DC–TC	2.2.2	15 000	25kV 50Hz	single rail	DC
	AC–TC	2.2.3	8000	750V DC	single and double rail	50Hz or $83\frac{1}{3}$Hz[2]
	AC on rail, DC track relay	2.2.3	4000	none	single and double rail	50Hz
	Audio frequency jointless TC (Aster[3] and TI21)	2.2.6	10 000	25kV 50Hz 750V DC	jointless	audio freq. 1.7kHz to 2.6kHz coded FSK
	High voltage impulse TC	2.2.4	1000	25kV 50Hz 750V DC	single rail	impulses 100V 3ms repeat freq. 3Hz
	Reed track circuits	2.2.3	2000	25kV 50Hz 750V DC	double rail	sinus 365 to 385Hz
FS	Coded current TC	2.2.3	4024	3kV DC	double rail	sinus 50Hz or 178Hz coded AM
	AC–TC	2.2.3	29 585	3kV DC	single and double rail	sinus 50Hz
	High voltage impulse TC	2.2.4	7	3kV DC	single and double rail	impulses 180V duration 10ms freq. 4Hz
	Audio frequency jointless TC	2.2.5	315	3kV DC	jointless	audio freq. 1000–3500Hz modulation AM 7.5–16.8Hz
CFL	AC–TC	2.2.3	998	25kV 50Hz	single and double rail	sinus 125Hz or $83\frac{1}{3}$Hz
	High voltage impulse TCs	2.2.4	48	25kV 50Hz 3kV DC	double rail	impulses 100V 3ms repeat freq. 3Hz
	DC–TC	2.2.2	353	25kV 50Hz	single rail	DC
	AC–TC	2.2.3	101	3kV DC	single and double rail	sinus 50Hz or $83\frac{1}{3}$Hz
	Short jointless TC	2.2.5	64	3kV DC and 25kV 50Hz	jointless	pulsed sinus base freq. 15kHz

Railway	Types used	Description section	Number of items installed on the network	Suited for electrified lines	Insulation of the track	Shape of signal in the track
NSB	DC–TC	2.2.2	220km lines	diesel	single rail	DC
	AC–TC	2.2.3	2400km lines	15kV 16⅔Hz	double rail	sinus 95Hz and 105Hz
	Audio frequency jointless TCs	2.2.5	3000	15kV 16⅔Hz	jointless	sinus 10kHz and 50kHz
NS	AC–TC 50Hz	–	1800[1]	1.5kV DC	single and double rail	sinus 50Hz
	AC–TC 75Hz	2.2.3	16 200	1.5kV DC	single and double rail	sinus 75Hz AM coded for ATP
	Overlay TC	2.2.5	<100	1.5kV DC	jointless	sinus 10kHz
	Jointless voice frequency TC (FTGS)	2.2.6	<100	1.5kV DC 15kV 16⅔Hz	jointless	audio freq. 4kHz to 17kHz coded FSK
	High voltage impulse TCs	2.2.4	<100	1.5kV DC double rail	single and 3ms and jointless	impulses 100V repeat freq. 3Hz
BV	DC–TC	2.2.2	25 000	15kV 16⅔Hz	single	DC

Notes:
[1]: No longer installed
[2]: 83⅓Hz track circuits are used in dual AC and DC and in AC only areas as well as DC
[3]: Aster TCs are not used on any type of electrified lines

In the following paragraphs, we deal with each of these sets by giving a description of a typical track circuit.

2.2.2 Direct current track circuits

From an historical point of view, direct current track circuits (DC–TCs) were the first in use. Nowadays, they are commonly used on BV, NSB, SBB and BR networks. In the other countries, they only have the legal status of a small minority.

Standard for all types of applications on BV is the direct current single track circuit. This solution is chosen first because it is cheap, simple and is widely available and secondly because it has only a few insulated joints that can cause faults. The basic diagram is shown in **Fig. 2.2**.

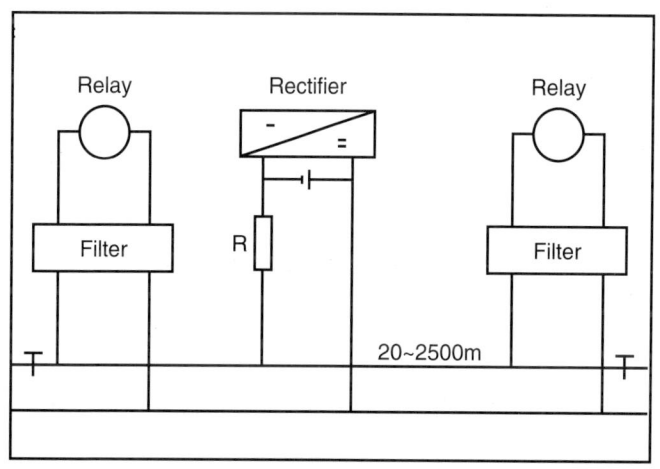

Fig. 2.2 Basic diagram: DC single track circuit

The track circuit is powered by 60Ah NiCd battery through a simple 7.0V rectifier. The relay is a biased, AC immune of the plug in type. The filter is a simple LR lowpass filter as protection against influence from traction currents (AC $16^2/_3$Hz traction). Track circuits longer than 200m have relays at both ends as protection against heavy geomagnetic currents.

On NSB, a similar type of track circuit is used.

The drop shunt and ballast resistance are the limiting factors of any track circuit. The drop shunt affects the safe operation whilst the ballast resistance affects the reliability. This implies varying maximum lengths according to the premises made for ballast resistance, drop shunt, relay and feed characteristics.

Fig. 2.3 gives a survey table of ballast resistance and drop shunt values.

On SBB's network DC track circuits are used in stations and on open lines, not in zones with DC traction or in interference areas with DC traction.

On BR's network, the track relay is preferably placed at the running on end of the track circuit to ensure maximum short circuiting. This relay is always followed by a slow to pick up repeat relay to guard against momentary loss of train shunt. This slow pick up feature operates over a time period of 570 milliseconds.

On all the networks mentioned, the polarity is staggered between adjacent track circuits, to avoid one track circuit feeding the other in the event of a block joint failure.

2.2.3 Alternating current track circuits

Almost all the European railway administrations use alternating current track circuits (AC–TCs). This type of TC consists of one single or both rails between insulated joints, a power supply and a receiver relay which are connected to the rail via feeding transformers on the one side and pick up transformers on the other side. Fig. 2.4 provides an illustration of the AC–TCs principle. It is a simplified outline of ÖBB's single rail insulated track section.

In order to provide constant supervision of the track, the circuit operates on the principle of the closed loop and in the condition of track free the track relay remains energized. The moment a railway vehicle enters the insulated track section, the closed

Railway	Maximum track circuit length	Ballast resistance	Drop shunt
SBB	1400m in future 800m	1.5 Ω.km on station areas 2.5 Ω.km on open line	1.0 Ω on points and track sections of up to 300m 0.5 Ω track sections over 300m
BR, diesel lines	1000m	2 Ω.km	0.5 Ω standard 1.5 Ω is accepted on heavily trafficked sections of electrified line
BR, electrified lines AC	650m		0.3 Ω is permissible on AC double rail to maintain working TCs
NSB	10km	0.6 Ω.km on station areas 0.5 Ω.km elsewhere	0.5 Ω on station areas 0.1 Ω elsewhere
BV	2500m	1.0 Ω.km gravel 2.0 Ω.km macadam	0.3 Ω[1]

Note:
[1]: With respect to geomagnetic currents, a voltage drop, of at least 2V and 4V pick up voltage over the rails is a better safety measure for a DC track circuit

Fig. 2.3 Ballast resistance and drop shunt values

loop is interrupted through a short circuit caused by the low axle resistance. The energized track relay will drop and indicate the condition of track occupied. In addition, a break of the rails or an additional bridging of the insulated joints or the rails by an obstacle can also be identified. Double rail insulation in connection with electric traction also requires the installation of impedance bonds at each insulated rail joint. Neighbouring sections are always supplied with different phases to detect a bridging of insulated joints, unless other specific techniques are used.

On the feeder side, different frequencies are encountered throughout Europe, ranging between 50Hz and 125Hz. On the receiver side, there exist intensity relays, phase shift relays, motor relays and electronic converters controlling one or two separate relays.

Minimum length of a track circuit depends on two parameters: it should be longer than the longest distance between two axles and it depends on the ratio line speed versus reaction time of the receiver relay and its surrounding circuitry.

The maximum length of the track circuit depends on:

- the ballast resistance, which varies from country to country;
- the drop shunt, which varies from country to country;
- whether there is single or double rail insulation;
- the frequency used;
- the protective resistors on feeder and receiver side, placed to prevent the traction return current from intruding into the weak current circuitry of the signalling equipment;
- the sensitivity of the receiver;
- the amplitude of the signal injected in the track.

Fig. 2.6 summarizes the majority of these conditions.

Special case: reed track circuits

BR use jointed reed track circuits in areas where dual immunity is required. Mechanically coupled reed filters are provided with the same frequency at the feed and relay ends. The filters are stable in frequency and have a bandwidth of less than 1Hz. Track frequencies can therefore be only 3Hz apart and range from 365 to 385Hz. Different frequencies for adjacent tracks guard against faulty operation due to block joint failure. A track circuit length of up to 600m is normal.

Reed track circuits are not compatible with three phase traction systems.

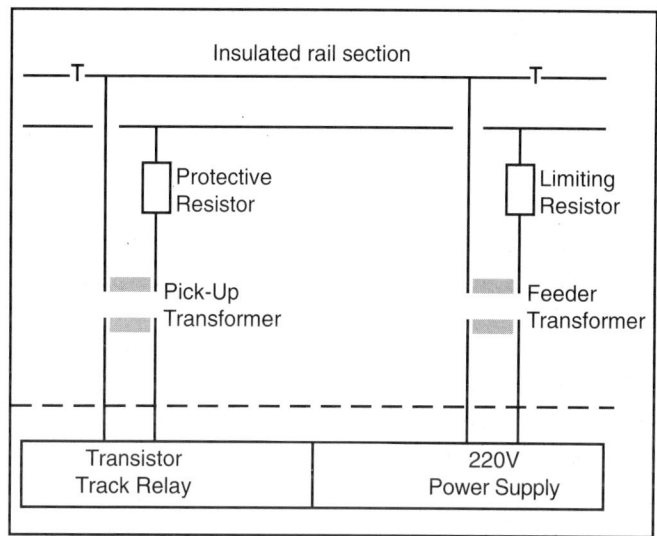

Fig. 2.4 AC track circuit principle

Fig. 2.5 Impedance bond – BR Southern Network

Fig. 2.6 Characteristics of the AC track circuits throughout Europe

Railway	Track circuit name	Insulation of the track	Frequency used	Maximum length	Type of receiver	Where applied
ÖBB	Electronic track relay	single rail	50Hz	2300m	Electronic circuitry controlling two inverse working relays; the first is energized if the track is free, the other one is energized if the track is occupied. That means in the case of a disturbance or a power failure the disturbed state is detected, as both relays are de-energized. If both relays should be energized, this is of course also detected as a wrong (distrubed) state	In all stations where geographical or electronic interlockings of Alcatel Austria are installed (about 60 stations in 1991)
		single rail	100Hz	1400m		
		double rail	100Hz	2300m		
		single rail	$106\frac{2}{3}$Hz	1400m		
		double rail	$106\frac{2}{3}$Hz	2300m		
	Motor track relay	single rail	100Hz	1400m	Motor relay driven by a two phase asynchronous motor with squirrel cage armature. The motor gets its turning moment from two energized coils: the track phase coil is energized via the track, the auxiliary phase coil from a local power supply. The turning moment is proportional to the product of the two currents and to the sinus of the phase angle between those two currents	In all stations where electromechanical, relay or electronic interlockings of Siemens are used (about 250 installations)
		double rail	100Hz	2300m		
NMBS/ SNCB	Birail	double rail	50Hz	2000m	Either a relay with two coils, one of them the track phase coil is energized via the track, the auxiliary phase coil from a local power supply. The relay is activated if the phase angle between the two currents is 90° and the amplitude of the current in the track phase coil is high enough, during a time of at least 1.2s. Or an electronic circuitry, controlling a DC relay. This circuitry implements the above described criteria with more accuracy	On electrified lines
	Monorail for electrified lines	single rail	50Hz	500m		
	Monorail for diesel lines	single rail	50Hz	1500m		On diesel lines

Railway	Track circuit name	Insulation of the track	Frequency used	Maximum length	Type of receiver	Where applied
RENFE	Phase relay	double rail	50Hz	2000m	Relay with two phase circuitry	In stations, in automatic blocks and in centralized control point systems, but not on high speed lines
		single rail	50Hz	800m	Relay with two phase circuitry	In recessing siding and switch zone
BR	AC track circuit	single rail	50Hz	600m	A double element two position vane relay requiring two power supplies, one from a local supply, the other from the track circuit	Areas with DC traction
		double rail	50Hz	2700m		
		single rail	$83\frac{1}{3}$Hz	600m		AC or AC/DC tractions
FS	Steady current	single rail	50Hz	700m	Vane relay, two elements, two positions	Station track circuits and non-coded current automatic block
		double rail	50Hz	2000m		
	Coded current	single rail	50+178Hz	900m	Relay-based or solid state control	Automatic block with code transmission for cab signalling
		double rail	50+178Hz	2000m	Logic equipment	
CFL	Siemens	single rail	125Hz	300m	Relay with asynchronous motor	Electrified lines 25kV, 50Hz
		double rail	125Hz	1200m		
	SEL	single rail	$83\frac{1}{3}$Hz	600m	Relay with cathode ray tube	
	Siemens	single rail	50Hz	250m	Relay with asynchronous motor	Electrified lines 3kV DC
		double rail	50Hz	1200m		
	SEL	single rail	$83\frac{1}{3}$Hz	600m	Relay with cathode ray tube	
NSB	AC track circuit	double rail	95Hz / 105Hz	1000m	Two phase motor relay; detection is based on frequency, phaseangle and energy levels	Electrified lines and all new lines
NS	AC–TC	single rail	75Hz	600m	A double element two position vane relay, requiring two power supplies, one from a local supply current, the other from the track circuit	Almost the complete NS network is equipped with it
		double rail	75Hz	1200m		

2.2.4 Track circuits with high voltage impulses (HVI–TCs)

High voltage impulse track circuits (HVI–TCs) have been designed in order to solve, in addition to the basic requirements, the problems that occur with the difficult problem of wheel–rail contact, the worst case being rusty rails. They are also used in dual traction zones.

They occur on the networks of NMBS/SNCB, SNCF, BR, CFL, FS, NS and RENFE.

An HVI–TC includes:

- a transmitter which generates impulses with a given shape, amplitude and recurrence frequency injected into the tracks by means of a matching insulating transformer or impedance bond.
- a receiver which detects the specific signal at the other end of the track section and operates the fail-safe relay if the specific signal responds to the correct positive and negative amplitude.

Whenever the specific signal is not received, is deformed or is too weak, the track circuit fail-safe relay is de-energized.

Impulses are delivered by the transmitter on the basis of capacitor discharges controlled by a thyristor at recurrent intervals. The capacitor discharges are operated in a circuit including the primary of the matching insulating transformer to the track. The receiver includes a transformer rectifier filter circuit without external power supply.

The energy received from the track is stored by the receiver for operating the fail-safe relay. This relay is fitted with differential windings designed to select only the asymmetric waveform delivered by the transmitter, thus eliminating the noise symmetrical waves mainly created by the rolling stock return traction currents.

Two adjacent track circuits can be insulated from each other by physical insulated joints. With insulated joints, in an electrified territory, the return traction current flows through impedance bonds. Jointless TC are available for short lengths up to 150m, as an overlay track circuit, mostly used on level crossing applications. These track circuits are centre fed.

Just to have an idea of what it looks like, **Fig. 2.7** shows data of the HVI–TC, as in use by SNCF.

- Impulse recurrence frequency: 3Hz
- Impulse width: 3 milliseconds
- Voltage peak value: 100V
- Current peak value: 100A
- With a ballast insulation equal to 2 Ω.km, the standard length of HVI–TC is as follows:

	single rail	double rail
non-electrified	600m	3500m
DC electrified (1500V)	400m	2000m
AC electrified (25kV)	200m	2000m

Fig. 2.7 HVI-TC data

2.2.5 Short jointless track circuits

To control the automatic barriers and lights at level crossings, supplementary track circuits have to be installed on the tracks. With these tracks that have been previously equipped with track circuits for automatic block for instance, a problem arises. Should all the work that has been carried out previously to install all the track circuits for automatic block become useless everytime a new automated level crossing is inserted in a line? To find a way round this, some railway companies introduced short jointless track circuits (ShTCs) which are superimposed on the existing AC–TCs. These ShTCs control the level crossing – both the closing and the opening of the barriers. This principle is applied at NMBS/SNCB and NSB. With SBB, FS and CFL, however, only the opening of the barriers is done by ShTC.

SBB use ShTCs for special applications where a second independent system is required, whilst NS use overlay track circuits to release points operated on site and also to release block in simplified block systems.

In all of these cases, ShTCs are superimposed on both AC–TCs and DC–TCs.

As in use by NMBS/SNCB, the ShTC applies in the following way. A 15kHz high frequency signal is generated in an emitter/receiver unit. This signal is injected in the track with bursts (repeat frequency 856Hz). The same unit picks up the signal again, amplifies it and controls a relay, which indicates whether the track is free or not. This ShTC is used on diesel lines as well as on electrified lines. Its length is somewhere between 27m minimum and 150m maximum.

2.2.6 Audio frequency jointless track circuits

The parameters of train detection have changed with the passage of time. These include:

- continuously welded rails;
- higher axle loads;
- GTO controlled traction motors;
- more powerful vehicles/higher traction return currents;
- higher train speeds;
- eddy current brakes;
- two-way working signalled tracks.

Conventional track circuits cannot work safely and reliably under all circumstances in this environment. As a result the audio frequency jointless track circuits were invented (AF–TCs).

Throughout Europe, three types of AF–TCs are in use:

- FTGS, as used by SBB, DB AG, RENFE and NS;
- UM71, as used by SNCF;
- TI21, as used by BR.

FTGS

The operating principle of the FTGS is the same as for all similar track circuits based on a short circuit bond between the rails with the adjacent sections of rail tuned to the allocated audio frequency.

The short circuit bond between the rails is S-shaped, the geometrical centre defining the boundary between adjacent sections. It ensures overlapping sensibility at the boundary. Lengths of the bond and thus range of sensibility depend on the chosen frequencies and vary between 7 and 19m. **Fig. 2.8** shows a standard AC–TC with S-bond joints.

The lower frequencies are preferred on open lines, whereas the higher frequencies are more suitable for station areas. Points and crossings can be covered. Centre feeding and cascading is possible. FTGS track circuits are used mainly on high speed lines on SBB, DB AG and RENFE or in dual voltage areas on NS.

The length of a standard FTGS track circuit, which is a maximum of 750m, is doubled by centre feeding. Trainshunt on open lines is maintained at ≤0.5 Ω, in station areas at ≤1.0 Ω.

UM71

A UM71 track circuit is generally terminated at each end by an electrical joint and linked to a remote transmitter or receiver via a matching unit and a twisted and shielded line cable. An insulating joint may also be used in place of the electrical joint at each end of the track circuit.

(a) Electric joint

The length of track included in the electric joint is terminated at each end by a tuning unit. Each tuning unit consists of an inductor and capacitors for obtaining:

- parallel resonance with the track inductance (impedance pole) at the frequency of the track circuit involved.
- serial resonance at the frequency of the adjacent track circuit. This serial circuit shunts the track, thereby preventing propagation of the signal beyond the limits of the track circuit.

A track air core inductor connected to the mid-point of the electrical joint optimizes performance and rebalances the return traction current between the two rails.

(b) Termination on an insulating joint

This is provided by the connection in parallel of an air core conductor and a tuning unit.

The continuity of the traction current return circuit is provided by:

- the mid-point of the track air core inductor, up to a permanent load of 300A in the case of AC electrification.
- a tuned impedance bond in the case of DC electrification.

(c) Alternation of frequencies

The track circuit uses different carrier frequencies to avoid crosstalk between adjacent sections of track. Four carrier frequencies are used:

- 1700Hz and 2300Hz on one track.
- 2000Hz and 2600Hz on the parallel track.

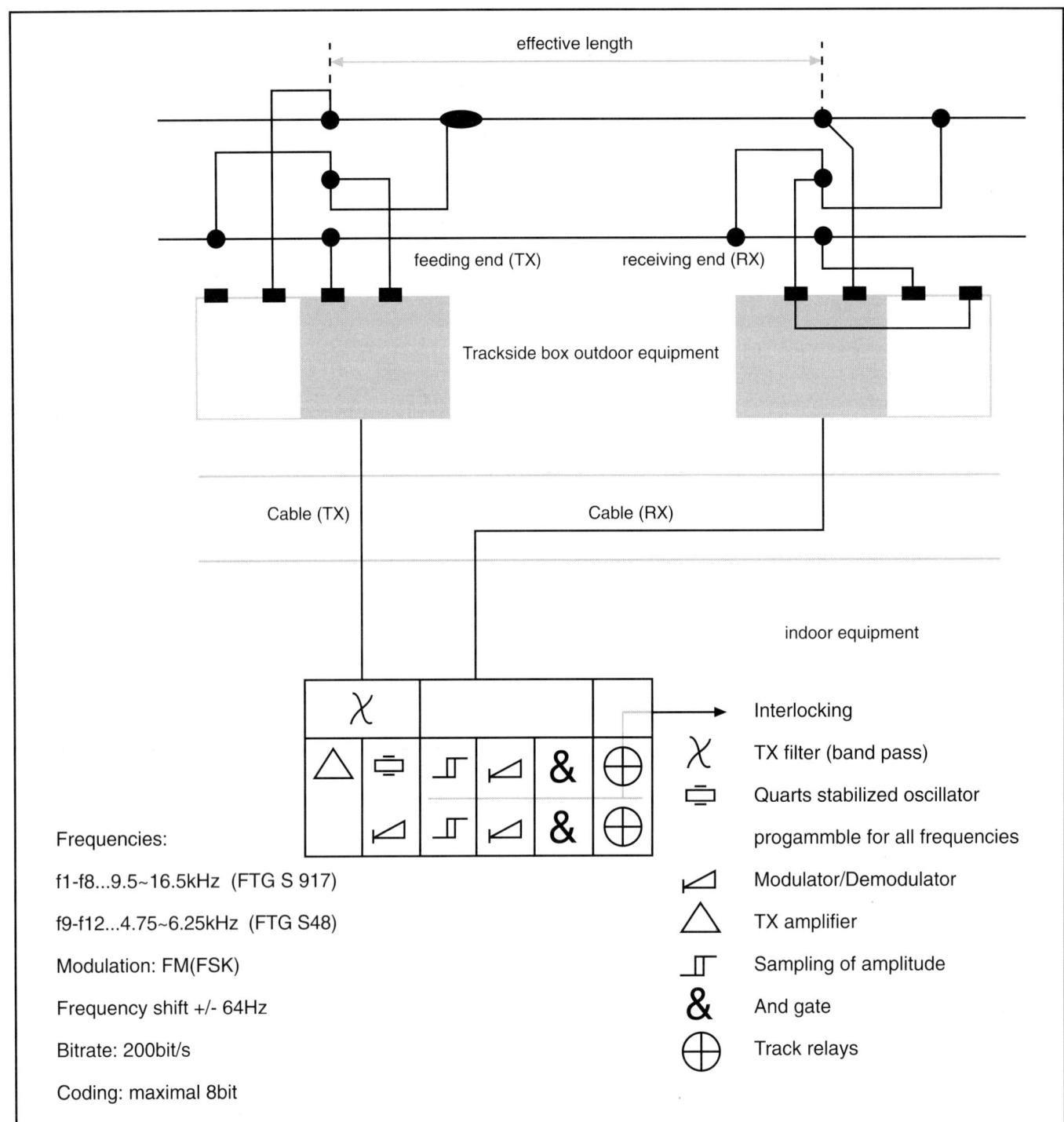

Fig. 2.8 Standard AF-TC with S-bond joints

The sine wave carrier frequency is modulated by a very low frequency square signal giving frequency shift of ±11Hz. This modulation ensures a very high immunity against traction current harmonics.

(d) Compensation

Compensation consists of improving transmission attenuation along the track by connecting capacitors of appropriate values between the two rails and buried in the ballast at given intervals. This compensation increases the maximum length of the track circuit.

The drop shunt value is 0.25 Ohm in station areas and 0.15 Ohm on main lines. This value is 0.5 Ohm on the complete North TGV line.

The insulation resistance is 2 Ohms per km in stations and on main lines, and 8 Ohms per km on all TGV lines. Those values are always taken whatever might be the length and type of track circuits.

This type of track circuit is also used to transmit ATP and cab signalling information to the TGV trainsets and Eurotunnel locomotives.

TI21 & Co

AF–TCs in use on BR are all of the voltage operated type with characteristics as defined in **Fig. 2.9**.

The Aster 1 Watt version employs a Z-bond layout forming two inductive loops tuned to resonate at particular frequencies. The band arrangement determines that adjacent track circuit overlap thereby ensuring that each TC is shunted before its neighbour clears. A continuous audio frequency current is fed to the track via the bond.

Likewise, a standard signalling relay is connected to the bond at the opposite extremity of the track circuit.

The Aster U version uses a frequency dependent short circuit to determine each boundary. These are overlapped for adjacent track circuits. Signals are injected and received from the rails via a secondary coil of the track transformer. Centre feeding is possible for a distance of 1000m in each direction to enable lengths of up to 2000m.

	Type		
	TI21	Aster U	Aster 1 Watt
Mode of operation	voltage	voltage	voltage
Signal frequency range (Hz)	1700–2600	1700–2600	1600–2800
Number of frequencies available	8	4	6
Type of modulation	4.8Hz FSK	–	–
Suitable for lines electrified at 25kV AC and 750V DC	yes	no	no
Electrical separation	tuned area	tuned area	tuned area
Length at 2 Ω/km ballast resistance	50–200m (low power) 200–2000m	50–2000m	50–2000m
End definition	±5m	±10m	±10m
Dead zone	no	no	no
Track to train information	no	no	no

Fig. 2.9 AF-TCs – BR

2.3 Axle counters

2.3.1 Survey of the European situation

Although axle counters have a similar function as track circuits in signalling practice, their use is not so widespread.

Looking at the current European situation, four groups of countries can be distinguished, reflecting the varying degrees of use of axle counters in their national railways.

The first category are the railway administrations that have a tradition in the application of axle counters: ÖBB, SBB, DB AG and CFL. The axle counters used in those countries are dealt with in more detail in **2.3.2**.

In the second category are those countries where axle counters are considered as a fringe activity, such as RENFE, SNCF, BR and FS. Here, they are considered as an alternative to track circuits and are used where cost or local conditions render the use of track circuits inappropriate. For a long multiple track circuited section of line, axle counters offer a cheaper and safer alternative solution if the line is lightly used and carries lightweight vehicles. Where the rails are bolted directly to bridge girders, leading to zero ballast resistance, axle counters have been used. They are also applied in areas prone to surface water such as tunnels and where lines cross at right angles at grade.

The third category of administrations don't use axle counters at all: NSB and BV.

The fourth and last category is new in the business. NS is experimenting with axle counters on lines only used by light diesel trainsets. NMBS/SNCB have already commissioned some axle counters on single track lines, in conjunction with a block system and are now introducing axle counters in freight stations and sidings. Here axle counters offer a higher degree of safety, and because they are rust resistant they are more economic in use. Indeed, in the points the rods do not need an insulation as in the case when track circuits are applied.

A supplementary problem arises when axle counters are installed at the head of a set of sorting sidings. The counter, which controls the detectors, has to know the correct position of the points, for example a close link with the interlocking equipment is a prerequisite. NMBS/SNCB solved this problem in a practical way. In stations with an electronic interlocking, the SSI computers are used to control the detectors, whilst in an all-relay installation a set of two, mutually coupled, industrial microcomputers read the relay contacts of the interlocking and record the axles present in the head of the set of sorting sidings. Their outputs are then fed back into the relay interlocking.

Some railways avoid the above described close link between the detectors and the interlocking equipment by accepting a simplified programme.

2.3.2 Axle counters in their traditional countries

Since not all tracks can be insulated sufficiently to permit the use of track circuits, the axle counter was developed for train detection independent of track condition. The philosophy is based on the assumption that a section is clear when the number of wheels that left a defined track section equals (not more and not less) the number of wheels that previously entered this section, provided the section was clear initially.

Results of earlier tests provided encouragement to proceed with development after 1945. The task was to design a sensor or wheel detector, able to distinguish individual wheels passing at any permissible speed, from any other object in the vicinity, and to detect its direction of travel. Another device had to be designed, which in collaboration with the wheel detectors was fast enough and had sufficient storage capacity for any number of wheels detected. This is known as a counter. To complicate the task, counting in and out can take place simultaneously and occasional miscounting has to be provided for, both technically and operationally. The distance between wheel detectors and counter should be as high as possible. Although axle counters are also used in stations, the main area of application is for open lines as an integral part of automatic block systems.

With the technological means available at the time, the first axle counter system was approved and accepted for commercial service in 1955. It consisted of wheel detectors with biased, magnetically operated contacts and electromechanical rotary selector counters, later motor counters, linked by signal cables. Output to the interlocking was via contact combinations of two relays never in concordance. In spite of many improvements, the magnetically operated contacts proved to be the limiting factor.

Higher speeds, longer trains, smaller wheels closer together, magnetic rail brakes, higher traction return

Fig. 2.10 Axle counter AZL70 type

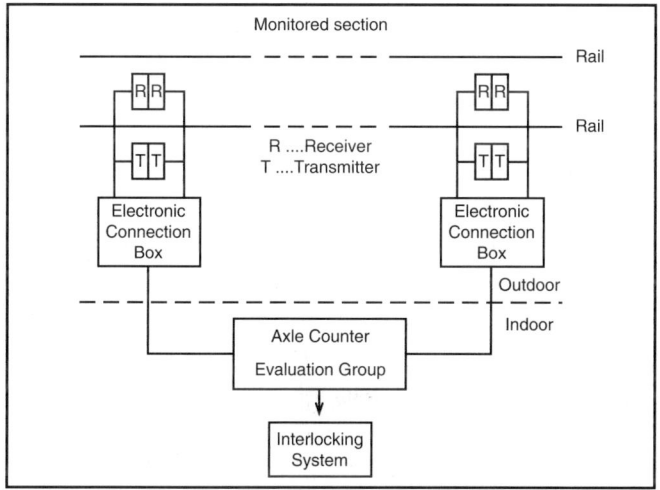

Fig. 2.11 Simplified outline of electronic axle counters

Railway	Type of Az	Number of sections equipped with axle counters
ÖBB	Electronic Az	850
	Motor Az	no longer installed
SBB	Integra	750
	Siemens	10
	Alcatel	10
DB AG	Motor Az	10 000
	AzS70 and AzL70	18 000
CFL	AzS70 and AzL70	17
FS	Siemens ITT SILIANI	414

Fig. 2.12 Survey of the degree of application of axle counters in Europe

currents, stray magnetic fields, and flangeless wheels could all only be handled by radically different wheel detectors and counters with increased capacity. New technology such as fail-safe discrete electronics resulted in new generations of axle counter systems, represented by AZL70 and AZS70. Both systems use electronic wheel detectors on the rail in which the coupling factor between the transmitting and receiving resonance circuit is affected by the passing of a wheel.

The resulting change is transmitted to the counter, where it is electronically evaluated. Wheel detectors may be mounted in a staggered way on different rails or in pairs on one rail. There is a minimal distance required between the sensors to derive the direction of travel. Electronic wheel sensors can be connected to only one counter, or can control both counters of adjacent track sections. The output interface remains as a contact combination of two antivalent relays. With the aid of an electronic converter, electronic wheel sensors can also be combined with a conventional motor counter.

Counters evaluate each complete overlapping pulse pair as a passing wheel. However, even an incomplete pulse pair, for example a wheel not fully passing the detector, results in one of the output relays dropping without the other one picking. This is interpreted as a fault.

A new generation on the basis of vital microcomputers is presently to be introduced. **Fig. 2.11** shows a simplified outline of the electronic axle counters.

In the future, DB AG plan the use of multiple axle counters with failure redundancy. They will offer a better cost efficiency and a very high reliability.

Fig. 2.12 gives a survey of the degree of application of axle counters on several European railway administrations.

The latest trend in axle counter technology is to immunize them against the influences of electromagnetic and eddy current brakes.

2.4 Treadles

2.4.1 General

For some applications neither track circuits, nor axle counters are well suited, for example when automatic level crossing are installed on secondary lines. In these instances there was a need for another category of train detection devices – treadles.

The main task of treadles is to signal the passage or the local presence of a rail vehicle.

Intrinsically they are not fail-safe, so for non-safety related applications, they can be used without supplementary precautions, whilst for those applications where safety is a prerequisite, they are doubled or completed by a TC.

2.4.2 Survey of the European situation

The first type of treadle in use was the mechanically actuated treadle with one or two wheel operated arms. The version with two arms allows discrimination of the running direction.

Basically the mechanical treadle consists of a cast iron housing carrying a high tensile steel bar. This bar absorbs the energy produced by the shock of the wheel on the arm and controls the reversal of electric contacts. The return of the arm is delayed by an oil dashpot. The latter ensures an adequate period of closure or opening of the contacts, independent of the time constant of the circuits.

The second type of treadle is the hydraulic one. It is activated by rail bending under passing wheels. The mechanical bending is amplified and damped by an hydraulic transmission. The fluid in it is either oil or mercury.

The pneumatic and magnetic treadles were invented at a later date.

The pneumatic treadle is based on the same principle as the hydraulic, in which the fluid is replaced by air.

There is also the magnetic wheel detector. Fundamentally the iron mass of the wheel flange deforms the magnetic field of a permanent magnet inside the detector. Due to this change in the magnetic field, electric contacts switch over.

Fig. 2.13 Mechanical treadle: two-way working track Forfex type

Later the permanent magnet was replaced by an oscillator which generates a magnetic field. If a wheel passes, the field is influenced, and subsequently the current in the oscillator circuit is reduced. A final relay drops, which indicates that the sensor is occupied. This magnetic detector is now often called a wheel sensor. By installing two detectors of differing frequencies one after the other, the passing of each train wheel produces a sequence of signals, which allow the electronic circuitry to determine the running direction. Sometimes the two detectors are installed on a different rail, ensuring a very high reliability, even if one of the two detectors detaches from the rail to which it has been secured.

Fig. 2.14 gives a survey of all the different types of treadles in use throughout Europe.

Fig. 2.14 Characteristics of treadles throughout Europe

Railway	Treadle type	Treadle's name	Applications
ÖBB	pneumatic	S44	– for initiation of automatic route setting – for release of train routes
	magnetic	A111	– to open automatically the barriers (& lights) of a level crossing
	magnetic	RSE45 wheel sensor	– to indicate a wheel which is trailing the rail frog of a point
NMBS/SNCB	mechanical	no longer installed	– to control level crossings on secondary lines
	pneumatic	no longer installed	– second condition to release a route in a signal box – second condition to open automatically the barriers and switch off the flashing lights of a level crossing
	magnetic		– in both applications the first condition is a track circuit
SBB	mechanical	–	– warning equipment for track works
	magnetic	inductive treadle	– second element in vital applications for clearing signals, opening of level crossings and switching off its flashing lights – switching element for triggering of automatic signals, automatic closing barriers – information for clearing call-on signals
DB AG	pneumatic	S44	– for announcements, initiation of level crossing protection systems – proving the presence of a train as condition for block normalization – route release in older interlockings
	magnetic	magnetic wheel detector	– for initiation of level crossing protection systems
	electronic	wheel sensor	– for detection of trailing moves on points equipped with non-trailable movable frog – for initiation of level crossing protection systems
RENFE	mechanical	–	– to give information to the level crossings – as indication of proximity circuits to the stations – as auxiliary element of the short track circuits of the interlocked level crossings
	magnetic		– to give information to the level crossing

Railway	Treadle type	Treadle's name	Applications
SNCF	mechanical	–	– to control particular functions connected with the accurate localization of train, eg: level crossings (control of gates and lights), control in marshalling yards
	magnetic	electronic	Same functions as the mechanical ones Are mainly used in snow region and where external undue operation may occur (spiteful acts) on mechanical treadles
BR	mechanical	–	– each initiating TC for an automated level crossing is cut by a treadle at the running-on end to give positive operation and guard against the non-detection of a lightweight vehicle
	magnetic	–	– similarly, reopening of the crossing requires operation of a treadle at the running outside – also provided at any signals within the approach area of the crossing to detect a train over-running a red signal; thereby ensuring the operation of crossing sequence
FS	mechanical	Forfex and Cautor	– to close automated level crossings for this application they are doubled and tested by a system which checks for correct operation of the arms – to release automated level crossings
	hydraulic	P70	– to release through route locking in interlockings – to release sections of manual block – to release level crossings
CFL	mechanical	–	– to initiate automated level crossing announcement – to neutralize temporarily a Crocodile fed by a local battery in case of a vehicle coming in contact with it from the wrong direction
NS	pneumatic	S44	– to mark the end of a block section – to initiate automated level crossing announcement – both applications only exceptionally
	magnetic	RSE45 electronic	– to mark the end of a block section – to initiate automated level crossing announcement – both applications only exceptionally

BV	pneumatic	–	– in combination with track circuits to:
			  – measure train length
			  – measure speed level
			  – lower barriers at platform level crossings

2.5 Vehicle sensors

2.5.1 General

Unlike treadles, which detect the wheel flange, vehicle sensors detect the iron mass of a vehicle. They also detect rail and road vehicles. In addition this type of sensor is used in the domain of road transport, for instance to control the red/green cycle of road traffic lights. The vehicle sensor, as used by some railway administrations, is usually used in pairs to detect the presence and/or the direction of a train.

Functionally a vehicle sensor acts rather like a treadle, in the detection of the passing of a train and, like a track circuit, in the detection of the presence or absence of a rail vehicle.

A vehicle sensor is composed of an 8-shaped loop. The iron mass of a rail vehicle above this loop causes the oscillating frequency to shift, due to the inductivity change of the loop circuit. Depending on the application of loops, they can be used in a single form as well as in pairs. Every loop has a separated output. In addition the output of both loops can be combined to close the detection gap at the intersection.

2.5.2 Survey of the European situation

Fig. 2.15 gives a survey of the use of vehicle sensors throughout Europe.

Railway	Type	Name	Applications
ÖBB	vehicle sensor	induction loop	– to open automatically the barriers (and lights) of a level crossing – to detect the presence of a rail vehicle at the level crossing
SBB	vehicle sensor	induction loop	– supervision of area at level crossing in order to prevent road vehicle caught between barriers – switching element for triggering automatic signals – automatic closing of barriers
DB AG	vehicle sensor	induction loop	– to detect the presence of rail vehicles at level crossings – to normalize the level crossing system
CFL	vehicle sensor	induction loop	– as strike in and out device at level crossing protections
NS	vehicle sensor	induction loop	– to mark the end of a block section, in special cases
BV	vehicle sensor	induction loop	– to detect the presence of *road vehicles* at level crossings

Fig. 2.15 Characteristics of vehicle sensors throughout Europe

2.6 Miscellaneous

2.6.1 Magnetic relay
The various private Swiss railways use a magnetic relay as second element in the vital applications for replacing signals, opening of level crossings and switching off flashing lights at level crossings. It is also used as switching element for triggering automatic signals and the automatic closing of barriers.

The magnetic relay requires there to be a permanent magnet on locomotives.

Alternatively the permanent magnet can be replaced by an electromagnet in cases where more than one set of information between train and track is required, for example, points and signal control.

2.6.2 Tail detectors
Currently, few railway companies use tail detectors, with the exception of ÖBB, DB AG and NSB.

On ÖBB there is only one application, and it is unlikely that any more will follow because there are other new solutions for side lines. The tail detector is intended for side lines without a train detection system, to have an indication that a train has completely left a section. From the point of view of equipment, it consists of a permanent magnet that acts as a sender and is attached on the coupling hook of the last car. A fixed receiver, situated between the rails, generates an impulse whenever a sender passes by. This pulse is stored in an interface until it is released by the signalman. The equipment works at speeds of up to 150km/h.

NSB use similar tail detectors in a configuration for simplified block operation without track circuits.

On DB AG tail detectors have been in use since about 1985. DB AG install special and simple signalling apparatus on certain lines with little traffic.

For automatic track free detection these lines will be equipped with axle counters or alternatively tail detectors.

For this purpose the principal way to apply tail detectors is to combine them with vehicle sensors (loops) in order to guarantee safety demand. Tail detectors on the above mentioned lines are applied for track free detection at stations as well as on open lines. Fundamentally the tail detector works on the previously described principle. DB AG have chosen tail detectors because they are widely available and cost effective. Since their introduction, DB AG have experienced a long period of positive results.

3 Switch Operating and Proving Systems

M. VALLEZ

Contents

3.1 General		144
3.1.1 Definition		144
3.1.2 Basic concepts:composition		144
3.2 Points		145
3.2.1 History		145
3.2.2 Development of track equipment		146
3.2.3 Development of the fittings		146
3.3 Control, locking and detection devices		148
3.3.1 Mechanical points operation		148
3.3.2 Electrically powered point operation		149
3.3.3 Electrohydraulic point machine		150
3.3.4 Electrohydraulic point machine, with rail clamp locks		151
3.3.5 Electrohydraulic point machine, with hydraulic power unit with accumulators		151
3.4 Locking devices		151
3.4.1 Clamp locking with horizontal hooks		152
3.4.2 Claw locking		152
3.4.3 Locking by clamp lock/housing/slide-plate (verrous-carter-coussinet:VCC)		152
3.5 Point detection		153
3.6 Derailer		154
3.7 Command and control circuits		154
Summary of the safety conditions		154

Illustrations

Fig. 3.1	*Composition of crossing apparatus*	144
Fig. 3.2	*Composition of points*	145
Fig. 3.3	*Operation phases of points movement*	149
Fig. 3.4	*Electric switch machine with throw and detector slides*	150
Fig. 3.5	*Clamp locking with horizontal hooks*	152
Fig. 3.6	*Locking by clamp lock/housing/slide plates (VCC)*	153
Fig. 3.7	*Electric point machine application*	155
Fig. 3.8	*Electrohydraulic point machine application*	158

3.1 General

3.1.1 Definition

Switch equipment represents a single point on the track which permits track continuity for a selected route. The multiple parts that make up the equipment cannot be common to those of the track.

It is important that switch equipment be inserted into the track as unobtrusively as possible and retain its main arming characteristics, whilst providing the rolling stock with continuous guidance.

All switch points must be capable of being *operated* but in addition safe traffic conditions demand that points are *immobilized* in the position selected and that this position is *detected*; ie, that one can be certain that fully applied switch blades are correctly set, that open blades are correctly open and that the position in the field corresponds to that of the signalling interlocking equipment.

The three functions – control, locking and detection – may be carried out by separate units or they may be incorporated within a single apparatus.

3.1.2 Basic concepts: composition

Points and crossing apparatus are composed of the following elements:

- the point, at the origin of the divergence;
- the crossing providing continuity at right angles to the intersection of the two diverging routes;
- the intermediate tracks, elements of continuous track, which provide the link between the aforementioned parts.

3.1.2.1 Description of the points

The points are composed principally of two half points each made up of:

- one *switch rail* which is the movable part. The ends of the switch rail are called the *heel* and the *point of switch tongue*. A distinction is made between two types of switch point:
  (a) *articulated points* are dimensionally stable and are fixed to the heel by means of an articulated joint;
  (b) *elastic points*, on the contrary, are rigidly fixed to the heel and move only by elasticity.
- a *stock rail*.
- *slide plates* fixed to the sleepers and lubricated.
- an assembly of *throw and control rods* connected to the switching and control devices.

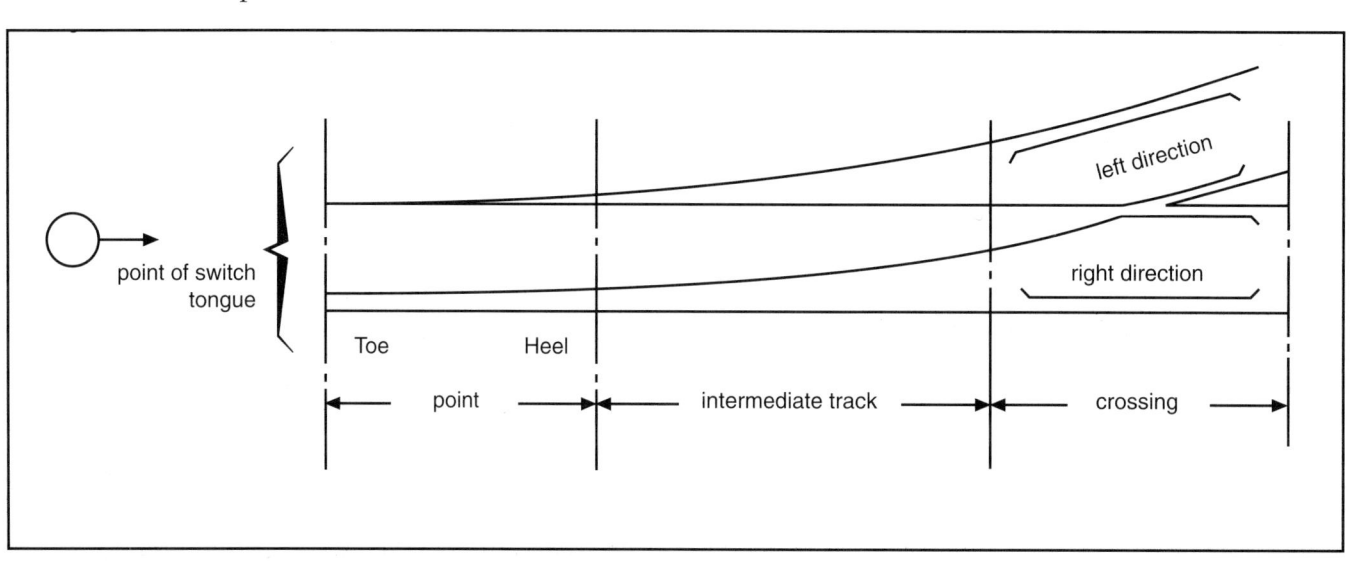

Fig. 3.1 Composition of crossing apparatus

3.1.2.2 Switching the points

The points may be crossed by traffic travelling:

- from the point towards the heel (passing the point facing);
- from the heel towards the point (trail the point).

Points are said to be:

- to the left when an observer facing the point sees the left direction given;
- to the right in the contrary case.

The *switching gap* is the distance between the switch rail and the stock rail at the site of the throw rod when the other switch rail is applied. With the points locked in one position, the *operating time* is the time necessary to unlock them, change their position and lock them in the opposite position.

The maximum *throwing force* is the maximum force that must be deployed at the axis of the rack rail in order to accomplish a complete throw over of the apparatus, ie:

- unlocking the closed blade and, for some systems, performing the first phase of the movement of the open blade;
- simultaneous movement of both blades;
- locking the new closed blade and for some systems performing the last phase of the movement of the new open blade.

3.1.2.3 Angle of crossing

Track apparatus is characterized principally by its angle of crossing expressed by the value of the tangent corresponding to the angle on leaving the diverted track. In fact, it is the radius that is its main characteristic; it is dependent on the operating speed selected on the diverted track. The higher the speed, the greater will be the radius (and therefore the smaller the angle).

These very small angle turnouts, allowing high speeds, are installed on high speed lines (HST). In contrast to other track apparatus, these turnouts are equipped with crossings comprising a movable frog. The operating principle of the movable frog is analogous to that of points with a spring switch blade.

The two basic elements forming the frog are:

- the *cradle*, which plays the part of the left stock rail and the right stock rail with their slide plates;
- the *movable point*, which plays the part of the right facing point or the left facing point depending on the position applied.

Throwing over the movable point is similar to that of the switch point.

3.2 Points

3.2.1 History

It is remarkable to note that the first known track equipment or points date from 1796. These were designed by a young English engineer called John Curr. On the basis of these designs Curr created a complete mining network in the county of Norfolk, England. An analysis of his drawings shows that the equipment that we still use was designed on the same principles.

While the basic principles of Curr's design remain the same, points, as well as their control, locking and

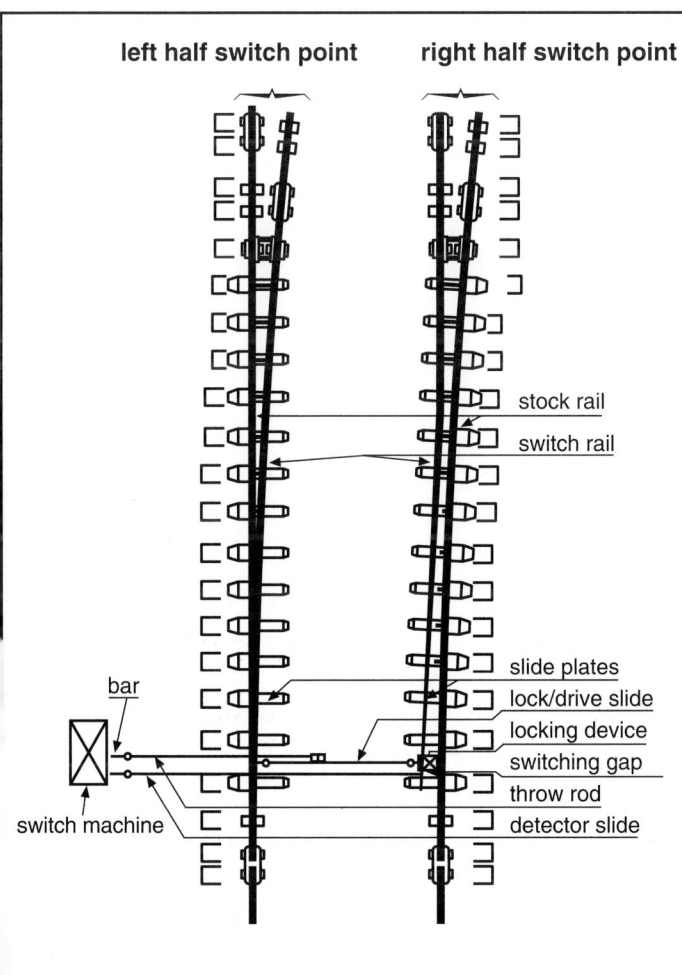

Fig. 3.2 Composition of points

detecting equipment, have developed considerably, and can now be crossed at speeds of more than 300km/h on straight track and 230km/h on curved.

The technological development of points and their equipment are discussed below.

3.2.2 Development of track equipment

3.2.2.1 Articulated points – flexible points

The remarkable development of the railway depended, to a great extent, on the ease of building up a route from relatively simple elements of construction – *articulated points*.

With the gradual increase in the speed of crossing over points and the growing need for passenger comfort, *flexible points* then became necessary.

The rigid assembly of the heel of the points to the continuous track has certain advantages:

- displacement of the point of the blade is done by bending it, which allows a smooth connection of the blade to the continuous track;
- elimination of the heel articulation reduces jolting, wear and the risk of breakage whilst increasing comfort.

In contrast, elastic (flexible) points require greater operating forces, and precautions must be taken so that the gap presented to the wheels by the open blade is adequate.

3.2.2.2 Profiles of rails and points

The development of the railways has led to a continual increase in the loads carried by axles and points have had to be adapted to take account of this.

Hence rail profiles have succeeded with many intermediate rail profiles, such as 40kg and 50kg and subsequently to UIC 60.

3.2.2.3 Sleepers

For many years wooden sleepers were practically the only ones used by European networks, but nowadays they tend to be supplanted by concrete sleepers over complete tracks.

The practical life of concrete is in fact much longer than that of wooden sleepers; they need less maintenance and, being heavier, they give the track greater stability.

Their integration with track equipment, however, is much more difficult as they require the use of powerful handling equipment. It should be understood that at present when renewing a set of points, the unit is assembled in its entirety at the side of the track and, when ready, is shifted into position. If the sleepers are made of concrete, this technique requires very powerful and very precise handling equipment.

Nevertheless, it would appear that in the long term, track equipment will be laid on concrete sleepers.

3.2.3 Development of the fittings

In parallel to the technological development of track equipment, their fittings (ie, the control, locking and detection devices have become gradually more sophisticated allowing track equipment to be crossed with complete safety at ever increasing speeds.

3.2.3.1 Local control – remote control

Individual local control of points by counterweights was followed by the appearance of signalboxes, giving concentrated control by means of bifilar connections or rod transmissions terminating at a mechanical operating unit. The field of action of this bifilar control is relatively limited (500m). Beyond that, electrical control is necessary. Nowadays, the latter is becoming more general.

3.2.3.2 Trailability

Points are trailed when a vehicle passes the heel when the blades are not positioned for the wave in question.

Trailing the points generally occurs when the signals have been jumped or when executing movements without signals.

Railways very quickly experienced the need to have track equipment that could be trailed, particularly in zones with frequent, low speed manoeuvring.

As a result, trailable control equipment and locking and detection devices were developed. With the increase in speeds, however, and the lengthening of points, the effectiveness of this trailable equipment was greatly reduced. At present, it is accepted that beyond 30–40km/h, irrespective of the trailable devices used, trailing the points *ipso facto* causes damage to the equipment. This is the reason why some networks have abandoned the trailing principle as soon as the normal points crossing speed exceeds a given threshold.

The situation amongst the European Networks

Set out below are the principles of the points switching equipment and the trailability adopted by some of the European railway administrations.

Austria (ÖBB)

ÖBB use the same technology as DB AG – ie, trailable switching equipment with integral locking control and external locking of the points by claw locks.

Belgium (NMBS/SNCB)

The switching equipment
This simultaneously provides the functions of position command and control of the points. An internal device immobilizes the equipment at the end of its travel. Mechanical locking of the points is done by horizontal hooks. Only the closed blade is locked.

The switching equipment and the hook locking device are trailable.

Controlled trailability
Some switch machines are equipped with an additional device (electromechanical actuator) which renders non-trailable the point equipment locked by the working of a route.

The additional electromechanical actuator drives home a clamp lock in the manoeuvring bar. The utilization and functioning criteria of the actuator are as follows:

- it is used on track apparatus crossed at more than 90km/h on straight track;
- it clamps the bar when the work of a route triggers the point machine.

Switzerland (SBB)

SBB have opted for trailability of the points switching equipment in the same way as DB AG. External locking of the points is done by a special ratchet clamp lock.

Germany (DB AG)

DB AG have opted for trailability of the command and locking equipment. The facing points are or are not locked by claw locks, depending on the section speed of the track apparatus.

The points switching equipment provides both the command and control functions for the locking of the points.

France (SNCF)

With regard to the trailability of the track apparatus equipment, SNCF have adopted the following principles:

- the points of switching zones travelled at low speed are equipped with trailable and reversible switching machines; the points are not at the point;
- on the points of the main track and tracks other than those already mentioned: non-trailable switching equipment immobilized at end of travel:
  - without locking of the point if the point is only crossed at a speed below or equal to 40km/h;
  - locked by clamp lock/housing/slide plate if the points are crossed facing at a speed in excess of 40km/h.

At speeds below 40km/h points controllers check the closing and opening of the facing points.

At speeds in excess of 40km/h position control of the points is provided by the clamp lock/housing/slide plate monitor.

However, on single track lines, in order to case the traffic operation, the switches are locked and trailable at a speed of 60km/h.

Great Britain (BR)

On BR switching of the traffic type of points is done by non-trailable apparatus and the points are locked by an external device. The trailable equipment is used on points sets in marshalling yards; the closed point is simply held by a spring or by pneumatic pressure.

Italy (FS)

FS use two types of switching equipment:

- point machines with internal locking;
- point machines with external locking.

Both point machines are trailable.

For both, temporary non-trailability can be achieved by means of electromagnetic clamps installed on the machine itself. Permanent non-trailability can be achieved by the addition of an internal mechanical device in the first case and by the use of non-trailable point locks in the second.

Non-trailability is required when the points are passed facing at 60km/h or more.

Luxembourg (CFL)

CFL have also opted for DB AG technology.

Netherlands (NS)
NS have adopted the following principles:

- for special sets of points (catch points, . . ., movable scotch block non-trailable switching equipment incorporating point position controller;
- for other points trailable switching equipment with two command bars (one for each point rail) incorporating point position controller;
- no external point locking is used; only internal locking of the switching apparatus is used.

Note

At high speeds (for example, in excess of 200km/h), the trailability of switching points is no longer acceptable as it would constitute a danger to the safety of traffic. In fact, the vibrational stresses in all directions created by the passage of trains along the tracks are then such that there is a risk of causing untimely changes in the position of the points.

On the SNCF TGV network, for high speed track apparatus, switch blade application controllers are installed between the additional unlocked engagements.

On DB AG, several switching machines are necessary to command the points crossed at high speed. This equipment develops an augmented throwing force. All the additional engagements of the points are locked and controlled.

On FS for speeds above 180km/h, two systems with external locking are used. Circumstances dictate whether the electromechanical or the electrohydraulic type is installed.

3.2.3.3 Clamping or locking of the points

To oppose any abnormal displacement of the points during the passage of traffic, the points are immobilized by permanently installed devices, the type of which is determined according to the crossing speed and direction, the situation of the equipment on the main tracks or on the service tracks and the points throwing method.

The concepts of clamping and locking the points throwing equipment are now discussed in further detail.

The clamping concept

The failure of just one of the constituent parts of the points immobilization system is sufficient to release it.

This mechanical protection against bursting acts as a kind of mechanical fuse between the external equipment and the actual switching apparatus.

The locking concept

The simultaneous failure of at least *two* constituent parts of the points immobilization system is *necessary* to release it.

Sundry clamping and locking equipment is described in **3.3**.

3.2.3.4 Permissive control

The permissive control of a points locking device is the maximum longitudinal displacement of the blade in relation to its stock rail that this device can tolerate whilst guaranteeing adequate locking.

The relative displacement of the blade in relation to its stock rail originates in:

- thermal expansion of the switch blade being greater at its facing point, the longer the point;
- machining and assembly play of the elements fixing the switch blade to the stock rail combined with the action of the train wheels.

Originally, the equipment used had a very low permissive control (non-permissive equipment). It was only suitable for short switch points. Gradually the equipment has been developed to tolerate a much greater permissive control; so much so that nowadays, equipment is described as permissive or highly permissive (for example, claw locks and clamp locks).

3.3 Control, locking and detection devices

The main features of points control are:

- throwing over the points
- holding the switch blades in their final positions;
- locking the switch blades in their final positions;
- electrical detection of the final position of the switch blades

The various devices in existence on European railway administrations are discussed below.

3.3.1 Mechanical points operation

Currently, mechanical systems are only in general use on some lines where traffic is not very frequent and on service tracks which are only used occasionally.

Apart from a few railways such as FS (60%), BR (45%), RENFE (39%) and NSB (25%), where the number of systems still in use is substantial, this type of operating device is tending to disappear in other countries (<10 to 15%).

Switch blades can then be moved:

- either actually on the spot, with an interlocked lever; or
- via a wire or rod transmission, a lever being still used at the command end to move the switch.

In both cases, a single or double keylock may have to be used before the signalman is able to use the lever.

3.3.1.1 Double key operated points
Instead of the points mechanism, an external box is connected. Locked in this box is a key. A second key is locked in the points – this inhibits throwing over the points.

If the position has to be changed the operator releases the key in the box by a command. The key can be taken out of the box and inserted into the points. As soon as both keys are in the points, the points can be thrown over manually by a lever. Depending on the actual position, only one of the two keys can be taken out and put back in the box again, the other one remains locked. The actual key in the box indicates the actual locked position of the points.

3.3.1.2 Single key operated points
Single key operated points are used where only one position of a set of points is needed for train routes.

An external box is situated near the points. Locked in this box is a key. This means that the points are locked in their normal position and cannot be thrown over.

If, for shunting purposes, the alternative position is needed, the operator releases the key in the box by a command. The key can be taken out of the box and inserted into the points. As soon as the key is in the points, the points can be thrown over manually by a lever. But the points must be returned to their normal position again in order to remove the key from the points and lock it in the box.

3.3.2 Electrically powered points operation
Switch point control is almost exclusively carried out by electrically operated units, powered either by direct current at 110–150V, 144V, 220V (NMBS/SNCB, BR, FS, CFL, NS, BV) or by alternating current at 150V, 220V – 3 × 380V (ÖBB, SBB, DB AG, RENFE, SNCF, CFL, NSB).

The choice of electrical power DC or AC is motivated by the need to have the necessary energy permanently available for this control.

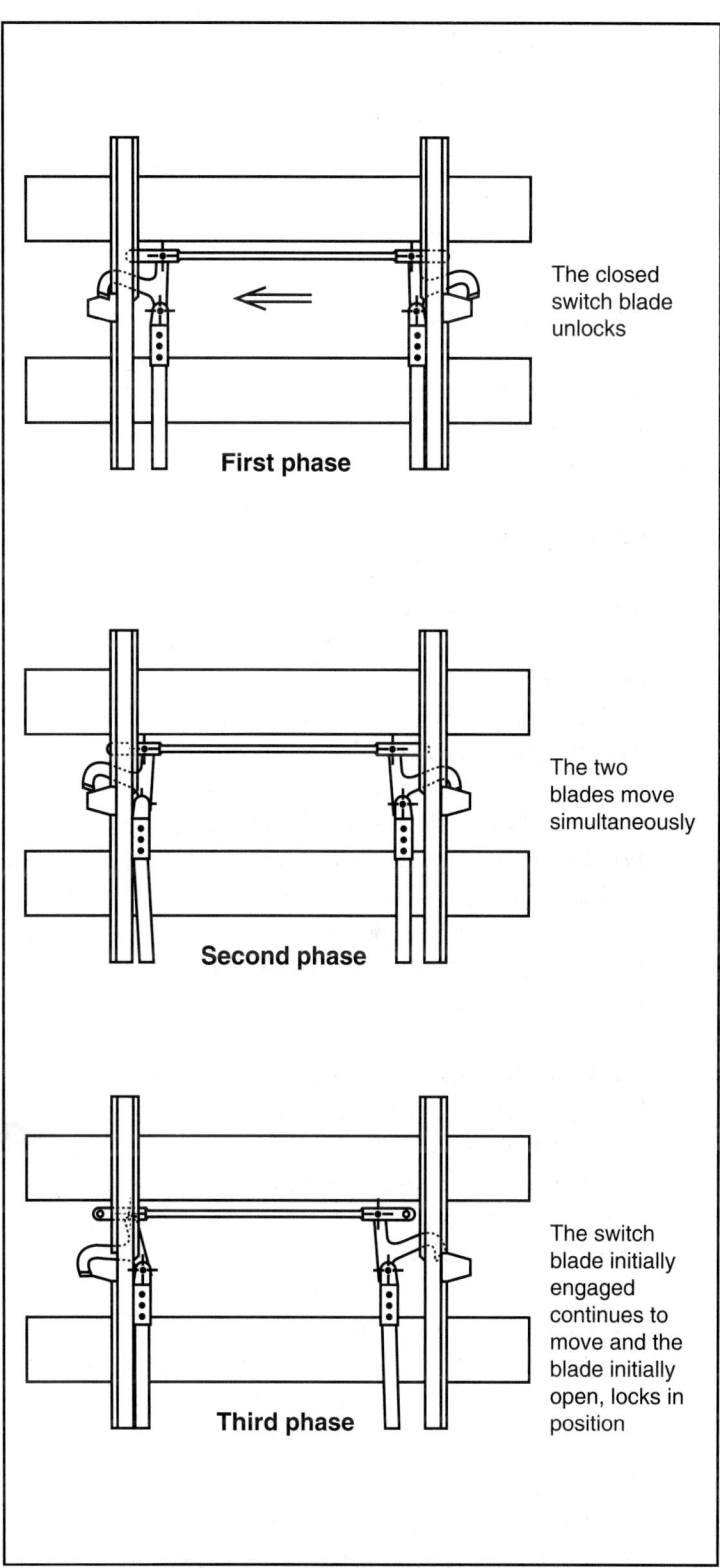

Fig. 3.3 Operation phases of points movement

3.3.2.1 Principle of operation
The operation of a points comprises of three phases:

First phase
The closed switch blade unlocks.

Second phase
The two blades move simultaneously.

Third phase
The switch blade initially engaged continues to move and the blade initially open, having completed its displacement, locks in position.

3.3.2.2 Note
On DB AG, CFL and FS, the principle of operation is:

First phase
The open blade moves first.

Second phase
As the open blade moves, the closed blade is unlocked: both blades are now moving simultaneously.

Third phase
The blade initially open is locked in the closed position while the other blade is still moving further on until reaching an open position which enables this blade to unlock the closed blade in case of being trailed.

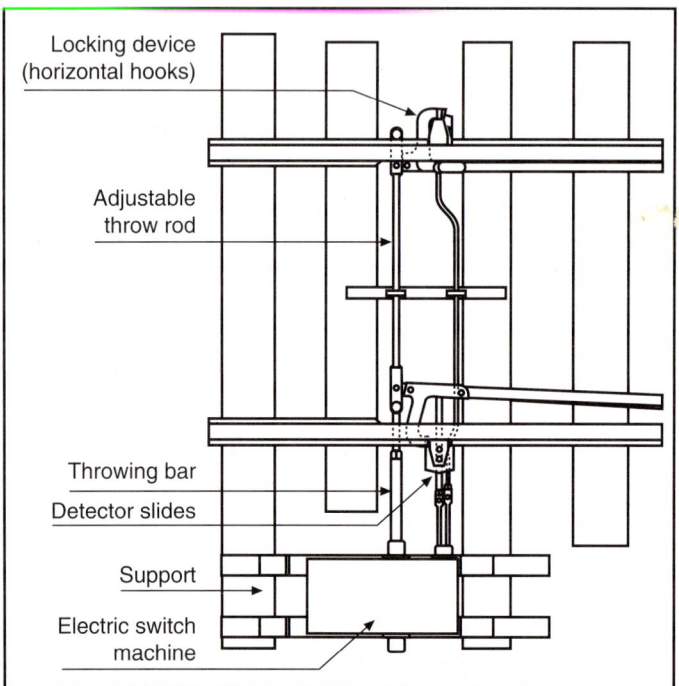

Fig. 3.4 Electric switch machine with throw and detector slides

3.3.2.3 Typical solution
The electric motor and its drive assembly are built into a cast iron housing equipped with a lockable sheet steel cover. The rotary motion of the motor is converted to the linear movement of the throwing bar by means of a reduction gear, a throwing and retaining clutch and a rack and pinion.

A built-in friction clutch containing a retaining and a throwing spring or being of an elastic type, prevents the motor from being damaged should the switch rails be obstructed. The friction clutch (retaining) keeps the facing point lock in position against the tension of the open switch rail.

Connection of the machine drive to the points is made from the throw bar to the stretcher rod by an adjustable rod. Each switch blade is fitted with a detector rod leading back to the machine to operate normal and reverse detection contacts.

Provision is made for hand cranking the machine in the event of a power failure or during installation and maintenance. The crank handle then inserted locates on a square extension of the ball screw shaft or directly on the motor axle.

Whilst hand cranking is in operation, the supply to the motor is isolated. Apart from the disconnection of the motor supply, all the normal functions of the machine – ie, locking and detection – are maintained during the hand cranking process, except on BR and FS where the point detection is not maintained at all during the hand cranking process.

3.3.3 Electrohydraulic point machine
A point machine, in which the throw bar is operated hydraulically rather than purely mechanically, is also used on ÖBB, DB AG, RENFE and CFL.

The electric motor, which is less powerful and more compact than those generally used in point machines, drives an oil pump. Fluid similar to that in aircraft hydraulic systems is used to maintain near constant characteristics in the varying temperatures encountered in those countries.

The drive motor, the hydraulics, the mechanical transmission links with the retaining device, the point detector locks, the detector bars with the monitoring devices and the oil storage container are built into a cast iron housing, equipped with a lockable sheet steel cover.

The force required for throwing the point is supplied by an electrically driven motor. The hydraulic modules consist of a reversible gear pump, two double-acting hydraulic cylinders, two adjustable relief valves, two check valves and the oil storage container.

Operation

With the operating command from the control equipment, the drive motor is started and the pump forces oil into the cylinder.

The cylinder moves a T–lever, whose rotary motion is converted into the required linear motion. When the end position is reached, the motor is switched off and the pump thus stopped.

In the final position, the switch blade is locked to the stock rail by the point lock, for example clamp locking device on the point, as the point machine does not itself have an internal locking device. The locked condition is maintained by the mechanical retaining device of the point machine.

Openings are provided for hand operation by inserting a crank handle which disconnects power operation.

3.3.4 Electrohydraulic point machine, with rail clamp locks

If the switch rail does not fit tightly up to the stock rail, the point detection circuit will not be completed and the signal will not clear. This common failure, experienced by BR, has led them to introduce the clamp lock, in which the switch rail on the closed side is clamped to the stock rail.

A clamp lock is provided for each switch blade driven electrohydraulically from two single-action actuators secured to a central thrust bracket fixed to the point soleplate. The two clamp lock units are connected by an insulated coupling bar. Hydraulic pressure is applied to either actuator from a power pack located outside the track. This contains a reservoir, an electric pump and normal and reverse control valves connected to the actuators by flexible hoses.

The device works on the principle of clamping the switch blade to its respective stock rail using a hook shaped lock arm pivoted from the switch blade. The actuator drives a lock slide which engages with the lock arm to lift it behind the stock rail thereby clamping and locking the blade firmly to the stock rail. At the same time, the coupling bar which links the two lock slides pulls on the opposite arm to release and open the opposite switch.

Detection is provided on each clamp mechanism by sealed limit switches.

Hand operation is effected from the power pack by disconnecting the electrical controls and inserting a crank handle.

3.3.5 Electrohydraulic point machine with hydraulic power unit with accumulators

For high speed lines point operation FS developed an electrohydraulic system consisting of:

- two hydraulic power units with accumulators, one for the blades and one for the movable frog. The units comprise an electric motor, a pump, two accumulators, a reservoir, non-return, safety and control valves. A complete throw over of the switching apparatus is accomplished in five seconds. Energy stored in the accumulators allows six consecutive operations;
- two single-action hydraulic cylinders for the switch tips which control two point locks;
- actuators whose number is different for points and movable frog secured to sleepers or the cradle of the frog. They are essentially double-action cylinders, internally locked by a non-trailable mechanical device and provide electrical position detection when operation is completed.

Emergency operation is carried out through the same hydraulic circuit used in normal operation and is controlled electrically by a three-position lever.

Simplified arrangements have also been designed for switches with intermediate tangents.

3.4 Locking devices

In order to prevent any abnormal displacement of the points during the passage of trains, the points switching mechanism must be immobilized.

Initially, clamp locking was ensured internally within the switch machine, thus immobilizing the points in relation to this apparatus fixed to the track sleepers. Clamping was therefore only totally efficient if each stock rail was itself absolutely solidly fixed to the sleepers.

In practice, on a track in service, the stock rails shift slightly and the equipment described above requires close monitoring. Under these conditions, it is a natural development to clamp each set of points in relation to its stock rail by an external device close to the tips of the points. This solution is used in the majority of railway networks.

At NS and FS (for speeds of up to 180km/h) however, *the tips* of the points are not equipped with any clamping device. The manoeuvring apparatus, robustly designed, alone provides the clamping function.

Several clamping devices are examined below.

Fig. 3.5 Clamp locking with horizontal hooks

3.4.1 Clamp locking with horizontal hooks
Clamp locking with horizontal hooks, widely used in Belgium, is a simple fitting. However, it does not provide for clamping the open point.

A bracket is bolted to the end of the point. It comprises a vertical pin around which pivots a hook which acts as the clamp lock by engaging with a pillow block fixed to the rail.

3.4.2 Claw locking
Claw locking has been adopted by ÖBB, SBB, DB AG, RENFE and CFL.

The essential characteristics of the claw lock are as follows:

- its robustness;
- its great intrinsic permissiveness;
- its wedging the open point in addition to locking the closed point;
- its trailability;
- its control over the position of the points must be guaranteed either indirectly in the manoeuvring equipment or directly by separated detectors.

It consists of:

- the lock chambers attached to stock rails;
- the lock/drive slide;
- the dovetail-shaped clamps bolted to the blades, movable within the constraints of the lock chamber and the matching recesses in the lock/drive slide.

The lock/drive slide is operated from the tumbler/point machine. During the first part of the operation the dovetail of the offlaying blade is kept in the lock chamber until it can escape into the recess of the lock/drive slide and thus unlock the blade. Simultaneously, the offlaying blade moves towards the stock rail, driven by the clamp via recess and engaged dovetail. When the onlaying blade is unlocked it joins the move, also driven by its clamp via recess and dovetail. This process continues until the new onlaying blade touches the stock rail. At this moment, the clamp's dovetail is forced out of the recess of the lock/drive slide to clamp behind the lock chamber, locking the blade securely to the stock rail. The lock/drive slide and the offlaying blade continue to move for a distance until sufficient overlap on the dovetail is provided.

FS have developed a similar system for speeds above 180km/h.

3.4.3 Locking by clamp lock/housing/slide plate (verrous–carter–coussinet:VCC)
Locking by clamp lock/housing/slide plate has been widely adopted by SNCF, in particular on all HST points. The system has also been installed by FS on the high speed line Roma–Firenze.

Differing greatly in design from that of the claw locks, it nevertheless has some similar characteristics such as robustness, adaptability to types of switch blade, permissiveness, etc.

This apparatus immobilizes a set of points by acting directly on the switch blades. The VCC are always used in pairs.

Fig. 3.6 Locking by clamp lock/housing/slide plates (VCC)

Apart from locking both blades, the two VCC internal controllers control the following functions: the position of each blade closed or open; the engagement of the lock of the closed blade; the effective opening of the open blade.

VCC are called dependent locks since they are moved together with the switch blades.

Of paramount importance is the fact that VCC controls are directly mounted on the pieces of equipment themselves that have to be controlled, without relying on any intermediate device.

3.5 Point detection

SNCF and FS

Besides the VCC-type locking equipment, it is necessary to monitor the position of a point individually.

Electrical control devices, also called blade controllers, are then used.

These are fitted to the tip of the point and made up of a supporting box fixed to the stock rail, on the outside of the track, containing electrical contacts and connected via a mechanical linkage to the mobile switch blade. Blade controllers are also used in Italy where points feature no internal point lock.

BV and many other European railways

On account of severe winter conditions, special precautions have to be taken.

The area between the switch rail and stock rail is heated (>7kW) to free those parts in the rig from snow and ice as much as possible.

No rail clamps are used at the points machines.

The point mechanism is a closed unit with internal points locking and detection contacts. There is also a small, 15W, heating element built in to prevent frost on the contacts.

In a rig with long switch blades (>10m), force has to be applied at more than one point on the blade. The distribution of the force with stretcher bars and links is impractical in Sweden.

Instead, two or more point machines are used on the same switch blade, one at the tip and the second, with a shorter stroke, halfway to the trailing point.

Intermediate simple, magnet operated contacts with no moving parts, except the blade itself, are installed outside the point machine. The contacts are checked in the detection circuit.

Depending on the type of switch and speed (>40km/h) through it, one or more contacts are sometimes necessary to check that the track gauge is correct and not too narrow. The contact part is mounted in a slide chair on the stock rail and the actuating permanent magnet is fixed to the switch blade.

3.6 Derailer

The derailer has the appearance of a half switch. It is intended to provide protection from an adjacent track, if necessary, by derailing a vehicle in an assigned direction; to do this, it creates discontinuity in a stretch of rails, thus directing the vehicle towards the ballasting.

Generally, the same type of electric point machine as for points are used for derailer stops. Sometimes they have a different gearing or different number of throw bar but basically they are the same.

FS do not use derailers but scotch blocks instead.

3.7 Command and control circuits

The electrical control of points may be done according to two essentially different techniques.

Under the technique called *direct command*, the power source is in the signalbox and the relay and contactors are concentrated in the box, as are the control circuits' source, as well as the control relay(s).

Under the *relayed command* technique, the power source is outside, close to a set of points. The same applies to the control source for the set of points in question.

As regards the cabling to be used, there are two distinct types of design: in the first, the wiring of the circuits connecting the switching apparatus and the control devices in the open to the signalbox are separate. Even different cables are used.

In the second design, the circuit diagram is designed in such a way as to allow the use of the *same wiring* for command and control.

The power source can use either direct or alternating current.

A non-presetting control diagram has been prepared.

Switching on the motor with a view to the displacement of the points is only done for the short period that follows its actual command; consequently, in the event of locking before the end of travel, the motor does not remain under power for a long period, since this could cause defects such as excessive frictional heat.

Any switching command that has not been carried out within 10 seconds is automatically cancelled by the triggering of a power contactor. The relay contacts are therefore no longer led to cut off the switching current.

Summary of the safety conditions

When a set of points is immobilized in the signalbox by a wave line, premature operation of its switching apparatus must be rendered impossible, even if a malfunction occurs or if the latter is already present on the command and control circuits.

The de-energized state of control must correspond to any position of the points outside the tolerances.

Fig. 3.7 Electric point machine application

Main Features	Units	AUSTRIA ÖBB	BELGIUM NMBS/SNCB	SWITZERLAND SBB	GERMANY DB AG	SPAIN RENFE	FRANCE SNCF
Number or % on the net		2500 (35%)	9500 (91%)	12 000 (99%)	85%	5800 (61%)	100%
Speed limit	km/h	200	160		200	200	230
Trailable		yes	yes	yes	yes	yes for 20%	no
Operating voltage range	Volt	3×380 AC	150 DC	3×380 AC	3×380 AC	2×220 AC	3×380 AC
Maximum line resistance	Ohm	39	18	40	70	8	50
Current absorption	Amp	2	5.8	1.2	2	5	1.5
Operating cycle duration	sec	2.5–5	5–8	4.5–7	5–6	6–7	4.2
Switch travel	mm	150–220–240	160–240	150–240	150–220	220	110–260
Throwing force	daN	500–800	400	580	5000	392–638	400
Trailing protection load	daN	700–1000	600–1000	980	7000	588–1472	–
Length/width/height	mm	1271/546/330	820/452/312	872/544/367	1260/434/290	1080/580/390	700/535/215
Weight	kg	110	150	180	110	190	91
Locking points:							
• inside of the point machine		–	–	–	–	yes	–
• separated facing point lock		yes	yes	yes	yes	–	yes
Number of lock slides		1	1	1	1	1	0

SWITCH OPERATING AND PROVING SYSTEMS

Main Features	Units	GREAT BRITAIN BR	LUXEMBOURG CFL	NORWAY NSB	NETHERLANDS NS	SWEDEN BV	ITALY FS		
Detector function									
• inside of the point machine		yes	yes	yes	yes	yes	—		
• inside locking device		—	—	—	—	—	yes		
Number of detector slides		2	2	2	1	2	—		
Emergency operation									
• hand crank		x	x	x	x	x	—		
• lever		—	—	—	—	—	x		
Number or % on the net		8500 (49%)	550	2200	4600 (87%)	4700 (90%)	40%		
Speed limit	km/h	200	140	130	140	130/200	300	250	180
Trailable		no	yes	yes	yes	yes/no	yes	no	yes
Operating voltage range	Volt	110 DC	3×380 AC	220 $16\frac{2}{3}$Hz	150 DC	220 DC	144 DC		
Maximum line resistance	Ohm	1	70	6	12	30	5	5	8
Current absorption	Amp	6	2	3–4	5	3.5	5	4	2.5–3.6
Operating cycle duration	sec	3.5	5–6	3	2	3.5	1.5	4	2.5
Switch travel	mm	108	220	60–160	130	170	150	115	150
Throwing force	daN	1000	700	1000	540	450/700	440	590	600
Trailing protection load	daN	8000	1000	1000	600		780	—	800
Length/width/height	mm	1540/460/260	1260/434/290	1260/434/290	880/670/460	774/905/313	913/565/334		
Weight	kg	220	110	110	175	160	150	320	240

SWITCH OPERATING AND PROVING SYSTEMS

Main Features	Units	GREAT BRITAIN	LUXEMBOURG	NORWAY	NETHERLANDS	SWEDEN	ITALY		
		BR	CFL	NSB	NS	BV	FS		
Locking points:									
• inside of the point machine		yes	–	yes	yes	yes	yes		
• separated facing point lock		–	yes	–	–	–	yes	yes	–
Number of lock slides		1	1 or 2	2	2	2	1	1	2
Detector function									
• inside of the point machine		yes	yes	yes	yes	yes	yes		
• inside locking device		–	–	–	–	–	yes	yes	–
Number of detector slides		2	1	2	2	2	–	–	2
Emergency operation									
• hand crank		x	x	x	x	x	x	x	x
• lever		–	–	–	–	–	–	–	–

[1] Limited by 200m maximum length of line circuit in electric traction areas.

Fig. 3.8 Electrohydraulic point machine application

Main Features	Units	AUSTRIA ÖBB	GERMANY DB AG	LUXEMBOURG CFL	GREAT BRITAIN BR	ITALY FS	
Number or % on the net		5000 (65%)	15%	292	8500 (49%)		
Speed limit	km/h	200	140	140	200	300	
Trailable		yes	yes	yes	no	no	
Operating voltage range	Volt	3×380 AC	3×380 AC	3×380 AC	110 DC	144 DC	150 AC
Maximum line resistance	Ohm	2×34	45	45	[1]	8	8
Current absorption	Amp	6	4	4	4	5	0.8
Operating cycle duration	sec	2.5–5	5–6	5–6	2.5	5	5
Switch travel	mm	140–240	150–220	220	108	115	115
Throwing force	daN	800	5000	500	430	1200	3600
Trailing protection load	daN	900	7000	700	8000	–	–
Length/width/height	mm	620/500/330	1300/548/290	1300/548/290	250/250/250[2]		
Weight	kg	160	160	160	25	320	960
Locking points:							
• inside of the point machine		–	–	–	–	yes	yes
• separated facing point lock		yes	yes	yes	yes	(5)	(13)
Number of lock slides		1	1	1 or 2	N/A		

Main Features	Units	AUSTRIA ÖBB	GERMANY DB AG	LUXEMBOURG CFL	GREAT BRITAIN BR	ITALY FS	
Detector function							
• inside of the point machine		yes	yes	yes	yes		
• inside locking device		–	–	–	–	yes	yes
Number of detector slides		2	2	2	2	(5)	(13)
Emergency operation							
• hand crank		x	x	x	[3]	x	–
• lever		–	–	–	N/A	–	Electric

[1] Limited by 200m maximum length of line circuit in electric traction areas.
[2] Plus pump unit 400/250/400.
[3] Hydraulic pump hand operated.

4 Signals

W. BÖHM

Contents

4.1	General	161
4.2	Sizes of signal lights	163
4.3	Types of lamp	167
4.4	Power supply, voltage rating, power of lamps and type of signal circuit	171
4.5	Control and proving systems	175
4.6	What to do in case of burnt out lamps of a main signal	176
4.7	Visibility distance	178
4.8	Distance between main signal and distant signal	178
4.9	Position of the signal posts	180
4.10	Distance between signals and control circuits	180

Illustrations

Fig. 4.1	Positioning of signals	161
Fig. 4.2	Dwarf signal (SBB) with three possible aspects, front side	163
Fig. 4.3	Dwarf signal (SBB) with three possible aspects, rear side	163
Fig. 4.4	Running signal light unit V136 from DB AG (section side view)	164
Fig. 4.5	Two lights with different lens diameters on one signal shield (ÖBB)	165
Fig. 4.6	Sizes of signal lamps	165
Fig. 4.7	Signal repeater (ÖBB) with three possible aspects, as an example of a matrix type signal controlled by halogen lamps and optical fibres	167
Fig. 4.8	Railway administrations where double filaments are in use	167
Fig. 4.9	Matrix type signal (speed indicator) from DB AG (section side view)	168
Fig. 4.10	Types of signal lamp	169
Fig. 4.11	Panel with signal transformers (with taps), including interface for INDUSI control (ÖBB)	171
Fig. 4.12	Typical signal lamp circuit (de-energized)	172
Fig. 4.13	Signal power supply	173
Fig. 4.14	Flashing frequency and ratio of mark/space period for flashing signal lamps	175
Fig. 4.15	Multi-aspect signal with junction indicator	181
Fig. 4.16	The new German running signal 'KS' type in situ	181
Fig. 4.17	ÖBB: main signal (showing free) with distant signal (indicating next main signal free)	181

4.1 General

Multi-aspect light signals with separate lenses for each of the aspects are used by most European railway administrations. The exception is Italy, where searchlight type signals are the most numerous in use on the network, together with dichroic mirrors signals.

In several countries searchlight signals, though still in use as an older application, are not considered standard. For example:

- Great Britain (BR);
- Netherlands (NS), defined as colour changer;
- Sweden (BV), for level crossing signals.

Basic signal aspects are indicated by one lamp per visible light and consist of one or several lamps. Additional information or special signals have a certain shape consisting of a matrix of light points. Examples include:

- signal repeaters on ÖBB, and NMBS/SNCB;
- special shunting signals and supplementary signals to main signals on SBB;
- numbers for speed or track indication, letters and other indicators used in the majority of countries.

In most cases a defined matrix of light points is controlled by one halogen lamp, the light being conducted to the light points along optical fibres.

Fig. 4.1 Positioning of signals

Railway	Height of lamps on signal posts	Aspects displayed at the same time on the same mast
ÖBB	between 3.7–6.5m above top of rail	maximum 3: • main signal with 1 or 2 lamps (additional speed indicator possible) • distant signal with 2 or 3 lamps (additional distant speed indicator possible) • departure signal
NMBS/SNCB	Lamp height of the red lamp normally 3.6m above top of rail with exceptions (lack of space) at 4.67 or even 5.35m	maximum 3: • main signal • speed indicator • indicator for wrong track or for dead end track
SBB	between 4.2–5.5m above top of rail	maximum 3: • main signal • distant signal • track occupancy signal or departure signal
DB AG	between 3.8–6.2m above top of rail	maximum 2: • main signal with 2 lamps at maximum (3 additional indicators for speed, direction and changing of track possible, or an additional subsidiary signal) • distant signal with 2 lamps (2 additional distant indicators for speed and direction, 1 additional lamp for short route possible) For the new 'Ks' signals: • only 1 aspect (indicators possible as above)

Railway	Height of lamps on signal posts	Aspects displayed at the same time on the same mast
RENFE	between 4.2–6.2m above top of rail	maximum 3: • main signal • speed indicator • direction indicator
SNCF	generally on level with driver's eye	maximum 2: • 'A' and 'R', or • 'A' and 'flashing R' Third light indication can be added: • route indicator lamp(s)
BR	the red aspect lamp is 3.7m above the top of rail, on the driver's eye level with 279mm spacing between each aspect	maximum 3: • main signal • route indication • position light subsidiary or 'Right away' indicator
FS	between 3.9–7.5m above top of rail	maximum 2: • main signal • distant signal, additional, speed indicator and directions signal
CFL	between 3.8–6.2m above top of rail	maximum 2: • main signal, additional speed indicator and route indicator possible • distant signal, additional distant speed indicator • route signalling indicator possible
NSB	between 3.5–4.9m above top of rail, except dwarf signals, repeater signals: 1–2m	maximum 3: • main signal, additional speed indicator and route indicator possible • distant signal • dwarf signal
NS	between 4.5–6.2m above top of rail	Maximum of one signal, additional speed indicator possible
BV	centre of signal head in line with the driver's eye level, approximately 3m above top of rail Signals on signal bridges may have the top lamp fitted 6.2m above top of rail	Only 1: • main signal, including built-in distant signal aspects if any • freestanding distant signal

In the older installations multi-lamp indicators are used, partly controlled by two different circuits, so that in the event of a one lamp failure it is only light intensity that is reduced.

Multi-lamp indicators on BR and NS are arranged with the lamps in several circuits to enable different characters to be shown using various combinations of lamps.

Signal lamps for the majority of signals are mounted on extended height signal posts, or alternatively, should this arrangement not be possible, they are mounted on signal bridges erected over the track.

BV additionally requires that simultaneously lit lamps are spaced at least 700m apart for main and distant signals.

On most networks, except NS, dwarf signals are generally used as shunting signals. On NS shunting signals do not exist, but dwarf signals are used as main signals in areas with a maximum speed not exceeding 40km/h.

On ÖBB dwarf signals are also used for main signals serving as start signals, starting from dead end tracks. On RENFE, dwarf signals can act as exit signals from the recessing siding and as entry or exit repeating signals. On NSB a dwarf signal can, in addition, be used in conjunction with a main signal.

4.2 Sizes of signal lights

The size of a lens and of the dispersion angle differs widely, partly dependent on the required visibility distance (and therefore on the type of signal), and partly dependent on whether the signal is located on a signal post or a signal bridge, or alternatively if it is a dwarf signal.

Heating of the front panes of selected important lights is sometimes provided, for example:

- On DB AG, for red at automatic block signals, yellow at certain signal aspects and on matrix signals.
- On SBB, for critical signals on mountain railways and generally associated with the new signal system 'N'.
- On ÖBB it will be provided in the future for those light signals that are to be replaced by halogen lamps and optical fibres.

Fig. 4.2 Dwarf signal (SBB) with three possible aspects, front side

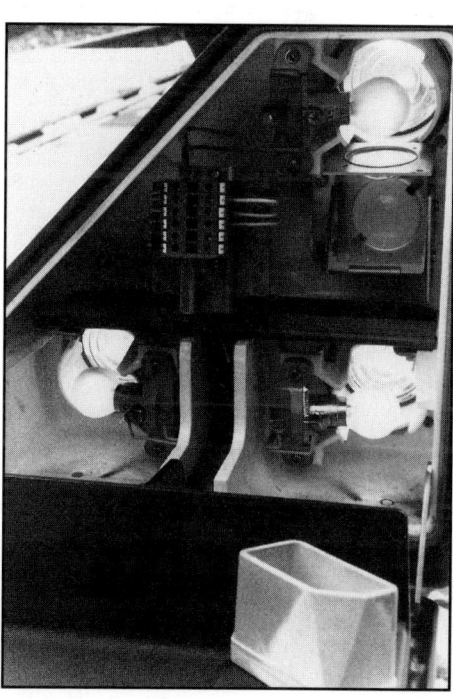

Fig. 4.3 Dwarf signal (SBB) with three possible aspects, rear side open

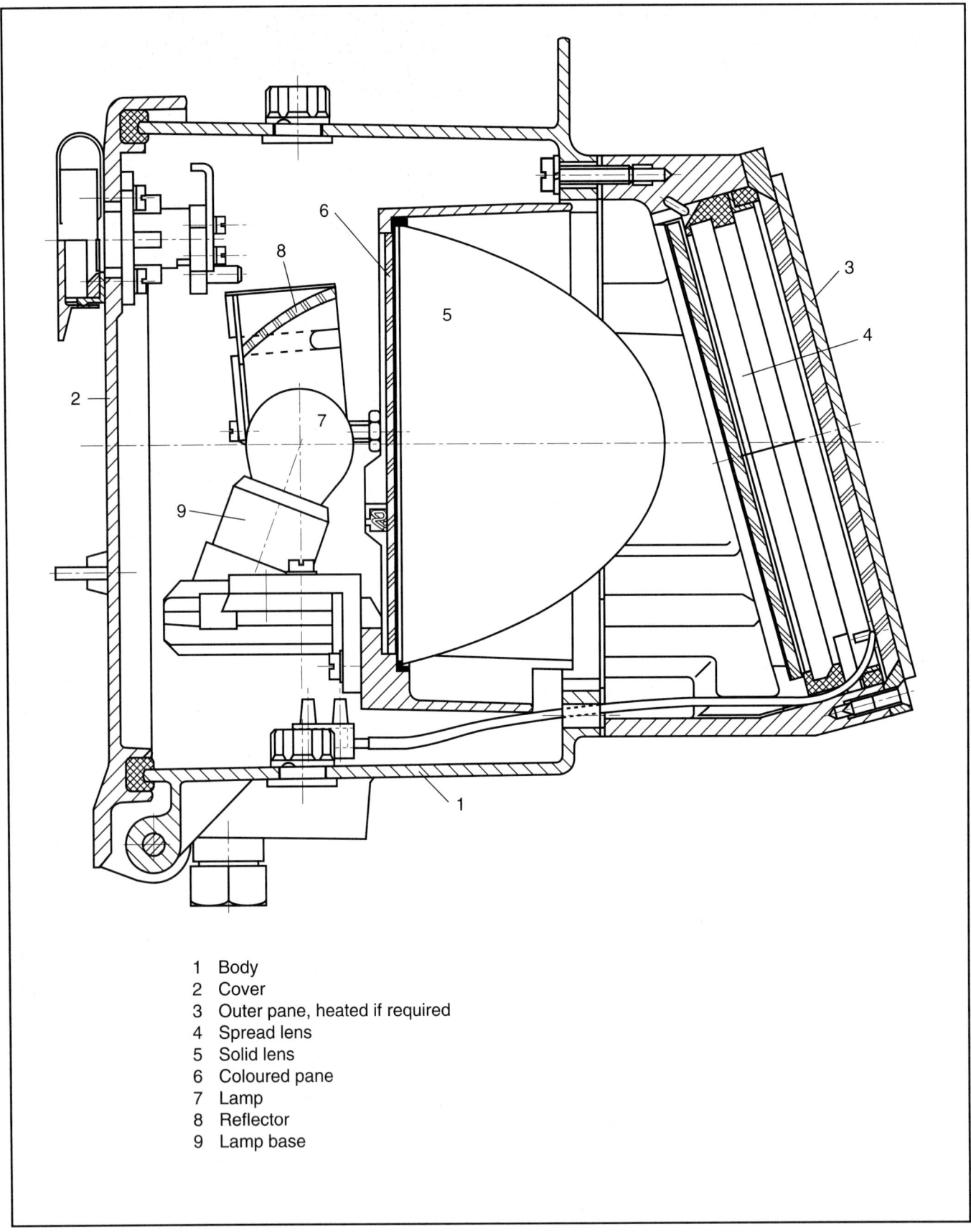

Fig. 4.4 Running signal light unit V136 from DB AG (section side view)

Fig. 4.5 Two lights with different lens diameters on one signal shield (ÖBB) (right)

Fig. 4.6 Sizes of signal lamps (below)

Railway	Lens diameter	Horizontal angle of dispersion	Type of signal, dependency
ÖBB	200mm		new types of main signals and distant signals
		4° or 6°	far visibility distances
		16°	close visibility distances and in curves
	70mm	not defined	shunting, subsidiary signal
	112 or 136mm	spread lens, angle not defined	older types of main signals, and distant signals, all other signals
NMBS/SNCB	160mm	4°, 10°, 20°	main signals: horizontal, 10° always for white
		1° 30' symmetrical	vertical on signal posts
		4° 45' asymmetrical	vertical on signal bridges
	160mm	60° symmetrical	shunting signals: horizontal and vertical
SBB	190mm	8°	visibility distance 350m
		20°	visibility distance 240–300m
		8° T45	visibility distance 260–280m
		8° T70	visibility distance 280–310m
DB AG	136mm	spread lenses with different lobe appliable and adjustable in the frame	main and distant signals
	70mm		shunting, subsidiary signal

166 SIGNALS

Railway	Lens diameter	Horizontal angle of dispersion	Type of signal, dependency
RENFE	210mm translucent step lens 125mm colour lens baffle lens in curves		all signals for better visibility
SNCF	160mm	10°	main signals
	70mm	8°	auxiliary absolute stop signal crossing light ('*oeilleton*')
BR	doublet lens: clear outer lens 213mm, coloured inner lens 140mm	5° 30°	main signals main aspects have a sector in outer lens to deflect part of the beam towards a driver of a train stopped close to the signal range of 1500m range of 600m
	doublet lens: clear outer and coloured inner lens both 92mm		dwarf signals, position light, shunting and subsidiary signal
	doublet lens: outer lens contains a prism section	reflector to broaden the beam for short range visibility	position light junction indicators, visibility distance 738m
FS	up to 149mm (working length 137mm)	7° symmetrical	curve radius >7500m
		14° symmetrical	7500m > radius >1500m
		20° (right or left)	1500m > radius >300m
CFL	136mm	8° and 20° symmetrical or asymmetrical	main and distant signals and most other signals
	70mm	not defined	white aspects in some cases
NSB	278mm (210 or 150mm in tunnels)	6° 16°	horizontal – general horizontal in curves
NS	213mm inner lens for colour, outer lens for beam, additional spread lens	20° asymmetrical	on high posts
	162mm colour lens, inner spread lens outer spread lens	40° 20°	dwarf signals
BV	210mm outer lens, 140mm inner lens + colour filter	6°	main and distant signals: two clear step lenses without any close up sector
	91mm	4°–20°	dwarf signals: clear single step lens

4.3 Types of lamp

Most countries currently use single filament lamps, whilst the general tendency is to change to double filament (except on NS). For some administrations double filament lamps are already standard (DB AG, CFL, for new signal system of SBB, BR). It is only partly introduced on main tracks of main lines (ÖBB).

A second tendency is to change to halogen lamps with conduction of the light by optical fibres. In many countries this has already been carried out, not only for signals or indicators which consist of a matrix of light points, but also for light signals, on signals in stations of high traffic routes, on block signals, on signal bridges and in tunnels, etc. There are no double filament halogen lamps, but in some cases two halogen lamps have been used giving the same effect.

On NS halogen lamps and optical fibres are only installed with signals of difficult access in order to facilitate the exchange of lamps.

Railway	Position of filament	Switch over and supervision
ÖBB	horizontal, parallel, both filaments with the same power	switch over and indication outside
SBB	horizontal, parallel, both filaments with the same power	switch over outside, indication to the signalman
DB AG	horizontal, parallel, main filament in the focus, both filaments with the same power	switch over outside, indication to the signalman
BR	main filament horizontal, auxiliary filament vertical, 3mm behind main, both filaments with the same power	switch over outside, switching relay located inside the signal head, indication to the signalman
CFL	horizontal, parallel, both filaments with the same power – main filament in the focus	switch over outside, indication to the signalman
NSB	horizontal, parallel, both filaments with the same power	switch over and supervision integrated in the geographic interlocking relay group

Fig. 4.7 Signal repeater (ÖBB) with three possible aspects, as an example of a matrix type signal controlled by halogen lamps and optical fibres

Fig. 4.8 Railway administrations where double filaments are in use

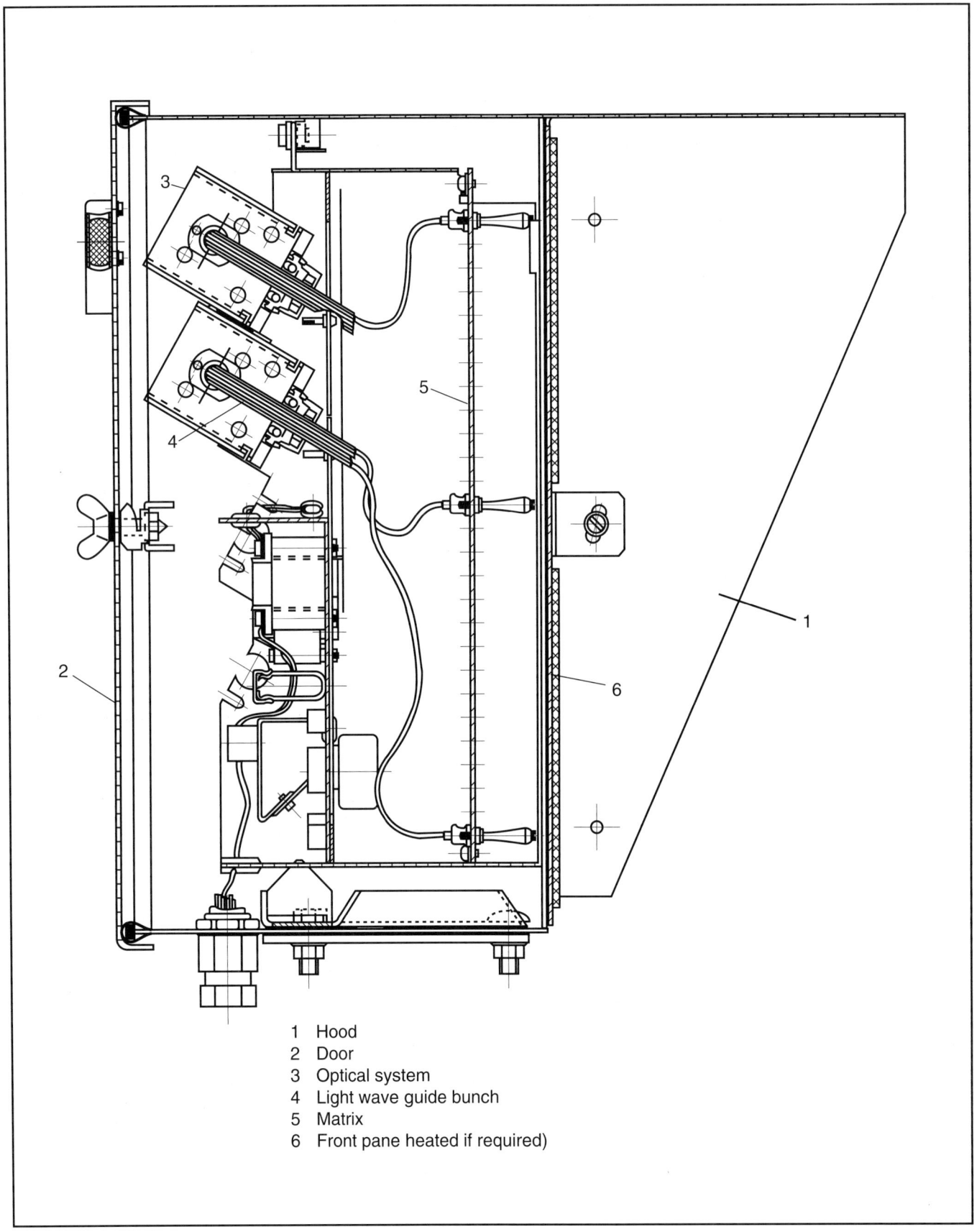

Fig. 4.9 *Matrix type signal (speed indicator) from DB AG (section side view)*

Fig. 4.10 Types of signal lamp

Railway	Changing of lamps/average lifetime requested	Standard lamps single/double filament	Halogen lamps and optical fibres
ÖBB	Replacement only if a lamp is burnt out Requested lifetime 12 months	single generally, double on main tracks of main lines	• signal repeater • speed indicator • distant speed indicator • signals on signal bridges, in tunnels • in the future block signals and main signals with distant signals in stations of high traffic routes
NMBS/SNCB	Permanently illuminated lamps: replacement periodically every 6 months Other lamps: every 24 months Requested lifetime: 4000 hours for 6.5V/7.2V/24V/40V lamps 1000 hours for 10V/110V lamps	single generally	• speed indicator • indication for changing of tracks • halogen lamps with either dichroic mirror or with an incorporated mirror
SBB	Replacement only if a lamp is burnt out Average life time: 6485 hours at day voltage for 40V lamps; 1787 hours at day voltage for 12V double filament lamps	single for old installations of signal system L double for new installations of signal system L, generally for signal system N	• special signals as addition to main signals (speed indication in new signal system, signals with shape of a bar)
DB AG	Replacement only if a lamp is burnt out Lifetime for double filament lamps, 2400 hours, for halogen lamps, 4000–5000 hours	double generally	• single filament/for speed indication, direction indication, change of track, and so on
RENFE	Routine replacement every 5000 hours Average lifetime 5000 hours	single generally	• speed indication • diverging junction

Railway	Changing of lamps/average lifetime requested	Standard lamps single/double filament	Halogen lamps and optical fibres
SNCF	Routine replacement every 4000 hours a lamp was lit Guaranteed lifetime 4000 hours	single generally	• speed indicators (halogen lamps, current study for use of optical fibres)
BR	Replacement only when main filament fails – lamp lifetime 1000 hours	double generally (auxiliary filament: reduced intensity)	• for letters and numbers behind a diffusion screen (matrix plate characters 400mm or 170mm high)
FS	Permanently illuminated lamps on tracks important for traffic: replacement every 2500 hours. Other lamps replaced when burnt out. Average lifetime required on acceptance tests is 2000 hours when fed at 12V, and 3800 hours when fed at 11.5V	single	• halogen lamps on dichroic signals only • no optical fibre on main signals
CFL	Permanently illuminated lamps: replacement periodically every 12 months Other lamps if burnt out	double generally since 1983 (single still in old equipment)	• indicators of theatre type, optical fibre, double filament lamps or two distinct lamp bundle systems with halogen lamps • speed indications • route indications • wrong track traffic indications
NSB	Replacement only if a lamp is burnt out, expected lifetime: 8000 hours	single generally except 'Oslo S'	no
NS	Replacement period 12 months Rated lifetime 16000 hours	single only	signals and speed indicators
BV	Normally lit lamps: routine replacement 12 months Other lamps: 24 months Expected life time: 8000 hours	single generally	• letters • figures and • other signs (matrix 5 × 7) • on departure signals • and on brake test signals

4.4 Power supply, voltage rating, power of lamps and type of signal circuit

The power supply is in most cases an alternating current of 50Hz, with the exception of SNCF where 400Hz is used because of smaller transformers, and NS where 75Hz is used, because a 75Hz power supply is needed for the track circuits.

Most countries use external type signal transformers, located at the signal, with a 110V or 220V primary voltage and transformation to a low voltage, in many cases 12V nominal. During the night in most countries the voltage is reduced to between 60 to 70% in order to prevent undue glare.

The tendency is to lower the power of lamps and to improve the optic system.

Generally lamps are not used at nominal values, but with a voltage reduced by some percentage in order to increase the working life span. For example on DB AG and BV the reduction is to 95%. However, in most cases the voltage is reduced to values between 9.7 and 11.5V instead of 12V nominal.

Fig. 4.11 Panel with signal transformers (with taps), including interface for INDUSI control (ÖBB)

172 SIGNALS

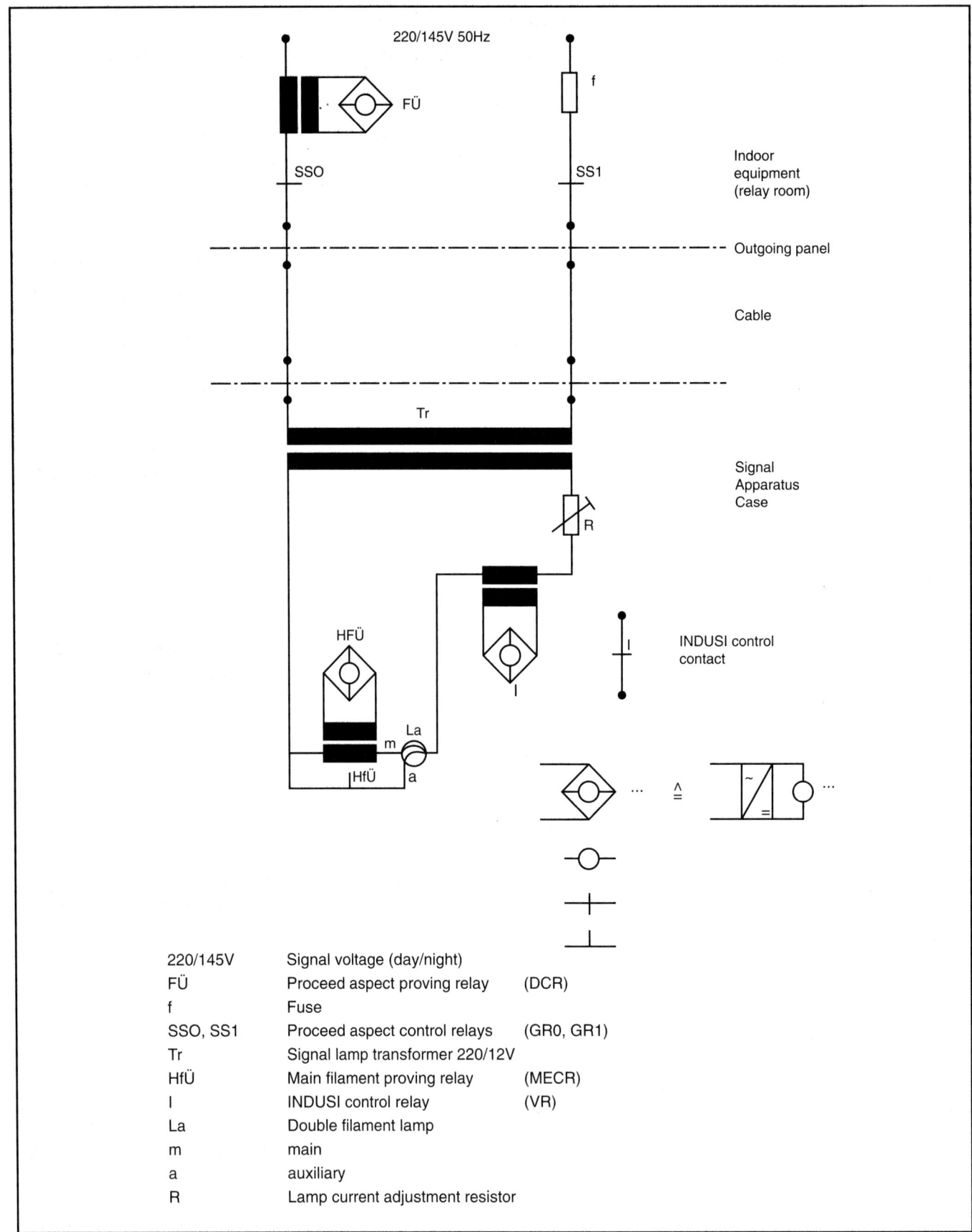

Fig. 4.12 Typical signal lamp circuit (de-energized)

Fig. 4.13 Signal power supply

Railway	Voltage supply	Reduction during night	Nominal voltage rating – power of lamps	Type of signal circuit
ÖBB	220V/50Hz +10/–15%	150–160V	12V/50W (old), 12V/35W (actual), 12V/20W (future) halogen lamps: 12V/42W (old) 12V/35W (actual)	Signal transformer, secondary adjustable with taps, 220V/12V for one, 220V/24V for two lamps in series
NMBS/ SNCB	110V/50Hz +/–5%	only for shunting signals	Main signals: 7.2V/15W Shunting signals: 40V/20W Halogen lamps: 24V/70W (old) 13.8V/30W (old) 12V/20W (new)	Lineside signalbox, lamps directly connected without transformers

separated dichroic mirror incorporated mirror incorporated mirror |
| SBB | Conventional signal system ('L'):

220V/50Hz or 220V/16⅔Hz +10/–15%

New signal system ('N')

220V/50Hz +/–10% | 80% for coloured lenses, 64% for clear lenses

64% | 40V/20W

12V/20W halogen lamps: 12V/20W | taps on secondary side of a central transformer or adjustable by resistors

Signal transformer 220V/12V, taps on secondary side |
DB AG	220V/50Hz	145V	12V/30W (actual), 12V/20W (future) shunting signals: 12V/15W or 10W halogen lamps: 12V/20W	Signal transformer, secondary adjustment by taps
RENFE	220V or 110V/50Hz +20%	no except for fibre optics	10V/19.4W (difference in charge <+/–5%)	Signal transformer 220V or 110V/10V, secondary adjustment by taps 0.5, 0.8, 9, 10, 11, 12, 13V
SNCF	127V/400Hz (alternating from 24V DC)	no (only for special light assemblies 19V/40W 10V drop)	Main signals: 7.2V/15W Level crossings: 6.5V/25W	Signal transformer, secondary adjustable with taps
BR	110V/50Hz	no	12V/24W

position light signal: 110V/35W | Signal transformer 110V/12V in signal head (one for each aspect) with main/auxiliary filament change over relay, taps on secondary side. From apparatus housing to signal length <200m |

Railway	Voltage supply	Reduction during night	Nominal voltage rating – power of lamps	Type of signal circuit
FS	Searchlight type signals:			Signal transformer
	150V/50Hz, 120V in tunnels	no	12V/20W, 9V in tunnels	Separate position and illumination check
	48V DC for position check circuit of the movable vane and for movable vane control			
	Dichroic mirrors signals:			
	150V/50Hz, 120V in tunnels	no	11.5V/20W, 9V in tunnels	Signal transformer
CFL	Old: 35–140V/50Hz	no	30V/15W, 12V/30W	Old installations: fed from the signalbox with 35–140V, adjustable by resistors
	New: 220V/50Hz	145V		New istallations: signal transformers 220V/12V or 220V/30V
NSB	220V/50Hz	190V	12V/24W	Signal transformer 220V/12V
NS	110V/75Hz	80V for signals, only in station areas. 50V for all speed indicators	Signals: Filament lamps 10.2V/30W Halogen lamps 10V/50W Speed indicators: Filament lamps 12V/3W Halogen lamps 10V/50W	Signal transformer 110V/12V, secondary adjustable by resistors
BV	Main and distant signals: 110V/50Hz	70% (77V)	12V/24W	Signal transformer 110V/12V, secondary adjustable by taps
	Dwarf signals: 45–55V	35V	55V/24W	Fed from relay room – no separate transformer
	Halogen lamps: (for matrix indicators)		70W	
	Solid state interlocking: Current for a lamp 1.9A	1.3A	12V/24W	Microprocessor controller, maximum 50m from the signal

Railway	Flashing frequency	Ratio of mark/space
ÖBB	60 pulses/minute	ratio 1 : 1
NMBS/SNCB	60 pulses/minute (+/−4%) except road signals at level crossings: 0.92Hz = 65 pulses/minute 1.71Hz = 35 pulses/minute	ratio 1 : 1 ratio 1 : 1 red lights ratio 1 : 1 white lights
SBB	60 pulses/minute	ratio 1 : 1
DB AG	60 pulses/minute	ratio 1.5 : 1 to 1 : 1
RENFE	60 pulses/minute	ratio 1 : 1
SNCF	70 pulses/minute	ratio 1 : 1
BR	60 pulses/minute	ratio 2 : 1 achieved by switching power 670ms on/330ms off
FS	60 pulses/minute	ratio 1 : 1
CFL	no flashing lights in the future (*by the middle of 1994*)	
NSB	60 pulses/minute except level crossings with two frequencies: 45 pulses/minute 90 pulses/minute	ratio 60%:40% ratio 25%:75% normal ratio 1 : 1 for warning
NS	75 pulses/minute	ratio 1 : 1
BV	80 pulses/minute	ratio 1 : 1

Fig. 4.14 Flashing frequency and ratio of mark/space period for flashing signal lamps

4.5 Control and proving systems

In conventional relay interlocking systems, control and proving of the lamp circuits is also effected with relays. In the new electronic systems the proving is partly done with electronic devices. It can be assumed that this method represents the future for all systems. In case of double filament lamps the proving of the main filament and switch over to the auxiliary filament is usually carried out externally at the signal. The proving of the lamp current, from the main or the auxiliary filament, is carried out by separate equipment (inside or outside).

ÖBB
Relay circuits for connection of the power to the lamps are currently in general use.

Central supervision of the lamp current of connected lamps is done by relays in conventional systems, by electronic circuits in one type of electronic system. All lamp circuits are able to detect short circuits and broken filaments.

NMBS/SNCB
Fail-safe relay circuits are used. An intensity relay checks the current through the lamp filament,

allowing the detection of a short circuit or a broken filament.

In computerized interlockings, this relay is replaced by an electronic circuit performing the same functions.

SBB

Relay circuits are used. In electronic interlockings partly electronic devices capable of detecting short circuits and interruption of filaments are used.

DB AG

Relay circuits are used in relay interlockings; in electronic interlockings relay circuits are used as well as semi and fully electronic circuits. All of these can detect interruption to main and auxiliary filament and of short circuits.

RENFE

Control is carried out by light relays, which pass the signal to the following and more restrictive indication, when for example the fuse has blown.

SNCF

SNCF have a local lamp monitoring block ('BKF' – *Bloc de Contrôle de Feu*), a signal light checking device for the supervision of the correct operation of signal lamps. This continuously checks the current value circulating through the lamp filament and controls a transfer relay, which in case of a burnt out lamp will control a more restrictive indication for this specific signal.

BR

Lamp proving is performed on the currently lit aspect except for the second yellow of the double yellow indication. Functions offered by lamp proving are broken filament detection and transfer.

FS

In the case of searchlight type signals, lighting checks are achieved by means of a current transformer located in the signal and connected to the lamp circuit. The secondary voltage is then rectified in the signalbox in order to feed a neutral check relay. For dichroic mirrors signals the lighting check is carried out by means of electronic devices fed by the secondary voltage of a current transformer, proportional to the feeding current of the lamp.

CFL

Pre-wired relay sets are used.

NSB

Electronic circuits are used in electronic interlockings, which provide the function for the detection of short circuits and broken filaments. Relay circuits with current relays for the supervision of the lamp current are used in relay interlockings.

NS

Relay circuits are used in special situations where lamp proving is necessary to avoid misinterpretation. This is only the case when two signals are placed next to each other between two tracks. The neighbouring signal will stay red if the signal is dark.

BV

In relay interlockings, safety current relays are used for lamp proving. In computer interlockings, electronic circuits are used for lamp current control and lamp proving.

Special features include detection of broken filaments and short circuits.

4.6 What to do in the case of burnt out lamps on a main signal

It is a general rule that a dark signal means stop. Exceptions sometimes exist for block signals that allow the permissive proceeding of a train.

In case of burn out of the main filament of a double filament lamp, an automatic commutation to the auxiliary filament is operated and in most cases an optical alarm is given to the operator.

There are further regulations and indications relating to burnt out lamps that exist in different countries. These are as follows:

ÖBB

Stop lamp: the dark signal means stop and has no influences on other operational conditions. For detection during night there is a white–red–white reflecting bar at the signal post.

Free lamp: if a lamp is burnt out the signal falls back to stop.

NMBS/SNCB

Stop lamp: dark means stop. The failure of the signal

conditions the aspect of the preceding signal. For signals controlled by a signalbox, an acoustic alarm is given to the operator, who acknowledges the alarm. For the other signals (for example, automatic signals), the driver reports the unlit signal to the operator.

Free lamp: dark means stop. Here too the failure of the signal conditions the aspect of the preceding signal. It is the driver who reports the unlit or incomplete signal to the operator.

SBB

Old signal system (L): if a free aspect lamp burns out the signal falls back to stop.

New signal system (N): in case of a fault a more restrictive aspect is displayed.

If a signal is dark: the signal in the rear is not cleared.

DB AG

In case of a fault a more restrictive aspect is displayed, in the worst case the signal is dark. If this happens the signal in the rear will not clear or return to danger.

Faulty or dark main and distant signals can be identified at night by reflecting identification panels at or before the signal post. Dark signals that may be passed at the driver's discretion if the signalman cannot be contacted and the rules are obeyed are specially marked with white/yellow panel.

RENFE

More restrictive indication in the signal, when the fuse of the lamp has blown, together with repercussions on the preceding signal.

SNCF

Free aspect lamp: a more restrictive aspect is displayed.

Stop lamp of a block signal: two types of block are used –

- the permissive block – the stop and proceed signal is a single red light or a dark signal. The signal is marked with the letter F (*franchissable*) meaning a stop and proceed point.
- the absolute stop block, which is used mainly in point and crossing lines, where the absolute stop signal takes the form of two red lights. The signal is marked with the letters NF or *non-franchissable*, meaning an absolute stop point.

BR

If a lamp fails completely, the main signal in rear which authorizes movement up to the failed aspect is maintained at red.

FS

Dark means stop. Wayside indication tables help the driver in locating signals. The lamp may be either failed for filament failure or switched off as a measure against the irregular position of the movable vane (mechanical failure). In both cases, in multiple light signals top light lamp failure cuts out the other lamps. The failure of a signal will condition the aspect of the preceding signal and provides an alarm to the maintenance staff.

CFL

If the auxiliary filament of a double filament lamp also fails, an acoustic alarm is given to the operator which has to be acknowledged. A free signal automatically returns to its most restrictive aspect. If the burn out affects the most restrictive aspect, it also becomes impossible to clear the signal and in respect of recent installations, also to clear the signal in rear.

A fall back solution is guaranteed by red and white reflecting bars on a main signal or reflecting symbol panels on a distant signal indicating the position of a totally dark signal.

NSB

If a red signal lamp burns out, dark means stop. If a free aspect signal lamp burns out and another legal aspect is indicated, it remains. If an illegal aspect is indicated, it drops to stop.

NS

Dark signal means stop. To mark the location of the signals on the open line reflective beacons are located 180, 120 and 60m respectively before the signal. The driver must also report an unlit signal to the operator.

BV

Stop lamp: dark means stop (and ATC has stop function). To mark the position of a dark signal there is a high intensity reflecting border surrounding the backboard of the signal.

Other lamps: fall back to stop.

4.7 Visibility distance

Visibility distance has to be fully understood by operating people as well as in terms of track layout. The required visibility distance is generally dependent on the maximum speed, but distances even for identical speeds are widely different.

ÖBB
Only defined for main signals, distant signals and protection signals according to the formula:
$$2.5 \times \text{max. speed in km/h} = \text{sighting distance in metres}$$
It must, however, be at least 100m.

NMBS/SNCB
150m for low speeds up to 60km/h. Up to 300m for high speed > 60km/h.

SBB
The regulation is that according to the maximum speed the engine driver must see the signal at least 10 seconds before arrival at the signal.

DB AG
Main signals:

- 300m for speeds up to 100km/h
- 400m for speeds 100–120km/h
- 500m for speeds >120km/h

If these cannot be provided, the provision of a distant signal repeater becomes necessary.
Distant signals: 150m announced by beacons.
Shunting signals: 100–300m.

RENFE
Main signals (entrance and exit): 200m.
   Block and distant signals: 300m.

SNCF
- 100m for speeds under 60km/h
- 200m for speeds of 60–120km/h
- 300m for speeds over 120km/h

If these distances cannot be achieved, the signals are preceded by signalling panels consisting of black bars on a white background.

BR
The standard for long range signals is 1500m, for short range signals, 600m.

Minimum permissible sighting distance in clear weather is 300m, but generally visible at longer distances – several hundred metres.

FS
- 150m for speeds up to 90km/h
- 200m for speeds more than 90km/h

Four wayside indication tables are located every 100m before each signal in order to attract the attention of the driver.

CFL
200–300m for main signals and distant signals, locally variable for shunting signals.

NSB
Required visibility distance is 200m for speeds below 130km/h, 8 seconds before arrival at the signal for speeds between 130 and 160km/h.

NS
The regulation is that according to the maximum speed the signal must be visible at a minimum of nine seconds before arrival at the signal in clear weather.

BV
At maximum speed < or = 140km/h the driver must see the signal at least 8 to 10 seconds before arrival at the signal.
   For high speed trains >140km/h, 400m is required.
   If these distances cannot be provided, a distant signal repeater is necessary.

4.8 Distance between main signal and distant signal

The distance between the main and distant signal usually depends on the maximum admissible speed, although even for identical speeds the distances are widely different.
   Generally on suburban routes with only passenger trains, the distance between signals is shorter than the normal maximum distance between a main signal and a distant signal. This is achieved by either special signalling or by ATC.

ÖBB
Distance varies between 400 and 1500m, in special cases up to 2000m. It depends on the maximum speed and on the gradient.

NMBS/SNCB
Varies between 370 and 2000m.

SBB
Old signal system (L): varies between 800 and 1600m dependent on the allowed speed limit.
New signal system (N): minimum can go down to 150m (short block system).

DB AG
Distance is in accordance with the braking distance of the line up to 1000m (700m, 400m). For improved positioning of signal posts or a combination of a distant signal and a rear main signal on one post, the distance may be increased by a maximum of 50%.

A shorter distance is only admissible in special cases. If it is shortened by more than 5%, this is indicated by an additional signal light, meaning reduced speed.

RENFE
Distance 1500m on flat ground with corrections depending on the gradient or up grade for a maximum speed of 160km/h.

For speeds up to 200km/h: a fourth winking green indication is installed on the signals previous to those that protect these sections of 1500m. This indicates to the driver that he must pass through the next signal at a maximum speed of 160km/h.

Under normal block conditions the maximum allowed speed is up to 200km/h and the fourth indication shows flashing green.

SNCF
Varies between 500 and 2800m and must not be lower than the stopping distance. If the distant signal cannot at least be placed at the stopping distance, it must be preceded by a flashing yellow signal.

BR
Varies according to the line speed, traffic type and gradient up to 2300m at 160km/h on a 0.5% falling gradient.

Trains permitted to run up to 200km/h are equipped with improved brakes allowing them to conform to the same distances as other trains travelling at 160km/h. In multiple aspect signalled territory signals may be spaced apart to 1.5 times the braking distance.

FS
FS lines are classified in two categories – referred to as A and B, depending on the speed of the various categories of trains expected to run on the line. The standard warning distance is 1200m for category A lines and 1000m for category B lines. Both measures are reduced by 200m if the average gradient between the two signals is not less than 1 in 100 rising.

When cab signalling is installed the standard distance is always 1350m. If the warning distance is less than the normal one, the warning signal has to be preceded by an appropriate aspect (flashing yellow).

CFL
Usual values 400m, 700m, 1000m, 1200m, depending on the maximum permissible speed of the line. It must be at least equal to the braking distance of the line.

NSB
- ≤ 100km/h : 1000m
- ≤ 130km/h : 1200m
- ≤ 160km/h : 1800m
- ≤ 200km/h : 3000m

NS
Minimum distance between signals allowing a green–yellow–red sequence, varies between 400 and 1344m for speeds between 40 and 140km/h on falling gradients up to 0.5%. However, shorter distances than normal braking distances are possible, in which case the signal aspect sequence will be according to the rules given in Chapter 1.

The maximum distance between signals is 1800m.

BV
Nominal warning distance is 1000m, with 100–150m added for slope conditions.

This distance may be shortened to 600m on low speed county lines <80km/h.

The distance may be increased to 1500m to achieve better visibility to the distant signal.

4.9 Position of the signal posts

There is no common rule for positioning signal posts. One country will have the signal located on the right side, whilst another country will have it located on the left. Mostly this is due to historical reasons. The construction of the old steam engines was closely connected to the position of the signals. For example, in a country with left-hand traffic the engine driver was sitting on the left side of the engine and a long steam engine boiler would have blocked the sight towards the right side. Today, of course, this would not be the case.

Generally a signal post may also be on the opposite side, when it cannot be misinterpreted.

On double track lines with right or wrong track, or with both way, traffic signals are always outside. For example, on ÖBB, NMBS/SNCB, FS, CFL, NS and BV that means for one direction on the right side and for the other direction of the same track on the left side; if possible, the signals should not be between the tracks. On FS signals are also outside the two main tracks within a station.

Alternatively there may be an arrow on the signal post which indicates that the signal is valid for the other side, for example on ÖBB, NMBS/SNCB and FS.

On the following railway administrations signal posts are generally positioned on the right side, because the trains are generally passing on the right side:

- ÖBB
- DB AG
- RENFE
- CFL, except the line connecting Luxembourg City to Arlon–Belgium
- NSB
- NS

Railway administrations with the signal posts usually on the left side, because the trains are usually passing on the left side are:

- NMBS/SNCB
- SBB
- SNCF, except Alsace Lorraine for historical reasons
- BR
- FS
- BV

If the gauge does not allow the installation of posts, signals are positioned on signal bridges constructed over the tracks.

4.10 Distance between signals and control circuits

Currently the distance limits between signals and control circuits are determined by the line resistance of wires and by relay circuits. Future developments may reduce the problems of distances. With the technology of electronic interlocking it will be possible to use a local power supply, local current proving circuits and transmission of information by optical fibre line systems.

ÖBB, DB AG, CFL

Maximum 6500m. This limit is given by the voltage loss due to the resistance of the connecting cable, by the cable capacity and by the maximum induced voltage along the cable.

Longer distances are possible by cable with a reduction factor, that means a copper shielding of the cable reduces the induced voltage. Up to 17km can be reached, for example block signals.

On CFL with a special type of cable in use a distance of 10km is possible.

NMBS/SNCB

Up to 10km from line side signalbox to central signalbox, and up to 50m from line side signalbox to the signal.

SBB

Old installations: maximum 4000m, dependent on wire diameter, with a maximum wire resistance of 40 Ohms.
New installations: maximum 6500m.

RENFE

Maximum 6500m from apparatus housing to signal.

SNCF

Maximum 1200–3000m with stations controlled from the main signalbox. On the open line several signals are controlled from a single equipment cabinet.

BR

Maximum 200m from apparatus housing to signal as standard for electrified areas. In non-electrified areas it can be as long as the cable volt drop permits.

FS
The maximum permitted electrical resistance is 80 Ohms. With a wire section of 1 sq.mm the maximum distance is 2.2km.

NSB
For main and distant signals: distance 2–3km.

NS
Maximum wire resistance 32.2 Ohms. Distance 2–3km.

BV
For main and distant signals: maximum 4000m dependent on voltage drop in the cables and interference from other circuits in the same cable.
For dwarf signals: maximum 600m.

Fig. 4.15 Multi-aspect signal with junction indicator

Fig. 4.16 The new German running signal 'KS' type in situ

Fig. 4.17 ÖBB: main signal (showing free) with distant signal (indicating next main signal free)

5 Interlocking Cabin
H. LINDENBERG

Contents

5.1	**Introduction**	183	5.4 **Operation interface**	205
	Early interlockings	183	5.5 **Power supply**	213
5.2	**Modern interlockings**	183	5.6 **Standards and legal obligations**	219
	5.2.1 Early relay interlockings	184		
	5.2.2 Geographical circuitry interlockings	187		
	5.2.3 Computer interlockings	189		
5.3	**Routes and route conditions**	196		
	5.3.1 Train routes	196		
	5.3.2 Shunt routes	196		
	5.3.3 Route conditions	196		
	5.3.4 Interlocking functions	197		

Illustrations

- *Fig. 5.1* *Summary of interlockings* 186
- *Fig. 5.2* *Spoorplan interlocking SpDr S600 – (DB AG)* 188
- *Fig. 5.3* *Microcomputer interlocking ElS cubicles housing upper level and interlocking level computers (DB AG)* 191
- *Fig. 5.4* *Solid state interlocking trackside functional modules and housing (BR)* 193
- *Fig. 5.5* *Standardized man–machine interface (ÖBB)* 208
- *Fig. 5.6* *Interlocking and associated power supply* 216

5.1 Introduction

Because railways have a number of unique characteristics, namely:

- bound to a unidimensional right of way;
- great masses are moved with considerable speed;
- low friction between wheels and rail;

the combination of these factors ensures that trains cannot normally be brought to a standstill within the distance which can safely be observed by the driver. Thus potentially, trains are a danger to others and are themselves exposed to dangerous situations. From the very beginning others had to be protected from trains and in turn trains too had to be protected. Safeguarding trains from human failure is the goal of railway signalling in general and of the interlocking in particular. Ways and means to achieve this have been developed in many countries, and originally pioneered in Great Britain. Compared with British practice, current practices in the countries investigated differ considerably. This is a result of different philosophical approaches, national characteristics, geographical circumstances and technical developments.

Interlocking as the heart of railway signalling, developed over almost the same span of time as railways themselves. The task was to ensure the provision of appropriate interlocking between all elements constituting the route over which a train is to travel, together with the associated signals to give safe and reliable instructions to the train driver. This, together with the aim to exclude the consequences of human failures, has always been effected with the technical means available at the time.

Early interlockings

The first interlockings were of an all-mechanical design. Thanks to the ingenuity of the design, certain types survive into this computer age. However, all are particularly vulnerable in winter weather, and stretch the physical condition of the signalman to its limits. Already in the late 1800s, long before all-mechanical interlockings were fully developed, ways were being sought to overcome the limitations.

The first attempts at improvement concentrated on using artificial power to ease the signalman's work and working conditions. After experiments with hydraulic, pneumatic and electric transmission between interlocking cabin and controlled elements, not all of which were successful, in the long run the all-electric control/detection stood the test of time. In all early 'power interlockings', the actual interlocking functions were invariably executed mechanically. It took many years and a generation of signal engineers before this task was entrusted to electrical contacts and circuits. Likewise, the points and signals generally remained the same, although power operated and detected. It was only in the 1920s that colour light signals appeared among the semaphores in significant numbers on main lines. The operating range of such power interlockings remained much the same as that of the all-mechanical systems. Train detection by observing track and trains remained the limiting factor. It wasn't until the 1930s that train detection on railways became widely entrusted to track circuits, and later to axle counters.

With the progress of technology, mechanical interlocking designs with power interlockings eventually gave way to all-relay interlockings. This development improved safety and working conditions for both signalmen and drivers, together with scope, control range and economy.

5.2 Modern interlockings

In the first relay interlockings, operation and operation sequence closely followed previous practice. Individual element operation, however, soon gave way to entry(start)–exit operation for route setting in total.

Of the two types of interlocking relays conforming to UIC 736i, the inherently safe N–type relay is used for vital circuits by SNCF, BR, NSB and NS. ÖBB, SBB, DB AG and CFL use the C–type, needing to be proved. NMBS/SNCB, RENFE, FS and BV make use of both types of relay. The relay type, of necessity, reflects in the circuit design. It is estimated that about 5% of all C–type relay contracts and associated wiring is necessary for the proving process. This penalty is economically compensated by lower price, smaller size and dense packing of such relays.

At FS proving of regular working of C–type relays is considered unnecessary, because current through relay contacts is limited by an external trip switch and all other features of relays are inherently safe. Although N–type relays are used as well, they are not considered mandatory for crucial functions.

5.2.1 Early relay interlockings

The early relay interlockings were invariably of the individually free wired type. Soon afterwards repeatedly used functions were grouped and modularized. Examples currently in use are:

- ÖBB: '*Drucktasten Stellwerk*' DrS (pushbutton interlocking), '*Vereinfachtes Gleisbildstellwerk*' VGS (simplified panel interlocking), '*Kleinstellwerkstechnik*' KSW (small interlocking)
- NMBS/SNCB: relay interlocking (no special name)
- SBB: Domino 55, Domino 69, DrS
- DB AG: DrS, DrS2, '*Modulares Compaktstellwerk*' MC L 84 (modular compact interlocking)
- RENFE: relay interlocking (no special name)
- SNCF: '*Poste tout-Relais à transit Souple*' (PRS) (relay interlocking with sectional release route locking)
- BR: 'individual lever', 'one control switch' (OCS), 'entrance–exit working' (NX)
- FS: two types of '*Apparato Centrale Elettrico ad Itinerari*' ACEIs (centralized electric route relay interlocking). 'With three orders of wiring' ACEI3, 'simplified version' Acei [There are three versions of ACEI, two of which fall under this heading. All are of the relay unit type. The first type is characterized by wiring in three orders, for example inside relay units, inter relay unit wiring on the same rack and inter rack wiring; the name chosen for this book is 'ACEI3'. A simplified version for small stations on single track lines developed more recently for economic reasons, shall likewise be abbreviated 'Acei'. For the third version see **5.2.2**]
- CFL: DrS, Domino 58
- NSB: NSI63
- BV: SJ 59

It is not uncommon for some of the above interlocking systems to be applied for large stations. Some interlocking systems are no longer installed: for example, individual lever, OCS, Domino 55, Domino 58, DrS, DrS2, VGS and ACEI3; although, many of these will remain in service where already installed for some time to come. Other systems represent quite recent developments, for example MC L 84, KSW and Acei. They are often designed particularly for smaller stations and for lines where the train sequence is not too dense, where most trains pass straight through the station and where little shunting is required.

Characteristics of the above relay interlockings can be summarized as:

- routes are locked and released as an entity – MC L 84, SJ 59, NSI63, DrS2, individual lever, OCS, Domino 55, VGS, NMBS/SNCB, RENFE, Acei; or, where desired, in two parts – Domino 55 and VGS.
- points operation during route setting is sequential – MC L 84.
- points operation during route setting is simultaneous – NMBS/SNCB and ACEIs.
- routes are set and cancelled by a single dedicated push button – PRS and ACEIs.
- buttons can be pushed and pulled to select functions – NX and ACEIs.
- three-position switches in use for individual points operation – NX, OCS and ACEIs.
- individual point operation commands are not provided for – PRS.
- routes, even conflicting ones, can be stored – PRS.
- transit interlocking keeps the points locked when a route has been set. They are unlocked one after the other with sectional track circuit release. This function may be cancelled – PRS.
- points must be set individually – individual lever, VGS and DrS2.
- signalled shunt routes are not provided for – MC L 84, Domino 58, SJ 59, NSI63, VGS, DrS2 and Acei.
- block working on adjacent lines is included – PRS.
- block working and train crossing on single track lines is automatic – SJ 59, Domino 55, Domino 58 and Domino 69.
- fleeting may be provided for – OCS, PRS, DrS2, Domino 55, NX, MC L 84 and ACEIs.
- command initiation may be by lever type switches – individual lever.
- functions are modularized, connected by free wiring – DrS, DrS2, Domino 55, Domino 58, VGS, KSW and ACEI3.
- two-stage button operation, route first, then signal – DrS/CFL and NMBS/SNCB.
- multi-stage operation, points one by one, then route/signal – individual lever, VGS and DrS2.
- signal replacing and route normalizing by switch/button manual operation – individual lever.
- a single switch/button per entry signal for each route initiates route setting, locking and finally signal clearing – OCS, PRS and ACEIs.
- the design is tailored for small stations on branch lines – KSW.

- indications by means of light emitting diodes – VGS.
- use of computerized design aids/tools – NMBS/SNCB, BR and FS.
- relays inside the cabin are of the C–type and relays at the track side are of the N–type – NMBS/SNCB.
- safety and function of free wired interlockings is assured and tested as far as possible in the factory (relay units), but mainly on site. Modulization decreases the need for site testing.
- basis for circuit design is a control table, which may be dispensed with at small standard stations.

Fig. 5.1 (pages 186-187) provides general information about the signalling and interlocking situation in the countries covered by this book. Where appropriate, relay and computer interlockings shall henceforth be summarized as 'modern' interlockings.

For a typical NX interlocking which is also suitable for large stations, the function, operation and testing of the BR type is described as follows:

Route setting control depends upon the operation of two buttons, one at the entrance and one at the exit. The buttons are arranged geographically on a diagram of the layout adjacent to each signal position. Where a signal acts as the exit from one route and the entrance to the next, a single button suffices for both functions. Where the destination is at buffer stops or in a siding, a separate exit button is provided. If the class of route – either main, shunt or restricted overlap – can be selected by the signalman, separate exit buttons are provided for these. In other instances the appropriate class of route is selected automatically dependent upon track conditions. Some systems employ a switch and button arrangement.

The buttons are sprung to return to the centre position, their operation being stored by the interlocking. The entrance button will show a flashing white light until the exit button is pressed. Provided the two buttons identify the extremities of a valid route, and the route is free to be set, the entrance light will become steady. Once an entrance button has been pressed, no other button can be operated as an entrance button within a prescribed area and a second button will always act as an exit. The calling of the route will ensure that the points are first set and locked in the correct position before the signal is changed to proceed. White route lights are displayed on the diagram which confirm that the route is locked. The route is cancelled either by pulling the entrance button or, when automatic normalization occurs, with the passage of the train. 'On' and 'off' signal indications are displayed on the diagram according to the actual aspect of the signal. Red and green are used for main signals, red and white for position light signals.

If the entrance button to a route is pulled when the route is approach locked, the signal is replaced to danger but the route lights continue to exhibit, signifying that the route is locked. The red signal indication flashes until the route normalizes or is *reoperated* by the signalman.

The movement of a train through a route is depicted by red track occupancy lights which replace the white route lights.

To enable the individual operation of points for relief or testing purposes, separate three-position switches are provided. When in the centre position, route setting is enabled over them. The left and right positions are used to set and lock the points normal and reverse.

Pre-selection is not provided but certain signals can be placed into automatic mode where trains normally follow the same route. A separate auto button is provided adjacent to the signal entrance button. Likewise a facility is provided for the replacement of some purely automatic signals in the event of an emergency.

Connections to points and signals

With relay interlockings separate line circuits are used for each function. Controls and indications are also kept separate. To save on cable cores some functions are polarized over individual pairs of conductors, for example points drive and detection. In other instances, such as aspect controls and indications, the return conductor for a number of relays may be shared.

Checking and testing

Design checks are carried out before wiring begins in order to ensure correct circuit design, use of equipment type and allocation of contacts and terminals. Where CAD has been used batch programs are run to effect automatic allocation of contact analysis and to check for circuit continuity and documentary procedures.

Wiring may take place either in the factory or on site and when complete is checked for correctness by

Fig. 5.1 Summary of interlockings

	ÖBB	NMBS/SNCB	SBB	DB AG	RENFE	SNCF	BR	FS	CFL	NSB	NS	BV
Total number of: interlocking controlled points	10 200	11 412	15 000	63 500	6052	—	18 000	22 900	532	2050	5000	6500[1]
controlled main signals (excl. block signals)	7000[1]	5844	7000[1]	35 500	4782	—	13 500	15 000	418	3350	5000	9000[1]
shunt signals	14 400[1]	5657	11 900[1]	42 600	7128	—	8500	8635	619	1850	–	4000[1]
interlockings in use (any type)	1084	466	825	4080	1190	2205	1855	2287	47	280	305	830
Out of the above total in use are:												
all-mechanical interlockings	792	135	142	1900	275	740	700	560	6	30[2]	2	160[2]
electromechanical interlockings	70	30	210	500	240	350	10	560	4	–	2	35
other early type interlockings	–	4[3]	–	–	2	–	–	4[5]	–	–	–	135
MODERN relay interlockings (non-Spoorplan)	132	296	157	625	474	910	900	1120	26	250[1]	300	430
MODERN Spoorplan interlockings	88	–	314	1015	184	205	200	40	11	1	–	38
MODERN computer interlockings	2	1	2	40	15	–	45	3	–	1	1	32
Out of all modern interlockings are:												
locally operated	195	250	255	1380	558	753	195	733	18	25	61	185
remote controlled from a CTC centre	27	47	218	300	352	272	950	430	19	225	260	450[1]
remote controlled, but prepared for local operation	27	47	218	300	352	245	200	400[1]	19	225	0	445[1]
Number of CTC centres	20	15	93	250[6]	17	27	300[6]	80	5	8	25	11

Total number of:	ÖBB	NMBS/SNCB	SBB	DB AG	RENFE	SNCF	BR	FS	CFL	NSB	NS	BV
Type of safety relay in use for relay in terlockings as per UIC 736i:												
non-controlled (N)–type	–	yes	–	–	yes	yes	yes	yes[7]	–	yes	yes	yes
controlled (C)–type	yes	yes	yes	yes	yes	–	–	yes	yes	–	[4]	yes
other relays used for non-vital functions?	no	no	yes	yes	yes	yes	yes	yes	no	yes	yes	yes
Master TRs are of the N– or C– type	C	N	C	C	N	N	N	C/N[7]	C	N	N	N

[1] Estimate
[2] With colour light signals
[3] Magnetostatic interlocking
[4] Exceptionally yes
[5] Hydraulic (hydrodynamic)
[6] Including remote control of satellites
[7] Introduced recently

a series of independent tests, including the important wire count to ensure that the correct number of wires, and not more than two, are connected to a terminal. A continuity test using a bell or buzzer, is performed on each individual wire to verify completeness of the wiring. This is performed before the insertion of fuses and relays into their positions, followed by a functional test on site where a circuit is proved to operate correctly in accordance with the control tables. Through testing is carried out to verify correspondence between the control panel and actual equipment on the ground and finally a principles test is made of the complete installation in accordance with the scheme plan.

It is interesting to note that on BV bells are not allowed for testing, as transients from the coil may destroy solid state devices.

5.2.2 Geographical circuitry interlockings

Modular geographical circuitry interlocking systems (henceforth called 'Spoorplan' for short) are highly modularized all-relay interlockings which have a number of attractive advantages:

- safety and function are to a large extent assured and tested during manufacture. Tests during commissioning are limited.
- projecting/installation work and manufacturing can run simultaneously.
- modifications to the track layout can be implemented in the interlocking with little effort.
- detailed control tables are not required, they may be substituted by relatively simple route tables.
- all routes offered by the track layout can be utilized.
- projecting and installation work and time is comparatively little and short.
- faults can be speedily corrected by replacing plug in equipment.

Most railways have at least experimented with this type of interlocking, but not all make general use of it as standard systems.

Spoorplan interlocking systems are in use with:

- ÖBB: various types of '*Spurplan Drucktasten Stellwerk*' SpDr . . . (Spoorplan push button interlocking)
- SBB: Domino 67, SpDrS60
- DB AG: SpDr . . . (various types)
- RENFE: electric interlocking with geographical modules
- SNCF: *Poste tout Relais Géographique*, PRG
- BR: Westpac, GEC-GS Geographical
- FS: *Modular Apparato Centrale Elettrico ad Itinerari*, MACEI [This abbreviation is used in this book to distinguish the three versions of ACEI.]
- CFL: SpDr . . . (various types)
- NSB: SJ 65
- BV: SJ 65

As an example of a Spoorplan interlocking in use with ÖBB, SBB, DB AG and CFL, a typical SpDr... system is described as follows:

The hardware consists of modular plug in units equipped with C–type relays. Units are coded to permit interchangeability only with units of the same kind. Relay operation and contact condition can be observed through a front pane. A number of standardized relay units take care of all command, indication, interlocking, control, proving, locking and detection functions. The total number of units making up an interlocking depends on the number of elements involved, ie, points and signals. All units are factory tested for correct wiring, function and insulation (see **Fig. 5.2**).

Relay units may be of different sizes, depending on the scope of the function. For economic reasons, one relay unit may contain either:

- all functions for a single element or purpose.
- a number of identical functions for different elements of the same kind.
- functions relevant to the interlocking and control/detection of one element.
- functions relevant to interlocking of several identical elements.
- functions relevant to control/detection only of one element.
- functions relevant to control/detection only of several elements of the same kind.
- supervising functions for all or part of the interlocking.

Elements may be main, shunt, distant, starting, home or block signals, points (per point machine), scissors (diamond crossings), exits for dead end routes and overlaps.

Standardized Spoor cables, in various lengths to suit the relay room layout, rack and unit arrangement, interconnect all relay units relevant for interlocking functions, according to the track layout. Vital functions between units pass via these cables. Spoor cables are factory manufactured and tested.

The circuitry of all relay units comprising an interlocking system is designed to handle all functions of all versions of elements under all known or anticipated conditions of layout and operations specified at the time of development. This represents the customer specific part of the interlocking and thus usually restricts use of a particular design to one customer. For the relevant application task, internal functions of relay units need to be completed by strapping, or programming certain circuit gaps. Special or seldom required functions which are not inherently provided must be added individually as free wiring.

Supervision units organize and prove commands and operations. To this effect, ring cables link the supervision units with all relevant units in order to achieve the desired common functions. Proving of push buttons and supervision of commands is executed in the supervision units. Stuck push buttons raise an alarm and inhibit commands thereafter, incorrect commands are rejected and alarms are managed.

In general, simultaneous dual button operation is used for commands. Two-stage operation may be used to effect special commands within a set time, for

Fig. 5.2 Spoorplan interlocking SpDr S600 – (DB AG)

example setting alternative routes. Single button operation is used only for non-vital commands, such as the silencing of audible alarms.

The equipment is centralized, which includes block signals within a distance of 6.5km from the interlocking cabin. This requires various voltages and power systems, superimposed control and detection functions, and, for example, interface circuitry at signals to save cable cores.

Route release is sectional, thus making individual route elements available for re-use as soon as cleared by the passing train. The logical sequence of tracks occupied and cleared is proved and automatic route release is prevented during track circuit power supply failure, as incidentally the dropping and picking sequence of track relays could simulate a moving train.

Equipment location

Interlocking equipment can be centralized as much as possible in the protective environment of the cabin or it can be decentralized to place parts of the interlocking near the controlled elements in order to save on cabin space and cables. Most railways (ÖBB, SBB, DB AG, RENFE, SNCF, FS, CFL, NSB, NS, BV) favour the former on conventional interlockings.

As an example, RENFE explains its philosophy:

'The tendency in the interlocking installations is to get the highest possible centralization of hardware. It has (been) shown that:

- the number of failures decrease with fewer interconnections.
- the maintenance and repair times are less when most parts of the interlocking are concentrated in the cabin.'

NMBS/SNCB uses decentralized control of signals and signal aspect sequence.

Where the decentralized approach is favoured (for example by BR, BV and others), the following is quoted as an example:

'On large BR installations interlockings are distributed along the line convenient to the points and signals under control. These are connected to the control centre by non-vital circuits using either direct wire or a time division multiplex system (TDM). Over short distances the direct wire method is used to transfer individual controls and indications between the control centre and the satellite interlockings, using discrete cores for each function. Beyond a few miles it becomes economic to use TDM techniques to communicate between the control centre and each interlocking.'

5.2.3 Computer interlockings

With the advent of powerful, small and cheap computers, it was merely a matter of time before these were utilized for interlocking functions in railway signalling. Compared with the traditional relay interlocking, savings in space, cables and remote control systems were expected, combined with a widening of scope and the resultant reduction in costs.

Most of the railway administrations covered by this book have currently entered the computer interlocking age or are about to do so.

Computer interlockings permit long range direct control via vital fail-safe links to field stations (nodes) at the hub of controlled elements.

Generally, standard interlocking conditions remain unchanged for the railway concerned. Similarly, commands available to the signalman remain the same. Where deviations occur, these are mentioned.

Computer interlocking systems have been developed in Austria, Germany, Great Britain, Italy and Sweden, either as company designs or as joint ventures with the national railways. All designed systems differ in many respects and the ones in actual revenue service on 1 January 1993 are now described.

ÖBB

An important consideration was the need for computer interlockings to replace the indoor equipment of obsolete relay interlockings, making extensive use of the existing outdoor equipment and cable network. This resulted in similarities between relay and computer interlockings.

The hardware structure is arranged in levels. The upper level handles the man–machine interface. The main interlocking level is represented by the section computers. The element control computers communicate with the associated elements through interface modules.

In computer interlockings the general application software is separated from the individual software, the project relevant data. All data for hardware and software configuration of a station are generated with the aid of a special tool. The software data are stored in EPROMs.

In case of track layout changes, hardware modules for new elements are added and connected by cables. Additionally the data for the microprocessors have to be changed. There is a projecting tool which gets the new layout as input and generates the new data as output. After transferring the data into EPROMs by a PROM programmer, the new EPROMs are plugged into the printed circuit boards.

Software modules are structured in small packages and subject to extensive tests and detailed proving for absence of program faults.

The correct function of all modules, programs and elements is checked regularly during normal operation.

The man–machine interface is standardized and consists of VDUs, supplemented by keyboard, mouse and printers. Great emphasis is put upon dependable, safe presentation of the process status, requiring dual channel control of the display(s). Failure of one channel, and thus non-safe display, results in a flashing picture.

Technical approach to computer interlockings, assurance of safety and provision of required redundancy differs between the individual contractors' designs within the framework and standards laid down by the ÖBB. Two systems are in use.

SMC 86 system

The SMC 86 system is based on the geographical circuitry approach.

All vital hardware in the different levels is realized by SIMIS® (Safe Microcomputer System) modules. Three computer channels are used for redundancy reasons. The main interlocking level is connected to the element control computers by an interlocking bus which is the main highway for vital information exchange between all modules. In future this will be used to increase the operation range from a control centre to one or several field stations from where the elements in the vicinity will be directly controlled and fed as usual.

There are data modules for supervision, points, scissors, entry and exit of routes, overlaps and the like, comparable to the Spoorplan interlocking.

Modifications are carried out by adding peripherals including the necessary software module and the application data.

ELEKTRA system

The ELEKTRA system is based on the geographical approach.

The computer architecture is split into two channels with diverse software for safety reasons.

Computers in the upper level and the element control computers are duplicated in each of the two channels as hot stand by. The hardware in the main interlocking level in each channel consists of triplicated computers for reliability reasons, known as the two out of three principle. If one computer in a channel has a different output it is disconnected. The process continues with the two agreeing computers. The main interlocking level is linked to the element control computers by serial connections.

NMBS/SNCB (see BR)

While mostly pertaining to standard SSI, there are some features of computer interlocking that are peculiar to NMBS/SNCB:

- SSI installations are always connected to an integrated electronic control centre (*Elektronische BedienPost* EBP).
- data transmission to and from trackside apparatus is via fibre optic cable.
- trackside modules are of a different technique and technology.

SBB (see DB AG/EIS)

DB AG (derivatives also for SBB, RENFE, NS)

An important original consideration was the need for computer interlockings to replace the indoor equipment of obsolete relay interlockings, making use of the existing outdoor equipment and cable network. This resulted in similarities between relay and computer interlockings, particularly as far as centralization of computers is concerned.

Both the hardware and the software structure is of modular design. All imaginable functional and operational requirements ever to be encountered have been specified and are incorporated in the customer specific software controlling the elements. For each element there is a data set containing the application relevant data (static part) and the process controlled data (dynamic part). The application relevant data are relatively simple to implement and to change in case of track layout modification.

The architecture of all computer interlockings is similar in that:

- display and interface computers and associated hardware represent the upper level and control the monitors (VDUs), indication panel (if provided), keyboard(s), control board(s), printer(s), etc, and communicate with the interlocking bus.
  Information interchange with train describer, LZB (continuous automatic train control), etc, is also at this level.
- the interlocking level is represented by the section computers, each of which takes care of a part of the station. The number of section computers is governed by the station size. All section computers are linked to the interlocking bus.
- also linked to the interlocking bus are the element control computers (the element control level) which in turn communicate with the dedicated elements through interface modules. From here onwards the conventional energy levels are used via copper conductors. Alternatively, light wave guides may be used for control/detection of signals, with energy fed separately.
- independence of parallel hardware channels is ensured by technical means (separation, screening, isolation). In this way simultaneous influence caused by external effects on the different channels is excluded.

The main highway for vital control and information exchange between all levels and modules is the interlocking bus.

Software modules are structured in small packages and subject to extensive tests and detailed proving for absence of program faults.

The correct function of all modules, programs and elements is checked regularly during normal operation.

At the man–machine interface great emphasis is put upon dependable, safe presentation of the process status, requiring dual channel control of the display(s). Failure of one channel and thus non-safe display, results in a flashing picture.

For the technician a special console is provided in the equipment room, containing a system printer for logging and maintenance purposes. An event recorder covers the last 24 hours. All commands and all status information is recorded. The recordings can be listed on the printer or be replayed on the installation test system.

Systems of the *Elektronisches Stellwerk* (ESTW) computer interlocking family are applied on small as well as large stations on all types of lines, as individual interlockings or as line interlocking centres.

Technical approach to computer interlockings, assurance of safety and provision of required redundancy differs between the individual contractors' designs within the framework and standards laid down by the Deutsche Bundesbahn. In use are the ElS and the ELL 90 systems.

ElS system

The ElS system is based on the geographical circuitry approach.

All vital hardware in the different levels is realized by the SIMIS® (Safe Microcomputer System)

Fig. 5.3 Microcomputer interlocking ElS cubicles housing upper level and interlocking level computers (DB AG)

modules. SIMIS® is characterized by its dual hardware structure. Both channels check each other by cross comparison. Identical outputs after an operation process initiate the process periphery. Differing outputs shut down the process output and disconnect energy supply to the elements.

Hardware in the upper level is realized by a three channel computer for redundancy reasons. Redundancy, if required, is provided by one or more spare module(s) in the interlocking and element control level. As all are linked to the interlocking bus, exchange of relevant data to take over the function of a failed module is ensured. Functions of the interlocking and element control computers are combined in area control computers (see **Fig. 5.3**).

This opens a new way to increase the range within which elements can be operated safely. The bus – the safe level – can be extended to one or several external nodes (*Dezentrale Stellbereiche*), from where the elements in the vicinity are directly controlled and fed as usual. A single interlocking cabin can thus be considered a distributed interlocking system able to control a line section of 100km or more without needing a separate individual remote control system.

There are data modules for supervision, points, scissors, entry and exit of routes, overlaps, etc. The module connection diagram can be compared with the unit connection diagram of the Spoorplan interlocking. Each module is informed about its adjacent modules. The software contains all interlocking functions and is brought to life as a system by the project relevant data.

Modifications are carried out by adding peripherals including the necessary software module and the application data.

Safety and function of the interlocking system is assured partly in the workshop and partly on site. A special installation test system is used for functional pre-tests.

The above generally applies also to SBB computer interlockings. It should be noted that, for the first time as a field test, light wave guides are used to control and monitor individual signals, particularly those situated remote from the interlocking cabin.

ELL 90 system
The ELL 90 system is based on the route approach.

Vital hardware in the upper and interlocking levels is realized by a three channel computer. Redundancy is thus provided by the two out of three principle.

Different outputs result in shutting down the disagreeing channel, the process continues with the two agreeing channels providing the output.

In the element control level, SELMIS (SEL Microcomputer System) safety modules are used. SELMIS is characterized by its dual channel hardware structure. Both safety systems use repeated processing and intermediate and final software check for identical results. In case of disagreement the SELMIS system is shut down. If redundancy is required, SELMIS modules must be duplicated.

The aforementioned generally applies to RENFE computer interlockings as well.

RENFE (see DB AG/ELL 90)

BR (also NMBS/SNCB)
Solid State Interlocking (SSI) comprises two dependent subsystems which are the Central Interlocking – the trackside data link and the functional modules. Together they form a complete system of fail-safe control over lineside signalling equipment. The system is designed in such a way that the interlocking is concentrated at one location.

The safety functions of the Central Interlocking are carried out by three multiprocessor modules (MPM) which operate in triplicate. Failure of an MPM results in the system working with the remaining two processors in duplicate. Should two MPMs fail, safety by redundancy cannot be achieved and therefore the complete system shuts down. The system software is divided into a number of programs and is common to all SSIs except for the geographic data used to customize each interlocking to a particular location. When a track layout is altered the only change usually necessary to the interlocking is the generation and testing of the new geographic data.

Output from the interlocking processors is in the form of command telegrams which are transmitted over the trackside data link to each trackside module.

The non-vital functions forming the signalman's commands and indications are handled by duplicated panel processors (PPMs) which form part of the interlocking. The duplication is provided for availability only, the system continues to function at this level with one PPM. Connection to a control panel is via a serial to parallel multiplexer whilst connection to an Integrated Electronic Control Centre (IECC) is directly into the signalling network portion of that system.

Data transmission to and from trackside apparatus is over duplicated screened twisted pair cable. Connections are made via Data Link Modules (DLM), again duplicated at the interlocking and at each trackside location. Remote data links are accessed via telecommunication PCM techniques connected to the two SSI subsystems by Long Distance Terminals. At each location that the data link calls, the lineside signalling equipment is connected to Trackside Functional Modules (TFM). These convert the interlocking telegrams into command outputs and convert parallel inputs into indication telegrams back to the interlocking. In this way data communication to each trackside location supersedes the need for remote control facilities (see **Fig. 5.4**).

Fig. 5.4 Solid state interlocking trackside functional modules and housing (BR)

Maintenance and fault finding are improved with SSI by having comprehensive diagnostic information available on a technician's terminal. All events within the SSI are logged on tape, thereby allowing off line interrogation in the event of an incident. In addition, a technician's terminal and printer reports all faults, allows for interrogation of the system, and enables the application of technician's inhibits to particular controls.

As an additional safety requirement, a hard wired means of replacing all signals to danger is provided by interrupting the power supply to the interlocking.

Checking and testing
In the case of SSI a check of source data is performed by a second engineer prior to the data being compiled. Following this, functional testing is carried out in the office on the design workstation. This has the facility to simulate trackside activity. Testing is performed by an independent engineer and where more than one SSI controls a given area these are linked to allow testing of their interfaces. When testing is complete, EPROMs are blown and decompiled for checking against the original data. Once installed, on site through testing from the control centre to the trackside is carried out.

Compared with BR relay interlockings, SSI allows for a reduction in much of the hardware previously associated with an interlocking. This is particularly true of trackside cables and associated work. Building requirements are also reduced.

One disadvantage of SSI is its slower operation due to the cyclic nature of processing compared with the parallel operation of relay circuits.

FS
Two different computer interlocking approaches are pursued. They are based on essentially different theories and are both considered viable for safety purposes on railways. Both are summarized by the Italian acronym ACS for computer interlocking system.

ACC system
The ACC system is based on the adoption of the two out of three architecture with software diversity for the interlocking level and two out of two for the field equipment controller level. System architecture is distributed. Central section and peripheral equipment

can be concentrated in the same room. Depending on the station size, the latter, however, may also be distributed, connected by a fibre optic link. Fail-safe hardware circuits are provided for:

- central exclusion logic;
- peripheral watchdog;
- field equipment controller interface.

The interface to outdoor devices can be with both relays and electronic components. Connection to the outdoor devices is via copper conductors in signalling cables with the emphasis on a minimum of cores.

Software tools are available for:

- preparation of application data;
- programming of the safety logic;
- verification of correctness of the above operations.

ASCV system

The ASCV (an Italian acronym for vital computer based interlocking) concept employs a single processor with double processing channels. Outputs are generated in two different ways with safety comparison. Safety is based on the presence of a signal for each permissive action and the dynamic check of inputs and outputs. The interface to outdoor elements is with relays or electronic components.

The principles and techniques applied by FS for validation of safety related software seek to provide consistence between the various levels of design by means of functional analysis, structured analysis and timing analysis, when safety critical timing constraints exist.

A comprehensive set of diagnostic aids is built into the ACS systems to facilitate rapid location of any faults. The ACS test themselves very thoroughly during the machine cycle. There is a technician's terminal which receives information from the diagnostic process and monitors all messages sent by the interlocking and the man–machine interface to each other. It provides the following facilities:

- printing the fault report;
- keeping a log of all changes of state in each part of the interlocking;
- enabling the maintainer to interrogate the interlocking through the keyboard in order to check the state of the various elements (signals, points, etc);
- enabling the maintainer to apply certain restrictive technician's controls;
- enabling the maintainer to enter certain other commands.

The keyboard is normally locked and access is under strict control.

Both types of ACS use standard equipment, adapted where appropriate, for signalman interface. The interface handles all information flowing between the interlocking and the signalman's display system, which is based on VDUs. Controls are in the form of button operation, mouse operation, and the like and are processed to generate route calls and controls for specific functions.

The man–machine interface provides additional ports for general purpose communications between ACS and other management systems such as train describer and centralized traffic control as required.

NSB (see BV)

NS

It is envisaged that computer interlockings on bigger stations and control centres will use a suitably adapted ElS system. In addition to the conventional NS interlocking principles, some functions have been changed and/or added. These are mainly:

- a procedure to check the correct working of track circuits, taking care of occasional loss of shunt. Sectional route release is now by proved logic sequential track circuit operation.
- unlike relay interlockings, route locking for normal train routes requires track circuits clear. This prevents premature activation of level crossing protection.
- operations taking advantage of the safe process status display. These are:
  - partial route release in spite of track circuit failures;
  - individual point operation command with track circuit failure, permitting 'Drive on sight' route setting;
  - blocking/unblocking of individual route elements to safeguard against unwanted operation and/or moves.

Safety relevant commands are protected against accidental use and only possible after a first command is repeated with a random number procedure within 20 seconds. They are logged on a printer.

NS cites its policy for the future as follows:

'At smaller stations it is foreseen to replace vital relay circuits with Vital Processor Interlockings (VPI) to be controlled by the Electronic Control Post (EBP) which replaces the non-vital relay circuits. The VPI concept is based on a single processor with two processing channels. It evaluates a set of Boolean expressions representing primordial interlocking logic.

The main processing logic of the VPI system, including the interpretation of external inputs, expression evaluation and the determination and checking of the input states, is designed using the concept of NISAL (Numerically Integrated Safety Assurance Logic). Permissive decisions are represented by large unique numerical values, constructed by combining numerical values representing each of the critical constituents of a permissive decision. The presence of restrictive system states or permissive actions are verified numerically.

The numerical values are used to generate a dynamic signal, which should be of an exact form and frequency to keep the vital relay up. The Vital Relay Driver portion of the system, responsible for the verification of check data and generation of output power, is designed using the concept of SAL (Safety Assurance Logic).

Software tools for preparation and verification are available.

VPI takes over the existing interlocking functions, however, also completed with a procedure check the correct working of track circuits as mentioned for the ElS system.'

BV (also NSB)

The vital functions of the new interlockings are performed by computers. This type of interlocking (SJ 85), unlike its predecessor (SJ 75), is fully electronic and uses no electromechanical components other than track relays.

The central computer is capable of controlling about 400 elements. If more elements are to be controlled, more interlocking computers can be linked under a common control and supervisory system. Central equipment is duplicated for system availability. In order to be ready to take over immediately, the hot stand by computer is updated regularly. Fail-safe functions are realized with diversity programming. The interlocking function concept is geographical, with the interlocking conditions described by algebraic equations similar to Boolean algebra. However, each variable can normally take up more than two values. There is also an engineering work support system generating:

- object data for the site computer;
- track diagram for the VDU;
- installation documents;
- equipment list.

The hardware is decentralized. Solid state element controlling modules are distributed in the field, housed in cabinets close to the respective elements. Each cabinet can hold up to eight such modules and a transmission concentrator. The element control modules are completely independent of the track layout. Depending on the element to be controlled, there are different modules:

- the control module for point machines generates three phase current for the motor. This circuit is immune to influences from traction currents.
- signal control modules have phase angle lamp control and lamp current detection circuits.

All modules have inputs for sensing track relay, point detection, and other contacts.

Interlocking computer and element controller transmission concentrators communicate via a distributed network, using a HDLC protocol. The communication network consists of transmission loops in a telecommunication cable or an optical fibre cable. In the latter case very long distances can be covered. Even in case of a break in the loop, communication with the concentrators is maintained. With loss of power at an element control module cabinet, the concentrator is automatically bypassed to allow the remaining concentrators in the loop to continue to function.

Safety and function of the interlocking system is assured partly in the workshop and partly on site. A special installation test system is used for functional pre-tests. Track layout changes are easy to carry out. Such changes and new functions are run and tested at the installation test system before they are used in the actual interlocking. Additional elements due to track layout changes can be accommodated at a spare place in the nearest cabinet. Reconfiguration of the interlocking is done by the engineering work support system.

The man–machine interface in the operating room consists of a large back projection colour VDU screen for the general overview and of normal monitors for close up display. Operation is by a standard computer keyboard. A train describer is integrated and can be used for automatic route setting.

For the technician a special console is provided in the equipment room, containing a system printer for logging and maintenance purposes. An event recorder covers the past 24 hours. All commands and all status information is recorded. The recordings can be listed on the printer or be replayed on the installation test system.

The computer interlocking can be applied for small as well as large stations on all types of lines.

Codes for the control of ATC beacons are generated in the interlocking computer and transmitted to the signal controller.

5.3 Routes and route conditions

Any signalled movement inside a station traverses a route. Depending on the type of unit making the move and on the degree of safety required, a selection of route classes is available. In addition, for any class of route, only one track path may exist between start and exit or one or more alternative paths may be available, one may be a priority path. As a general rule, provision and use of an alternative route must be of practical use. It must permit bypassing a track element which, if faulty, could otherwise prevent setting the desired priority route between the same start and exit.

5.3.1 Train routes

Trains traverse train routes, providing the highest degree of safety. Such routes require points available, set and locked, all tracks clear beyond clearance markers (NS, RENFE and FS provide a drive on sight route, on which tracks are not proved), and flank protection (except NS; RENFE only provides flank protection in some cases). Tracks between route points and corresponding elements providing flank protection are proved clear by most railways. Those which do not do so (ÖBB, NMBS/SNCB, SNCF, BR and FS), sacrifice detection for availability. Most railways provide an area of protection to prevent immediate danger in case a train overruns the exit signal marginally (overlap). Exceptions are NMBS/SNCB, SNCF and NS.

Where trains are to be routed onto an occupied platform track, special routes with corresponding signal aspects may be available (NMBS/SNCB, SBB, BR, FS). NS use the drive on sight route. RENFE has something similar.

Depending on the traffic pattern, signals for trains may temporarily be automatic in operation to relieve the operator from routine work. Such facility may be, for example, fleeting, which will not release a route after a train. The signals act as automatic block signals, clearing once the train route conditions are met again. Alternatively, automatic route calling (ARC) may be used. Here, the only possible route is set automatically, triggered by an approaching train. The route is train released and remains so until the next train approaches. In larger stations or where otherwise desirable, automatic train routing (ATR), based on train describer or other destination criteria, sets the appropriate route towards the ultimate destination at the most appropriate time in order not to affect other routes too much. Where more sophisticated sources of data – for example, timetable and priority criteria - are available the same effect can be achieved more appropriately by automatic route setting (ARS).

Most railways do provide some sort of relief to permit traffic to continue when, due to technical failures, a normal train route and subsequently a main signal proceed aspect is not possible.

5.3.2 Shunt routes

Shunt units move along shunt routes. These routes generally require less conditions than train routes and are subsequently inferior from a safety point of view. This is acceptable, however, due to the generally low speed of shunt moves. As to the actual conditions for shunt routes, there is a great variety. Where all shunt routes are treated equally, usually one shunt signal proceed aspect suffices. Most railways do not prove the destination track clear, particularly if it is a platform or similar berth track. Likewise, most do not prove intermediate tracks clear nor do they require overlap, flank protection or exit signal alight. NS do not have or use shunt routes and shunt signals at all. All shunt movements are treated in the same way as a train move.

5.3.3 Route conditions

Trying to summarize all railway route conditions would be very confusing. For this reason, all individual railways' conditions are described separately, particularly where they differ from the summary in **5.3.1** and **5.3.2**.

The required interlocking conditions may be laid down in detail in control tables. Where standard conditions are generally realized in customer relevant circuits/software, route tables (*Fahrtentabellen*) or similar may suffice. For small standard stations individual tables may not be required at all.

As a typical example for detailed control tables, BR practice is now described.

The route control table is arranged in five sections:

- *Section 1* details the controls that allow the signal to show a proceed aspect for the particular route. These include: which track circuits must be clear and occupied, if any; the points that must be proved in the correct position; the next signal that must be proved alight; and which track circuits disengage the signal and whether first or last wheel replacement is required.
- *Section 2* relates to the interlocking determining: those opposing routes and other routes from the same signal which must be proved normal; track circuit controls where these are used to select a particular class of route, for example main or draw ahead; the points that must be set or free to move; and the conditions which must apply to release the route automatically as the train passes the signal.
- *Section 3* contains the conditions under which approach locking is effective, ie which track circuits are occupied, and the sequential track circuit occupation required to release the approach locking and the alternative timing release period.
- *Section 4* is used to assign the controls to each signal aspect available for the route by referring to the relevant aspects of other signals, junction indicators and track circuit occupancy.
- *Section 5* deals with any route locking imposed by the route on opposing signals, detailing which occupied track circuits maintain route locking, on which other routes, and how that locking might be released when a movement has come to rest within the route.

5.3.4 Interlocking functions

For manual route setting (route selection) entry signal, exit and class of route are defined as command(s) to the interlocking.

Selected as routes may be:

- train routes
- shunt routes
- special routes

and any of these as:

- alternative routes.

Route availability is checked (route proving) and, if positive, the call is stored where relevant (route marking).

Points in route and overlap (where required) are operated to the required position and locked (route setting). Available points/derailers/signals/level crossing barriers/gates closed provide flank protection (where required). Successful proving of all tracks (where required) clear, exit signal alight (where required), opposing signals locked, points set and locked, direction of route set, flank protection and overlap available (route proving), results in route locking.

Subsequent to route locking, the signal aspect is called (signal selection) and finally cleared (signal control). The proceed aspect (including subsidiary aspect where applicable) depends upon the route being straight or via a turnout branch of points, the length of overlap (protection distance) available, sometimes also upon the aspect displayed at the exit signal and probably also the distance to that signal. Point deflection (turnout angle) may have a direct influence on the main or a subsidiary signal aspect (for example, speed indication). In the cleared signal, all required conditions of a route are continuously monitored except with:

- SBB the tracks;
- BR the tracks where propelled trains are to be expected; and
- ÖBB the first track.

Where approach locking is provided, it is activated the moment the signal clears.

Until a route is locked, it can generally be cancelled manually without delay. Before a signal has cleared, an already locked route can also be cancelled manually. This may be either without delay, delayed, or without delay, recorded. After the signal has cleared, manual route cancellation replaces the signal immediately. The actual route cancellation is effected either immediately too, or delayed for some time. The actual time being either rigid or depending on maximum train speed, on type of line (mixed traffic or passenger only) and signal spacing. Only that part of the route not occupied after the elapsed time will be cancelled. Where approach locking is provided, immediate route cancellation is possible as long as the monitored approach tracks are clear.

It is general practice that the signalman can manually, selectively and immediately replace a proceed signal to danger, regardless of other conditions. This does not affect the route as such and, as long as the route conditions are available, the signal can be cleared again manually.

When a train passes the entry signal for a train route, the proceed aspect is replaced to danger either by the first axle occupying the first track section of the route, the first axle of the train occupying the second track section, or by the last axle leaving the berth track. Where the latter two methods are used, track clear proving in the proceed aspect must cease at an appropriate time, or certain tracks are not proved at all.

Where the master track relays (TRs) are of the C–type, proving is necessary, normally automatic sectional route release is by proved logical sequential occupation and subsequent clearing of track sections. Although TRs in Sweden are of the N–type, this proving is nevertheless carried out in interlockings of more recent vintage. In Italy, C–type TRs are appropriately protected by trip switches. Where TRs are of the N–type, logical sequential track proving is not necessary. Alternatively, signalman operated, or automatic route release *in total* may be used. On NMBS/SNCB rail treadles effect route release. Each released route section is immediately available for another task.

A number of railways detect the track circuit feeding power to prevent simulation of a train activated sequence of track relays dropping and picking again due to power failure and return.

Approach lock release is effected when the train has entered the route.

Where overlaps are required, several possibilities are available. The overlap is either treated like a train route complete with flank protection. Track proving may not be required, trailing points only may be locked and facing points may remain unlocked and free to operate. Alternatively, special interlocking conditions are specified and apply. If an overlap is not later integrated into (part of) a route, it is either cancelled automatically after a time delay, or it may be cancelled manually without delay, unrecorded, or recorded.

Generally, shunt routes require fewer conditions to permit moves towards occupied tracks via an undetected path. However, full flank protection is standard with SBB and BV. Some railways prove all except the destination track clear (NMBS/SNCB; RENFE (only if under signalmen's control, '*Maniobras Centralizadas*')). BV differentiate two types of shunt routes, depending on the route being clear throughout or not. Where additional protection is required – for example when passenger coaches or wagons with dangerous goods are shunted – flank protection may have to be provided.

Shunt signals usually have just one proceed aspect. Sometimes, however, more than one aspect is possible, as on SBB and BV. FS provide two aspects on tall (long range) shunt signals only. In that case the shunt signal aspect depends on the selected route's condition or on the exit signal aspect. With a shunting move passing a shunt signal/aspect, the proceed aspect is reset to danger with either the last axle leaving the berth track, the first axle on the first track section (on BV reset to the caution aspect), the first track section in the route cleared, or manually.

Generally, at computer interlockings, the functions are the same as in relay interlockings. Occasionally, additional commands are possible. Commands are also comparable where conventional control/operating panels are used. However, devices are often employed which interface more easily with computer interlockings.

Below, the practice in use on current interlockings is discussed.

ÖBB

Train routes initially require all tracks clear for the signal to clear. Thereafter the first track section is no longer proved. The standard overlap is 50m. Where the track layout permits, one of two different overlap alternatives may be selected. Points within the overlap are set and locked and flank protection is provided. Overlap tracks are not proved clear. If the standard overlap cannot be provided, reduction to 0m is permitted with a 40km/h entry speed. Tracks between route points and corresponding elements providing flank protection are not proved clear. Before route locking, the train route may be cancelled without delay. After route locking, manual route cancellation may be effected without delay, recorded.

The entry signal aspect depends on:

- points straight or turnout;
- the turnout angle;
- route length, for example less than normal signal spacing or braking distance; and
- the length of overlap.

Automatic train route release is effected by proved sequential occupation and subsequent clearing of tracks. Electronically controlled track relays assume a fault status with power supply failure, distinct from clear and occupied. The fault status does not cause route release. In case of other detection systems, train effected route release is prevented during track circuit power supply failure. The entry signal is replaced to danger with the first axle on the second track section. It is for this reason that the first track is proved only prior to signal clearing. Overlaps not converted into routes are manually cancelled without delay. Exceptionally, automatic delayed overlap release is used where appropriate. Train routes can also be set by ARC and/or ATR. If the main aspect of an entry signal does not clear due to conditions not being fully met, the signalman can operate a subsidiary relief aspect.

Shunt routes in need of special precaution may require flank protection.

With a shunting move passing a shunt signal, the proceed aspect is reset to danger with the first track section in the route cleared.

Route tables are the basis from which geographical dependencies for projecting work are derived, for example projecting tools for computer interlockings.

NMBS/SNCB

Train routes require flank protection, provided this can be achieved without conflicting flank protection calls on points. Tracks between route points and corresponding elements providing flank protection are not proved clear. Signals are generally situated 50m from danger points. Route locking proves direction of route normal or wrong and treadles normal. Before route locking, manual route cancellation may be effected without delay. After route locking, manual route cancellation may be effected without delay, recorded. The entry signal aspect depends on:

- points straight or turnout;
- the turnout angle;
- route length, for example less than normal signal spacing or braking distance;
- routing in track normal or wrong direction; and
- the exit signal aspect.

Where level crossings are situated such that the initiation distance would need to be longer than the actual distance between signal and level crossing, signal clearing is delayed to allow for appropriate warning time. Rail treadles situated at entry and exit effect route release in total in addition to the clear tracks. Where sectional route release is desired, additional treadles are needed. The entry signal is replaced to danger with the first axle on the first track section. Precautions are made to keep the signal closed. Train routes can also be set automatically by ARC and/or ATR/ARS. If the main aspect of an entry signal does not clear due to conditions not being fully met, the signalman can lock the route artificially by a recorded key operation. After this, he authorizes the driver by paper order or telephone to pass the danger signal.

Shunt routes are called *caution* routes. They prove all but the destination track clear. With a shunting move passing a shunt signal, the proceed aspect is reset to danger with the first axle on the first track section.

Disconnected sections of catenary can be the reason to prevent a signal clearing for a route under such a condition. In this way danger to persons and equipment is prevented as part of the interlocking system.

Route tables are the basis for projecting work.

SBB

Train routes require all tracks clear until the signal clears and not thereafter. Instead of an overlap as such, special interlocking conditions ensure that potentially endangered routes within a range of 100m beyond the exit signal cannot be set simultaneously. If 100m cannot be provided, reduction of the safeguarded area to 60 or 40m is permitted with 60 and 40km/h respective entry speed. On tracks leading to a buffer stop, drivers have to rely on route knowledge to adjust the speed to 30km/h. The exit signal is proved alight to ensure a distinct target. In areas where there are no shunt signals, main signals can provide flank protection for train routes, subject to the restrictions of the special interlocking conditions. Before route locking, the train route may be cancelled without delay. After route locking, manual route cancellation may be effected without delay, recorded. Subsequently, to prevent premature manual or automatic (from stored commands) undesired route setting or other operations, any commands are blocked for two minutes thereafter. The entry signal aspect depends on:

- points straight or turnout;
- the turnout angle;
- route length, for example less than normal signal spacing or braking distance; and
- the length of overlap protection.

If the destination track is an occupied platform track, a subsidiary aspect '*Besetztsignal*' is displayed. Automatic train route release is effected by proved sequential occupation and subsequent clearing of tracks. Train effected route release is prevented during track circuit power supply failure. Home signals are replaced to danger with the first axle on the second track section. Other signals are replaced with the last axle leaving the first section. It is for this reason that the tracks are only proved until the signal clears. Special interlocking conditions are automatically released some time after arrival at the destination track. The delay is not rigid and is part of the special interlocking conditions. Train route commands that cannot immediately be executed are stored until the route is available, whereupon the stored call becomes effective automatically.

Train routes can also be set by fleeting, ARC and/or ATR. Programmed crossing moves are also possible. If the main aspect of an entry signal does not clear due to track circuit failure, the signalman can bypass track proving, whereupon the signal clears. If a failure other than track failure is the reason for the entry signal not clearing, a relief aspect can be displayed.

Where different traction power supply systems meet, interlocking with signalling is provided so that locomotives for one system can only be routed onto tracks with the correct traction supply provided.

Shunt routes require flank protection. For the overlap, special interlocking conditions apply for 20m, variable depending on the destination track condition. Two different shunt aspects are provided, depending on the shunt route terminating at the next or beyond the next shunt signal. With a shunting move passing a shunt signal, the proceed aspect is reset to danger with the first track section in the route cleared.

All tracks are assumed to be used as potential routes, therefore only exclusions have to be dealt with. With the geographical system, all systematic locks are provided for. Non-systematic locks such as the special interlocking conditions governing the overlap have to be specified individually.

DB AG

Train routes require 200m overlap. Where the track layout provides more than one overlap alternative, the most suitable one may be selected. Points within the overlap are set and locked and flank protection is provided, tracks are proved clear. If the standard overlap cannot be provided, reduction to 100 or 50m is permitted with 60 and 40km/h respective entry speed. At terminal tracks or with overlap 0m, 30km/h entry speed is displayed. The exit signal is proved alight to ensure a distinct target. Where feasible, level crossings are treated as flank protection. In cases where flank protection provided by a signal (*Lichtschutz*) is not considered sufficient, remote protection must be provided from further away (*Fernschutz*). Before route locking, the train route may be cancelled without delay. After route locking, manual route cancellation may be effected without delay, recorded. The entry signal aspect depends on:

- points straight or turnout;
- the turnout angle;
- route length, for example less than normal signal spacing or braking distance; and
- the length of overlap available.

Automatic train route release is effected by proved sequential occupation and subsequent clearing of tracks. Train effected route release is prevented during track circuit power supply failure. The entry signal is replaced to danger with the first axle on a track section ≥50m beyond. Overlaps not converted into routes are manually cancelled, without delay, recorded. Automatic delayed overlap cancellation may be provided where appropriate. Train routes can also be set by ARC, and/or by ATR. If the main aspect of a signal does not clear due to an established cause, the signalman can operate a relief aspect for 90 seconds. During this time other route setting involving points operation is prevented (*Laufkettensperrung*).

Where level crossings are situated so that the initiation distance would need to be longer than the actual distance between signal and level crossing, signal clearing is delayed to allow for appropriate warning time.

Shunt routes in need of special precaution may require flank protection. With a shunting move passing a shunt signal, the proceed aspect is reset to danger with the first track section in the route cleared.

Route tables are the basis for projecting work.

RENFE

A train route on its own does not require specific flank protection or overlap. Level crossings, however, are activated. It is when additional routes in the area are set that special conditions must be met with regard to overlap and flank protection. Compatible routes and applicable conditions are laid down individually with the aim to balance safety of trains against efficient use of the station's track network. The exit signal is proved alight to ensure a distinct target. Fused signal lamps result in appropriate change of signal aspects in the rear. Before route locking, the train route may be cancelled without delay. After approach route locking, manual route cancellation may be effected with up to 3 minutes delay. The entry signal aspect depends on:

- the route being straight or turnout at entry and/or exit of the station;
- route length, for example less than normal signal spacing or braking distance; and
- the exit signal aspect.

Automatic train route release is effected by occupation and subsequent clearing of tracks and prevented during track circuit power supply failure. The entry signal is replaced to danger with the first axle on the first track section. Overlap release is effected automatically 30 to 60 seconds after clearing the last point track circuit. Approach lock is effected by occupation of the track in the rear of the distance signal and released with the first axle on the first track circuit in the route.

Train routes can also be set by ARC, fleeting or ATR. If the main aspect of an entry or exit signal does not clear due to track circuit failure in the route, the signalman can operate a (*Autorización de Rebase*) relief aspect.

Shunt routes set individually by a signalman (*Maniobras Centralizadas*) require all tracks clear except the destination track. With a shunting move passing a shunt signal, the proceed aspect is reset to danger with sequential occupation of the berth track and the first track circuit in the route and subsequent clearing of the berth track. Where there is no berth track circuit or when it is not cleared, the shunt signal aspect is reset with sequential occupation of the first and second track circuit in the route and subsequent clearing of the first track. Such shunt routes are also subject to special overlap conditions. Back shunt signals situated at facing entry points double as dwarf route indicators. Alternatively, a part of the station may be temporarily isolated by locked points and used for shunting under local control with all shunt signals clear (*Maniobras Locales*).

The running management provides, as basis for project work:

- several layout plans;
- movement list: route tables, incompatibilities table;
- alternative routes;
- priority routes.

Because of the complexity, some software has been developed to assist with the creation of operating programs.

SNCF

Train routes require flank protection. Tracks between route points and corresponding elements providing flank protection are not detected. Before route locking, the train route may be cancelled without delay. After the route is locked, manual route cancellation is effected with 3 minutes delay. The entry signal aspect depends on:

- points straight or turnout;
- the turnout angle; and
- the exit signal aspect.

Approach locking is activated the moment the signal clears. Where level crossings are situated so that the initiation distance would need to be longer than the actual distance between signal and level crossing, signal clearing is delayed until the barriers are closed. Automatic train route release on high speed lines is effected by sequential occupation and clearing of tracks. On other lines, automatic route release is achieved with the first track occupied, a treadle operated and the first track cleared. Alternatively, dead locking between points and signals is used. The entry signal is replaced to danger with the first axle on the first track section or by rail treadle. Train route commands which cannot immediately be executed are stored until the route is available, whereupon the stored call becomes automatically effective. Train routes can also be set by fleeting. The approach lock is released with the approach tracks all cleared.

With a shunting move passing a shunt signal, the proceed aspect is reset to danger most often with the first axle on the first track circuit in advance of the signal.

Control tables and route tables are the basis for projecting work.

BR

Train routes require 200m overlap. Where the track layout provides more than one overlap alternative, a suitable one may be selected. Trailing points within the overlap are set and locked, tracks are proved clear and controlled level crossings must be closed. Facing points within the overlap are not locked for flexibility of operation. If the standard overlap cannot be provided, reduction to 50m is permitted with the entry signal approach cleared to enforce reduced entry speed accordingly. At terminal tracks the drivers have to rely on route knowledge.

Tracks between route points and corresponding elements providing flank protection are not proved clear. The exit signal is proved alight to ensure a distinct target. Before route locking, the train route may be cancelled without delay. After route locking, manual route cancellation will be effected after up to 4 minutes delay, the time depending on the type of line and the signal spacing. The entry signal aspect depends on:

- points straight or turnout; and
- the exit signal aspect.

Where approach locking is provided, it is activated the moment the signal clears. If the destination track is an occupied platform track, a train route setting operation will result in a call on route (no overlap) and a draw ahead aspect. All other requirements are as for train routes. Automatic train route release is effected by occupation and subsequent clearing of tracks. The entry signal is replaced to danger with the first axle on the first track section. Where propelled trains are expected, a parallel path consisting of berth and first tracks occupied is made up to keep the signal off. The last axle leaving the berth track replaces the entry signal. Track detection ceases the moment the train enters the route. The approach lock is released with the first two track circuits in advance of the entry signal occupied successively and the first one cleared again. The overlap is automatically released 120 seconds after arrival at the destination track, the actual time depends on type of line, either mixed traffic or passenger only, and signal spacing. Train routes can also be set by fleeting (automatic mode), ARC and/or ATR/ARS.

Where automatic level crossings are situated so that the distance between signal and level crossing would need to be longer than the distance available, signal clearing may be delayed to give appropriate warning time.

With a shunting move passing a shunt signal, the proceed aspect is reset to danger with the last axle leaving the berth track. The shunt aspect may be supplemented by a route indication if additional information to the driver is essential.

Comprehensive control tables are prepared as the basis for controls and conditions applicable to a project.

FS

Train routes towards home signals and starting signals common to more than one track (group starting signals) require 100m overlap. 50m overlap suffices towards normal starting signals (relating to only one track). Points within the overlap are set and locked and flank protection is provided where possible. The entry signal aspect depends on:

- points straight or turnout;
- the turnout angle;
- route length, for example less than normal signal spacing or braking distance; and
- the exit signal aspect.

The exit signal is proved alight to ensure a distinct target. If the destination track is a very short track, or an occupied platform track, a restrictive aspect (double yellow) is displayed, automatically in the first case. Approach locking is activated the moment the route is locked, whether or not the signal clears. Before route locking, the train route may be cancelled without delay. After route locking, manual route cancellation may be effected without delay as long as the approach track is clear. With occupied approach track, the route may be cancelled without delay, recorded, at locally operated stations. At remote controlled stations, route cancellation will be effected after 5 minutes delay. The entry signal is replaced to danger with the first axle on the first track section. Automatic train route release is effected in different ways, depending on whether or not sectional release is provided. In the first case, route locking (stable locking) is cancelled together with the approach lock after the replacing track section is cleared. Sectional release continues by occupation and clearing of tracks. In simplified interlockings, automatic release is effected when the train clears the route. The approach lock is released after the replacing track is cleared. In older installations the overlap was cancelled manually. In installations that are or will be unmanned, overlap release is automatic, initiated with the release of route locking. Delay time is up to 5 minutes, depending on the circumstances, for example the expected train speed and the length of the berth track.

Train routes can also be set by fleeting, ARC and/or ATR. If the main aspect of an entry signal does not clear due to conditions not being fully met, the signalman can operate a relief aspect on home and starting signals for a drive on sight move.

Shunt routes require incompatible routes check. Two different types of shunt routes with relevant shunt signals/aspects exist. Normal (short range) shunt routes require neither tracks clear, overlap nor flank protection. Longe range shunt routes are provided in large stations to speed up longe range shunt movements. The special aspects at tall shunt signals require tracks clear and exit signal alight, but neither overlap nor flank protection. With a shunting move passing a shunt signal for a normal route, the proceed aspect is reset to danger with the last axle leaving the berth track. Longe range shunt signals are replaced with the first axle on the first track section.

CFL

Route conditions are as per DB AG with the following deviations: Train routes require 100 to 200m overlap. If the standard overlap cannot be provided, overlap 0m may be chosen, permitting 60km/h entry speed. The entry signal aspect depends on:

- points straight or turnout;
- the turnout angle;
- overlap length; and
- route length, for example less than normal signal spacing or braking distance.

Where the destination track is a split platform track, a subsidiary aspect is displayed automatically if the second part track is occupied and thus leaves only the first track clear with overlap 0m. Train routes can also be set by ARC.

Where different traction power supply systems meet, interlocking with signalling is provided so that locomotives for one system can only be routed onto tracks with the appropriate supply provided.

Signal layout plans, route tables, points tables and signal aspect tables are the basis for projecting work.

NSB

Route conditions are generally the same as those for BV. With regard to overlaps, the practice differs. The overlap length may vary between 0m at routes towards home signals and 200 to 400m at routes towards starting signals. Points in the overlap are locked and detected as for routes.

Sections of catenary can be disconnected by the signalman.

Signal layout plans, route tables, point tables and signal aspect tables are the basis for projecting work.

NS

A normal train route monitors all tracks clear and all points, movable bridges and other devices between entry and exit signals set, locked and detected in the correct position, and additionally, no route locking in the opposite direction. This requirement is to meet the fail-safe conditions which are a consequence of using non-vital circuitry for route proving and marking. Route locking is preceded by route locking proving, checking that all points in the route are in the correct position in order to prevent route locking in the case of points failure which would make signal clearing impossible. This is to prevent undue activation of level crossings and other warning devices. A possibly necessary delay time is initiated only after success, and must elapse in case of a level crossing or other warning device at short distance beyond the signal. Since the warning device is only activated after route locking, this time delay guarantees the required minimum warning time, as a train, waiting at/approaching the signal, could accelerate as soon as the signal clears. No delay is required if the route is locked before an approaching train has passed the Lx activation point. The signal clears immediately, at the latest after the previous train has left the route. As all trains traverse routes, no flank protection is required. No overlap is used. On open lines, however, a protection distance of 200m is exceptionally provided, for example at movable bridges.

Signals and replacing track are approximately 9 to

15m apart to prevent signal replacement being noticed by the driver in a locomotive with the cab at the rear end. A new route setting over a still locked part of a route to the same exit signal is already possible as soon as the points permit the new route. Route setting from the same entry signal is already possible as soon as the preceding train has replaced the signal. In both cases, the signal will clear only after the first train has fully passed the exit signal. In the case of drive on sight routes, the signal will clear immediately, permitting the second train on to the same route, still occupied by the first train. A platform track, split in two parts by placing signals for both directions half way on the track, can be used to admit arriving trains from both sides at the same time.

A not yet approach locked route can be cancelled without delay. Route cancelling after approach locking will hold the route for ≥120 seconds, on condition that an approaching train has not passed the signal within this time. After cancelling an approach locked route without a train approaching, it is immediately possible to call a new route. The entry signal aspect depends on:

- points straight or turnout;
- the turnout angle;
- route length, for example less than normal signal spacing or braking distance;
- condition of the destination (platform) track; and
- the aspect of the exit signal.

Approach locking is activated the moment the signal clears. The route is released with the first axle on the first track circuit. The approach locking will only release when the first and second TCs are occupied. As elements are henceforth maintained only by train occupancy in the route, this measure is taken to guarantee the presence of more than one axle in the route, in order to diminish the risk of loss of shunt. A new normal route can already be called again with a train still occupying part of it, provided decisive points are already cleared. Automatic sectional route release is effected by clearing of tracks. In case of a TC terminating within the clearance marker of a point or crossing, this element will only clear if the following TC is also cleared. This measure prevents the setting up of a new drive on sight route with the points in the other position whilst the preceding train could stand foul of that route, which would not be detected since track circuits will not be proved in such a route. Train effected route release is prevented during track circuit power supply failure. The entry signal is replaced to danger with the first axle on the first track section. Automatic train routes will not release and permit fleeting.

If a signal cannot be expected to clear for a train route due to occupied or faulty tracks, the signalman can select a drive on sight route instead, with a corresponding aspect. Where, in case of a drive on sight route, the first track circuit is occupied or faulty, the signal must be replaced by cancelling the route manually.

It should be mentioned here that the drive on sight route is seldom used. Typical application is to call trains onto occupied platform tracks, into undetected territory, or to keep traffic moving where train detection is faulty.

Where different traction power supply systems meet, interlocking with signalling is provided so that locomotives for one system can only be routed onto tracks with the correct traction supply provided.

In computer interlockings the following deviations from relay interlockings apply:

- in the route locking proving for a *normal* route it will be checked that *all* track circuits in the route are cleared. This will delay a new route locking until the preceding train has passed the exit signal. Activation of level crossings will thus only start when the signal is ready to clear.
- automatic train route release is now effected by *proved sequential* occupation and subsequent clearing of tracks.
- special route setting with all elements blocked to safeguard trains passing danger signals upon instruction.
- cancelling of a *special* route.

Signal layout plans, signal aspect sequence plans and the specification of operational requirements are the basis for projecting work.

BV

For train routes the overlap demand depends upon the situation, entry speed and whether or not ATC is provided. Standard with respect to conflicting train routes is 200m beyond the exit target, which can be a signal, stop board or clearance marker. At the end of a route permitting >40km/h, or >80km/h with ATC supervised entry speed, the overlap distance may be reduced to 100m in case of a conflicting shunt route and to 50m in case of fixed obstacles such as a terminal track. Reduction of the safeguarded area to 0m is permitted with a low speed route of ≤40km/h or ≤80km/h with ATC. Tracks between route points and corresponding elements providing flank protection are proved clear in new interlockings. Main signals as flank protection are acceptable, subject to the above overlap conditions, for example if they are exit to a set train route, 200m overlap must be provided. Before route locking, the train route may be cancelled without delay. After route locking, manual route cancellation will be effected after ≥90 seconds delay, the time is dependent on maximum train speed and the signal distances. The entry signal aspect depends on:

- points straight or turnout;
- route length, for example less than normal signal spacing or braking distance; and
- the length of overlap.

With ATC, permitted speeds are transmitted and supervised. Automatic train route release, sectional at newer interlocking systems, is effected by proved sequential occupation and subsequent clearing of tracks. Train effected route release is prevented during failure of the track circuit power supply (double pass function). The entry signal is replaced to danger with the first axle on the first track section. Special overlap conditions are automatically released 90 seconds after arrival at the destination track.

Train route commands that cannot immediately be executed are stored until the route is available, whereupon the stored call becomes effective automatically. Fleeting is a standard function in most installations. Stations on single track CTC lines have crossing automats for automatic route setting at the most appropriate time.

Where automatic level crossings are situated so that the distance between signal and level crossing is insufficient as an initiation distance, signal clearing may be delayed to provide appropriate warning time.

For shunt routes, two different shunt aspects are provided:

- the route being clear throughout; or
- if any part of the route is occupied.

Shunt routes require flank protection as standard. A third shunt aspect can be displayed for free shunting in an area temporarily isolated from the rest of the track network. With a shunting move passing a shunt signal, the proceed aspect is reset to caution with the first axle on the first track section. The shunt signal is reset to danger manually, with the shunt route released a short time thereafter. Automatic route release is effected when the shunting move proceeds past the next shunt signal.

Signal aspect tables are used as the basis for project design work.

5.4 Operation interface

All modern interlockings may be either operated locally or remote controlled from a centralized traffic control (CTC) centre.

In all relay interlocking systems non-vital and vital functions may be integrated or separated. If all functions are integrated, signalling relays are used throughout. Where the N–type relay is used, for cost reasons there is often a desire to separate vital and non-vital levels of function.

Non-vital operation levels with suitable good quality relays are used by RENFE, SNCF, BR, NSB, NS and BV. Lately, relays are giving way to programmable logic controllers (PLCs), for economic reasons. Computerized systems are also used, mainly to increase the operation aids available to the signalman and to confront him with identical command operations and indications, regardless of the type of interlocking controlled, for example relay and computer interlockings of various makes.

Depending upon the operational method and the extent of the station, control and indication part of the interface to the interlocking may be separated or combined.

Individual element commands from the panel require either simultaneous depression of a master (common) button situated separately from the track layout portion to define the task required, together with a button in the track to define the element addressed. Alternatively, dedicated single buttons or switches are used.

Route commands require start and exit definition, where applicable, including marking of alternative routes. The class of route desired must also be defined. Both buttons in the relevant position in the track layout on the operating panel are operated sequentially or simultaneously. On some installations, a single associated button or switch is operated, defining direction and class of route and overlap. Any route provided by the track layout, including alternative routes, may be used for maximum operational flexibility, or only selected routes may be available.

When the route cannot immediately be set, the route call may be stored, becoming effective when conditions permit; alternatively, it has to be repeated manually.

Button proving may be executed to safeguard against unintentional commands which may be caused by sticking buttons.

In modern interlockings the commands are initiated either:

- by individual push buttons/switch(es);
- via a key pad in coded form;
- through a computer interface;
- by fleeting (also called through routing, automatic operation, etc);
- by automatic route calling (ARC);
- by automatic train routing (ATR); or even
- by timetable/priority based automatic route setting (ARS).

Size of installation, work load for signalmen and desire for automation govern the method chosen. Various methods of operation may be at the signalman's disposal.

Visual indications are provided on the operating or indication panel in the appropriate place, steady or flashing, and if necessary supplemented by audible alarms. Colour, shape of aperture, mode of appearance, and type and pattern of sound distinguish type and importance of an indication, reminder or alarm. Where the signalman has to base his decisions on the displayed process status the indications assume vital importance. In this case, if areas beyond the direct operation range are controlled from a CTC centre, the remote control system used has to meet fail-safe or similar standards. FS rely on strict observation of the appropriate operation rules.

In computer driven interfaces where the process status is displayed on monitors or something similar, the indications may appear in different colours, shapes or modes, often supplemented by train number or other relevant information.

Generally, where conventional control/operating panels are used together with computer interlockings, the command functions and indications are practically the same as in relay interlockings. As the man–machine interface is often adapted to suit the computer – for example push buttons give way to keyboards, tracker balls, mouse, light pen, etc – and indication panels may be substituted or supplemented by VDUs or projection screens. The signalman must adapt to the machine-specific command inputs and display.

To minimize delays caused by malfunctions of technical equipment, most railways make use of technical means to bypass certain safety levels of the interlocking, for example track clear proving. Where such commands (henceforth called relief commands) are possible, they are usually recorded automatically and/or the respective buttons/levers/keys are sealed.

Two kinds of such relief commands can be distinguished:

- commands requiring just special attention to minimize the chance of inappropriate execution, for example cancellation of unused routes, part routes or overlaps; and
- commands enabling the signalman to bypass vital interlocking functions, for example points locking by track circuits.

In the latter case the driver may either be aware or unaware of this fact, depending on whether a special signal aspect is displayed, or the interlocking conditions are bypassed in a way permitting the normal proceed aspect to be shown. It is obvious, therefore, that where a driver is unaware of an irregularity, the responsibility rests entirely with the signalman, who must observe and follow regulations before executing the respective command.

The philosophy of attribution of responsibility may differ. Either the signalman is responsible for all commands including relief commands and the consequence, or the responsibility for safety is shared with or passed to the driver.

Where no relief commands are provided, instructions from the signalman must be transmitted to the driver in a dedicated manner, either personally, in writing, by telephone, by radio or other means. The driver must then proceed as per regulation. As an example, a Swedish driver may, after the signalman has convinced himself that there is no danger for the train, based upon recorded dialogue, '. . . continue with a maximum of 40km/h (supervised by ATC), and be prepared to stop within half the distance he can overlook. For any point in the route the driver must check the position himself'.

In Great Britain, to pass a controlled signal at danger, the driver requires verbal permission from the signalman or a hand signal from a handsignalman working under the signalman's instruction. Upon receipt of this, the driver has to proceed at caution, that is to run on sight. Speed is limited to 25km/h over facing points.

Before authorizing a movement to pass a signal at danger the signalman must ensure that:

- the portion of line is clear and safe for the movement;
- the barriers or gates at manned level crossings are closed to road traffic;
- the points in the route are operated to and locked in the required position and the corresponding normal or reversed indications are showing;
- any ground frame release giving access to the route is normal, unless it requires to be operated for the movement;
- reminder appliances are placed as appropriate.

When authorizing the movement the signalman must tell the driver:

- how far the movement may proceed;
- to proceed cautiously, prepared to stop short of any obstruction;
- to approach cautiously and not pass over any manned level crossing, or, when the crossing is automatic and will not operate normally, to approach cautiously and not pass over, until it is safe to do so.

Written permission is required when two or more consecutive stop signals are to be passed at danger. This is known as *temporary block working*, and the driver is given a paper ticket.

At automatic signals if the signal telephone has failed, the driver may proceed as at controlled signals, at caution. He must not enter a tunnel without first ensuring that the line is clear throughout.

At semi-automatic signals, the driver himself must additionally check the correct position of points before passing over.

The most typical and significant operation relevant characteristics of each railway's interlocking system are now described.

ÖBB

The classical interface with the interlocking consists of an operating desk with push buttons for route setting and individual element control in the appropriate place on the track layout. Groups of common master buttons with associated counters are situated separately and define the task required. Indications are integrated into the panel face at the appropriate place.

Commands are initiated by the simultaneous operation of two push buttons. A few commands require as a first phase the operation of a function selection button to start a timer, during which the second two button operation must be executed. Fringe commands may be effected by single button operation only. Individual element commands from the control panel require simultaneous depression of a master button together with a button in the track portion to define the element addressed. Route commands from the panel require start and exit as well as class of route and overlap exit definition. Alternative routes must be marked beforehand as a first phase. To prevent undesired commands, buttons are checked for sufficiently short operation and for plausible selection of type and number. Sticking buttons and wrong number (other than 2), or selection of non-logical pairs of buttons cause an alarm and rejection of the command. On large conventional installations commands are effected in coded form from a numerical keyboard (*Nummernstellpult*). After verifying the displayed code, it is transferred to the interlocking for execution, or alternatively, with push buttons from a condensed control panel. Indications are displayed on a vertical polygonal indication panel. More recently, a computerized aid *Videopult* is used where command input is by light pen on a colour monitor on which the process status is also displayed. The signalman has the option of selecting route setting by ARC and/or ATR.

The signalman's vital decisions for relief commands and the consequences depend upon fail-safe presentation of the process status.

Recorded commands available to the signalman, which bypass safety levels and may only be executed after the appropriate regulations are observed, are:

- display of the *Ersatzsignal* or the *Vorsichtsignal* aspect if the main aspect fails.
- overriding a disturbed block section.
- change of position of a point with TC occupied/faulty.
- restoring correspondence of a trailed point.
- cancellation of a locked train route.
- resetting of an axle counter.
- overriding a disturbed level crossing.
- opening the level crossing barriers.

Fig. 5.5 Standardized man-machine interface (ÖBB)

On new installations, the *Einheitliche Bedienoberfläche* (EBO – standardized man–machine interface) is common to both national suppliers and consists of up to five colour VDUs per signalman, supplemented by keyboard, mouse and printers. Remote control, train describer, automatic route setting, automated shunt route setting, train protocol, catenary switching and interfaces to continuous automatic train control (LZB) and to a train supervision centre can be integrated. Great emphasis is put upon safe process status display by dual channel control of the VDU(s). Failure of one channel, and thus non-safe display, results in a flashing picture. (See **Fig. 5.5**.)

NMBS/SNCB

The classical interface with the interlocking consists of a control desk with push buttons for route setting and related controls. Indications are displayed on an indication panel.

Commands are initiated by individual dedicated push buttons. Route commands from the control desk require start and exit as definition. Class of route is selected by pushing one out of four buttons (train/shunt route and both as in or out routes). Alternative routes are selected by dedicated individual route buttons. Individual element commands from the control desk require individual push button operation.

The station inspector has at his disposal relief commands which bypass safety levels and may only be executed after the appropriate regulations are observed:

- route release with track circuit or treadle malfunction.
- route locking although not all route conditions are satisfied.
- change of direction of working over an open line track.

On new installations a computer based operation interface similar to the integrated electronic control centre (IECC) of BR, is used. Overview and detail screens are available plus a general purpose display. Command inputs and output messages are handled by the latter. Route setting is carried out by the automatic route setting (ARS) module, receiving up to date information from the timetable processor (TTP), all remaining under the signalman's control.

SBB

The classical interface with the interlocking consists of an operating desk with push buttons for route setting and individual element control in the appropriate place on the track layout. Groups of common master buttons with associated counters are situated separately and define the task required. Indications are integrated into the panel face at the appropriate place. On large installations, indications are displayed on a separate indication panel.

Commands are initiated by the simultaneous operation of two push buttons. A few commands require as a first phase the operation of a function selection button to start a timer, during which the second two button operation must be executed. Fringe commands may be effected by single button operation only. Individual element commands from the control panel require simultaneous depression of a master button together with a button in the track portion to define the element addressed. Route commands from the panel require start and exit as well as class of route and area of special protection definition. Alternative routes are set with the aid of green push buttons.

To prevent undesired commands, buttons are checked for sufficiently short operation and for plausible selection of type and number. Sticking buttons and wrong number (other than 2) or selection of non-logical pairs of buttons, cause an alarm and rejection of the command. In larger conventional installations routine commands are executed by keying code numbers on a numerical keyboard. After verifying the displayed code, it is transferred to the interlocking for execution.

A train route call which cannot be executed directly may be stored, becoming effective when conditions permit. After manual route cancellation, all commands are blocked for 2 minutes to prevent untimely early setting of already stored conflicting routes. Programmed automatic crossings can be arranged at small remotely controlled stations. At larger stations the signalman has the option of selecting route setting by ARC and/or ATR, or as fleeting through the station.

Because the signalman is responsible for all commands including relief commands and the consequences, indications assume vital importance.

Recorded commands available to the signalman, which bypass safety levels and may only be executed after the appropriate regulations are observed, are:

- clearing the main signal (*Notfahrt*) in case of a faulty track circuit.
- display of the *Hilfssignal* aspect if the main aspect fails.
- overriding a disturbed block section.
- change of position of a point with TC occupied/faulty.
- restoring correspondence of a trailed point.
- cancellation of a locked train route.
- resetting of an axle counter.

Special commands are available to block and unblock individual elements and functions to prevent undesirable operation and release.

In computer interlockings the man–machine interface in the operating room consists of a common polygonal indication panel and/or up to five monochrome and colour VDUs per operator, supplemented by keyboard(s) and printer. Great emphasis is put upon safe process status display by dual channel control of the VDU(s). Failure of one channel, and thus non-safe display, results in a flashing picture.

DB AG

The classical interface with the interlocking consists of an operating desk with push buttons for route setting and individual element control in the appropriate place on the track layout. Groups of common master buttons with associated counters are situated separately and define the task required. Indications are integrated into the panel face at the appropriate place. On larger installations, indications are displayed on a separate indication panel.

Commands are initiated by the simultaneous operation of two push buttons. A few commands require as a first phase the operation of a function selection button to start a timer, during which the second two button operation must be executed. Fringe commands may be effected by a single button operation. Individual element commands from the control panel require simultaneous depression of a master button together with a button in the track portion to define the element addressed. Route commands from the panel require start and exit as well as class of route and overlap exit definition.

Alternative routes must be marked beforehand as a first phase. To prevent undesired commands, buttons are checked for sufficiently short operation and for plausible selection of type and number. Sticking buttons and wrong number (other than 2) or selection of non-logical pairs of buttons, cause an alarm and rejection of the command. In larger conventional installations routine commands are executed by keying code numbers on a numerical keyboard. After verifying the displayed code, it is transferred to the interlocking for execution. The signalman has the option of selecting route setting by ARC and/or ATR.

The signalman's vital decisions for all commands including relief commands (*Hilfsbedienungen*) and the consequences depend upon fail-safe presentation of the process status.

Recorded commands available to the signalman, which bypass safety levels and may only be executed after the appropriate regulations are observed, are:

- display of the *Ersatzsignal* or the *Vorsichtsignal* aspect if the main aspect fails.
- overriding a disturbed block section.
- change of position of a point with TC occupied/faulty.
- restoring correspondence of a trailed point.
- cancellation of a locked train route.
- cancellation of an overlap.
- resetting of an axle counter.

Special commands are available to block and unblock individual elements and functions to prevent undesired operation and release.

In computer interlockings the man–machine interface in the operating room consists of up to five monochrome and colour VDUs per operator, supplemented by keyboard(s), control board (digitizer), key button and printer(s) for recorded commands. An indication panel is not required in new installations. Great emphasis is put upon safe process status display by dual channel control of the VDU(s). Failure of one channel, and thus non-safe display, results in a flashing picture. Train describer, automatic route setting, train protocol, interfaces to continuous automatic train control (LZB) and to a train supervision centre can be integrated.

RENFE
The classical interface with the interlocking consists of an operating desk or panel with push buttons for route setting and individual element control in the appropriate place on the track layout, and common master buttons to define the task required. Indications are integrated into the panel face at the appropriate place. On larger installations indications are displayed on an indication panel or on VDUs.

Commands are initiated by the simultaneous operation of two push buttons or at other installations by successive depression of start and exit push buttons. Individual element commands from the control panel require simultaneous depression of a master push button to select the task together with a button in the track to define the element addressed. Route commands from the panel require start and exit definition. Alternative routes must be set with the aid of intermediate buttons. On larger installations commands are effected in coded form from a control panel with key pad after verification. The signalman has the option of selecting route setting by ARC, fleeting, or ATR based upon a train describer.

Recorded commands available to the signalman, which bypass safety levels and may only be executed after the appropriate regulations are observed, are:

- clearing a relief signal (*Autorización de Rebase*) in case of a faulty track circuit.
- change of position of a point with TC occupied/faulty.
- cancellation of a locked train route.
- resetting of an axle counter.

Special commands are available to block and unblock individual elements and functions to prevent undesired operation and release. Suspected faulty elements in a locked route may be operated for test purposes. Trailed points are documented by a counter.

In computer interlockings the man–machine interface in the operating room consists of an operation panel or, if required, of monochrome and colour VDUs, supplemented as required by keyboard, indication panel, control board, digitizer, and printer. Where safe process status display is required, this is achieved by dual channel control of the VDU(s). Failure of one channel, and thus non-safe display, results in a flashing picture. Automatic route setting and interface to a train supervision centre can be integrated.

SNCF
The classical interface with the PRS interlocking consists of a control desk with push buttons for route setting and other controls. Indications are displayed

on an indication panel. In PRG interlockings the interface consists of an operating desk with push buttons for route setting and individual element control in the appropriate place on the track layout. Indications are integrated into the panel face at the appropriate place.

Route commands are initiated by successive depression of exit and start push buttons in that order. Alternative routes can be set as successive part routes. In installations with a non-vital operating level, a train route call may be stored, becoming effective when conditions permit. The signalman has the option of selecting route setting by fleeting (permanent route setting).

As an example for a sophisticated non-vital operation level, the SNCF *Poste tout Relais à Commande Informatique* (PRCI) computer controlled relay interlocking is now described.

The operating flexibility desired by operators who greatly appreciate the advantages of programmed route setting, and the development of microprocessors, have resulted in a move towards computer control of signalling installations. The assumption that computer based applications would offer the greatest advantages at the interlocking/operator interface has proved correct in practice. This has led to the development of the PRCI, which makes widespread use of computerized technology to facilitate logic exchanges between man and machine.

The PRCI consists of two parts:

(1) a computer control system based on various computerized modules, such as SNCI for automatic/manual route control and SNTI for teletransmission.
(2) the fail-safe interlocking module consisting of relays arranged in geographical circuitry, as in the *Poste tout Relais Géographique* PRG signalbox.

One of the main advantages of the PRCI is its enormous operating flexibility. At Paris Montparnasse for example, a station with 196 sets of points and 156 signals is controlled by one to four operators depending on the volume of traffic.

Whilst PRCI is well adapted to larger stations, it can also be used for smaller stations and single track lines which can thus easily be remotely controlled from a central control panel in a district control office.

Some special features of PRCI include:

- route command through a dialogue with a keyboard and VDU;
- the possibility to store many commands in advance;
- ability to prepare the commands for several trains on various routes.

Recorded commands available to the signalman, which bypass safety levels and may only be executed after the appropriate regulations are observed, are:

- change of position of a point with TC occupied/faulty.

BR

The classical interface with the interlocking consists of an operating desk with push buttons and/or switches for route setting in the appropriate place on the track layout, and individual switches/buttons for individual element operation. Indications are integrated into the panel face at the appropriate place. On larger installations commands are also effected by push buttons and indications are displayed on a free standing operation panel.

Commands are initiated by the successive operation of start and exit push buttons in that order. Some systems employ a switch and button arrangement. Individual points commands from the control panel are effected by three position switches. In the centre position route setting is enabled over them, the left and right positions are to set and lock the points normal or reversed. Route commands from the panel require start and exit as well as class of route and overlap exit definition. In some instances the appropriate class of route is selected automatically dependent upon track conditions. Alternative routes must be marked by presetting and locking decisive points or by the use of intermediate push buttons. The signalman has the option of placing certain signals in automatic mode (fleeting), or to choose route setting by ARC and/or ATR.

No relief commands are available.

The integrated electronic control centre (IECC), represents the latest computer based style of operation for large schemes. It comprises two local area networks on which are configured a number of processor modules. One of the two networks deals primarily with signalling functions. The segregation is designed to maintain quick response times. It is the signalman's display system (SDS) module which forms the man–machine interface. A work station consists of up to five VDUs, a keyboard and a trackerball device. Overview and detail screens are available plus a general purpose display. Command inputs and output messages are handled by the latter. Route setting is carried out by the automatic route setting (ARS) module, receiving up to date information from the timetable processor (TTP), all remaining under the signalman's control. Trackerball and associated control buttons are used to set and cancel routes and to operate individual elements.

FS

The classical interface to the interlocking consists of a control desk from which all commands are effected and on which some indications are displayed. Route setting and cancellation is effected from route push/pull buttons arranged in logical order in rows and columns. An associated indication shows that the route has been marked. Three position switches associated with all main signals permit automatic operation, inhibit clearing and permit relief operation. Points and level crossings have switches for individual operation with associated status indications. Open lines can be blocked against undesired movements. Route commands are initiated by depression of one push button. Alternative routes are available only in larger stations and chosen by individual buttons. The other indications of all individual elements within the station (points, signals, track circuits, level crossings, treadles, block etc, as well as all set routes) are displayed on a separate indication panel. The signalman has the option of selecting signal operation by fleeting, or route setting by ARC or ATR. In the Genova CTC office the operator can call for the display of computer generated alternatives (*guide operatore*) to solve traffic conflict situations. He can establish and implement priorities and thus exercise conflict management.

On more recent installations beyond a certain size, commands are effected from a keyboard with start and exit definition.

Commands available to the signalman, which bypass safety levels and may be executed after the appropriate regulations are observed and seals are broken, are:

- display of the '*Segnale di avvio*' or the '*Segnale di chiamata*' aspect when conditions prevent the main signal from clearing.
- change of position of a point under various fault conditions.
- cancellation of a locked train route.
- locking of a train route.
- conclusion of incomplete sectional route release.
- operation of signals if a level crossing is faulty.

On electronic interlockings, the man–machine interface consists of fail-safe VDUs, supplemented by a keyboard.

CFL

The interface to the interlocking is generally as per the interface employed in Germany (DB AG); however, no relief aspects and associated commands are available. ATR is not used.

No computerized operation level is in use.

NSB

The interface to the interlocking is generally as per Sweden (BV), with the following deviations.

In ordinary sized interlockings, start and exit of routes is marked by two returning switches operated simultaneously. Other commands are given with returning and non-returning switches. Commands to change the area of responsibility, for example from CTC operator to local signalman, require simultaneous operation of two different switches or switch and push button. In larger interlockings, key pad control is used to define start and exit, together with a selective command button. At CTC centres, a keyboard is used for commands, consisting of a mnemonic code plus a three digit element number. When the remote control link is disrupted, the field stations automatically resort to automatic block posts.

Individual points may be operated from a local control panel, known as ground stand, and located nearby for shunting moves. In this case no track locking is included and in case of track failure relief commands can be issued from this panel.

NS

The typical operator interface with the relay interlocking consists of a control desk with push buttons for route setting in the appropriate place on the track layout and switches for individual points control and other controls outside the track layout. Alternative routes must be marked by presetting and locking decisive points or, in exceptional circumstances, by the use of intermediate push buttons. The indications are either integrated within the operating desk or displayed on a separate indication panel. The signalman can set automatic routes for fleeting.

An *Elektronische Bedien Post* (EBP) (electronic control post) was introduced in which the traditional non-vital relay logic is replaced by computers. The new interface consists of a keyboard and control monitor together with a colour VDU and/or indication panel.

In combination with EBP, a VPI electronic interlocking can be used to replace the traditional vital relay logic whilst retaining the same function.

For the ElS computer interlockings, the German philosophy was adopted, in which the indication functions are considered vital. In this context new relief functions became possible:

- partial route release (unlocking points locked by faulty TC).
- change of position of a point with TC occupied/faulty.

Safety relevant commands are protected against accidental use and only possible after a first command is repeated with a random number procedure within 20 seconds. They are logged on a printer.

Not directly bypassing safety levels are the commands:

- special route setting without signal clearing.
- cancelling of a special route.

Commands are available to block and unblock individual elements and functions to prevent undesired operation and release for protection of staff working in or near the track:

- blocking/unblocking of points.
- blocking/unblocking of route elements.

BV

The classical interface with the interlocking consists of an operating desk with push buttons for route settings and individual element control in the appropriate place on the track layout, and common master buttons to define the task required. Indications are integrated into the panel face at the appropriate place. On larger installations commands are effected in coded form from a control panel with key pad, verification plus master button operation for execution. Indications are displayed on an indication panel.

Individual element commands from the control panel require simultaneous operation of a master button to define the task required, together with a button in the track to define the element addressed (SJ 65). Alternatively, dedicated single buttons or switches are used (SJ 59). The route call may be stored, becoming effective when conditions permit. Fleeting is a standard function at most stations. At larger installations the signalman has the option of selecting route setting by ARC and/or ATR.

On single lines in CTC territory, small stations are equipped with crossing automats. These take information from the interlocking and from the line block system on both sides of the station. Should only one train be approaching, the automat sets a route straight through. When trains are approaching from both sides the automat issues a command sequence to arrange that crossing. If the transmission system between control centre and field station is interrupted for more than two minutes, the local station automat is activated as fall back level.

In newer CTC centres and at larger interlockings, the man–machine interface consists of a large back projection colour VDU screen for the general overview and normal monitors for close-up display. Operation is by a standard computer keyboard. The signalman can build his own automatic functions, thus replacing the crossing automats in conventional interlockings. A train describer is integrated and can be used for automatic route setting (ARS).

Individual points may be operated for shunting moves from a local ground stand control panel. In this case no track locking is included and in case of track failure relief commands can be issued from here. Where no such local control panels are provided, a track circuit bypass button is provided in a locked cabinet in the cabin some distance away from the signalman. In case of TC failure, signalman and dispatcher have to co-operate in order to throw the point.

5.5 Power supply

As interlockings are vital for railway operation, care must be taken to maintain the supply of electrical power at all times. Besides the actual interlocking, power must also be available for signals, train detection systems, point machines, process status display and fringe equipment. The primary power source is generally the public power supply or the railway's own power supply. In case of primary power supply failure, an alternative source of power is often available to take over for any length of time. The components of a signalling system can tolerate power fluctuation and interruption to varying degrees. Whereas relay interlockings are relatively tolerant, computer interlockings are highly sensitive. In conventional relay interlocked signalling systems one or more batteries are kept constantly charged in order to supply the DC consumers and take over to feed the AC consumers via motor generator sets or static invertors until the alternative supply either takes over or the primary supply is restored. SBB, BR, NSB and BV are exceptions. Computer systems invariably need uninterrupted power supply (UPS) systems to remain in operation.

The signal cabin's power supply invariably provides DC for the interlocking. Signals are AC or DC fed with either normal voltage around the clock, or with normal voltage during daytime and reduced voltage at night. This is to dim the signals at night and thus prevent glare from irritating the drivers. Track circuits are fed by DC or AC, power may be detected to prevent accidental route release which could occur when track relays happen to drop and pick in the logical sequence with track power failure and return. Point machines may be three phase AC, single phase AC, or DC operated, points operation may be simultaneous, alternatively staggered or successive to limit the peak load on the points supply network. Where three phase networks are used, the phase sequence must be monitored to assure correct rotation of motors. Indication lamps are generally AC fed, with the exception of NSB.

The power supply may be galvanically isolated from the primary power supply and floating against earth. For testing purposes, battery, AC supply and three phase supply networks may be suitably connected. Where the interlocking and/or associated circuit are earth free, checks for possible insulation faults may be carried out sporadically, regularly or continuously. This is to detect the first earth fault in a signalling installation before a second fault could provide a path via earth to bypass vital interlocking and thus create a potentially dangerous situation.

In case of prolonged primary power supply failure other sources of semi-permanent power may be used, for example diesel generators.

The individual railway's power supply philosophies differ and the more important figures are summarized in **Fig. 5.6**. The situation on the different railways will be described individually.

ÖBB

The primary power source is the public power supply. The DC for the interlocking is supplied from a trickle charged 60V battery. Signals are AC fed with normal voltage during daytime and reduced voltage at night. Track circuits are fed by AC. Where motor relays are used, they are fed from a three phase 50Hz, or 100Hz network to avoid $16\frac{2}{3}$Hz traction interference. Appropriate phase distribution ensures fail-safe operation and detection of faulty block joints. Point machines are three phase AC operated, points operation is staggered. The power supply is galvanically isolated from the primary power supply and earth free. Earth leakage detectors continuously check for insulation faults. For testing purposes, DC, AC and three phase points machine supply networks are earthed at monthly intervals.

In case of primary power supply failure the battery takes over for a limited time, also feeding motor generator sets or static invertors. In case of prolonged primary power supply failure a diesel generator takes over.

In computer interlocking systems the UPS consists of rectifier/inverter modules with a float charged 220V or 380V battery of about 15 to 30 minutes capacity as standby, and supplies the signalling installation with regulated 220/380V, three phase, 50Hz. This is transformed to all required voltages for signals, points, track circuits, etc. All DC consumers including the computers are fed via transformer/rectifier sets. Alternatively, only the computer interlocking itself is fed by a UPS, whereas signals, points, track circuits and the UPS itself are supplied as in conventional interlockings. The voltages/networks are earth free. The AC supply for ventilators, monitors, printers and other peripherals is earthed.

NMBS/SNCB

The primary power source is the public power supply. The DC for the interlocking is supplied from a trickle charged 48V battery, the positive pole is earthed. Signals are AC fed with normal voltage day and night. Track circuits are primarily fed by AC, point machines are DC operated, points operation is simultaneous. The DC power supply and the centre point of the AC transformer are earthed. For earth leakage detection, differential relays are used.

In case of primary power supply failure a second power supply, derived wherever possible from another high voltage transformer station, takes over. Where no second power supply is available the battery bridges the changeover gap for up to two hours until a diesel generator takes over.

SBB

The primary power source is either SBB's traction supply or the public power supply, whichever is more economic. Both can alternatively feed a rotary inverter, driving a three phase generator, which in turn provides a virtually uninterrupted 50Hz supply for the interlocking. From this, the DC and AC consumers are supplied without further standby. The 48V DC for the interlocking is supplied from a rectifier, exceptionally backed up by a trickle charged battery, the negative pole is connected with the star point of the three phase points supply network for functional reasons and earthed. Signals are AC fed, earth free, with normal voltage during daytime and reduced voltage at night. DC track circuits are primarily fed by AC. Point machines are three phase AC operated, points operation is staggered.

In case of failure of the actually feeding power, the alternative power source takes over feeding the inverter.

In computer interlocking systems the UPS consists of rectifier/inverter modules with a float charged 380V battery of about 30 minutes capacity as standby, and supplies the signalling installation with regulated 220/380V, three phase, 50Hz. This is transformed and rectified to meet all required voltages for signals, points, track circuits and DC consumers, including computers. The AC supply for ventilators, monitors, printers and similar equipment is earthed.

DB AG

The primary power source is the public power supply. The DC for the interlocking is supplied from a trickle charged 60V battery. Signals are AC fed with normal voltage during daytime and reduced voltage at night. Track circuits are primarily fed by AC. Where motor relays are used, they are fed from a three phase 50Hz, or 100Hz network to avoid $16\frac{2}{3}$Hz traction interference. Where both $16\frac{2}{3}$Hz and DC are present as traction supply, 42Hz is chosen. Appropriate phase distribution ensures fail-safe operation and detection of faulty block joints. Point machines are three phase AC operated, points operation is staggered. The power supply is galvanically isolated from the primary power supply and earth free. For testing purposes and functional reasons, battery negative pole, AC supply neutral and star point of the three phase points machine supply network are connected. An earth leakage detector checks for insulation faults continuously.

In case of primary power supply failure the battery takes over for a limited time, also feeding motor generator sets or static invertors. In case of prolonged primary power supply failure a diesel generator takes over. During times of primary power supply failure, points operation is sequential.

In computer interlocking systems the UPS consists of rectifier/inverter modules with a float charged 380V battery of about 30 minutes capacity as standby, and supplies the signalling installation with regulated 220/380V, three phase, 50Hz. This is transformed to all required voltages for signals, points, track circuits and the like. All DC consumers, including computers are fed via transformer/rectifier sets. The voltages/networks are earth free. The AC supply for ventilators, monitors, printers and other peripherals is earthed.

RENFE

The primary power source is the RENFE owned mains. The DC for the interlocking is supplied from a trickle charged battery of either 24, 48, 50 or 60V. Signals are AC fed with normal voltage during day and night. Track circuits are primarily fed by AC, point machines are AC single phase operated, points operation is mostly sequential. The power supply is galvanically isolated from the primary power supply and earth free. On the high speed line AVE an earth leakage detector checks continuously for insulation faults.

In case of primary power supply failure the battery bridges the changeover gap until the local public power supply takes over. In more important installations a diesel generator is provided.

In computer interlocking systems the UPS consists of rectifier/inverter modules with a float charged 380V battery of about 60 minutes capacity as standby, and supplies the signalling installation with regulated 220/380V, three phase, 50Hz. This is transformed to all required voltages for signals, points and track circuits. All DC consumers, including the computers are fed via transformer/rectifier sets. The

Fig. 5.6 Interlocking and associated power supply

	ÖBB	NMBS/SNCB	SBB	DB AG	RENFE	SNCF	BR	FS	CFL	NSB	NS	BV
DC voltage for interlocking (V)	60	48	48/60	60	12/24/48/60	24	50	50	48/60	40	12/28	24/48
Signal feeding voltage, daytime (V) / frequency (Hz)	220 / 50	110 / 50	90 / 50	220 / 50	220 / 50	127 / 400	110 / 50	150[16] / 50	220 / 50	250 / 95/105	110 / 75	110 / 50
Signal feeding voltage, night-time (V)	155	110	70	145	220	127	110	150	145	190	50/80[10]	75
Point machine operation voltage. If AC, 50Hz 1/3 ph = single/three phase	220/380 3ph	150DC	220/380 3ph	220/380 3ph	220[19] 1ph	220/380 3ph	120DC	144DC 150AC 1ph	220/380 3ph	220[12] 1ph	136DC	220DC[4]
Points operation staggered	yes	no	yes	yes	no	yes	no	no	yes	yes	yes	no
Track circuit feeding voltage, AC (V) / frequency (Hz)	3×380 3×220 / 100	110 / 50	35...90 / 50	3×220 3×180 / 100[5]	220 / 50	115 / 50	110 / 50	150 / 50	3×220 3×180 125/50	220 / 95/105	110 / 75	220 / 50
Other AC (V) / frequency (Hz)	–	–	65 / 125	220 / 50	110 / 50	–	110[2] / 83⅓	80 / 50	220 / 83⅓	–	110[2] / 50	–
Track circuit feeding voltage, DC (V)	–	–	–	–	yes	–	–	–	24	3...15	–	7
Nominal indication lamp voltage (V)	24AC	48AC	24AC	24AC	10AC	–	24AC	24AC	24AC	24/40=	24AC	24AC
Interlocking power supply floating?	yes	no	no[8]	yes	yes	yes	yes[2]	yes	yes	yes	yes	[6]
Check for earth leakage automatic?	yes	yes	no	yes	no	yes	yes	yes	yes	yes	yes	yes[13]
Check for earth leakage regularly at ... intervals c = continuous	monthly	c	[9]	c	c[1]	c	–	c	c	c	c[11]	c
Check for earth leakage (other, when?)	–	–	–	–	–	–	–	18	–	–	9	–

	ÖBB	NMBS/SNCB	SBB	DB AG	RENFE	SNCF	BR	FS	CFL	NSB	NS	BV
Interlocking power supply earthed? If interlocking battery earthed, pole	no	yes plus	yes minus	no	no	no	no	no	no	no	no	no
Primary power (P=public/R=railway owned)	P	P	P or R	P	R	P	P	P	P	P	P	R
Standby power provided by												
second power supply	no	yes	yes	no[14]	yes	yes	yes	yes	no	no	yes	yes
diesel generator	yes	no[14]	no	yes	yes[1]	yes[7]	yes	yes	yes	yes	no	yes
traction power	no[14]	no	yes	no	no	no	yes	yes[17]	no	yes	no	no[14]
battery only for ... h	1/4–1/2	2	no	1/2..2	1..3	6..24	no	6..8	1..3	2..12	3	no[15]

Notes:
[1] On high speed line (AVE).
[2] Obsolete.
[3] Also 220V DC.
[4] Also 220/380V 3 phase.
[5] At areas with AC 16⅔Hz traction; 42Hz at areas where AC and DC traction supply meet.
[6] External circuits yes, internal circuits no.
[7] Mobile.
[8] Signal power supply earth free.
[9] At regular intervals.
[10] 50V for speed indicators, 80V for main signals.
[11] 12V DC and 136V DC.
[12] 50Hz, also 3 phase, and 16⅔Hz.
[13] Main signal and point detection circuits.
[14] Exceptionally yes.
[15] At level crossings only, capacity depending on traffic.
[16] 48V DC for control and indication circuits.
[17] Indirectly.
[18] At regular intervals in older installations.
[19] Also 24V DC and 110V DC.

voltages/networks are earth free. The AC supply for ventilators, monitors, printers and the like is earthed.

SNCF

The primary power source is the public power supply. The DC for the interlocking is supplied from a trickle charged 24V battery. Signals are AC fed with 400Hz which permits the use of smaller transformers. Track circuits are primarily fed by AC 50Hz, point machines are three phase AC operated, points operation is staggered. The power supply is galvanically isolated from the primary power supply and earth free. Earth leakage detection inside the relay room is realized with an insulation fault detector connected between battery negative and earth. An earth leakage detector continuously checks for insulation faults.

In case of primary power supply failure the battery takes over for six to 24 hours, also feeding static invertors for point machines and signals.

BR

The primary power source is the public power supply. The DC for the interlocking is supplied directly from a rectifier. Signals are AC fed with normal voltage during day and night. The signal power supply is galvanically isolated from the primary power supply and floating against earth. Track circuits are primarily fed by AC. Point machines are DC operated. An earth leakage detector continuously checks for insulation faults.

In case of primary power supply failure the second power supply, a diesel generator set or the traction power supply, takes over.

In computer interlocking systems the equipment is designed to operate from a 110V AC supply, consuming 35W per cubicle. It can be housed in heated or non-air conditioned environments. To maintain continuity of operation during power changeover a UPS unit is usually installed. In earlier installations without UPS, other than very short power failures could result in a cold start with all associated delays.

FS

The primary power source is the public power supply. The interlocking cabins are fed, depending on the circumstances, by either:

- the primary power supply with automatic switch-over to a standby supply based on a substation owned by FS, supplemented by a generator set;
- a rectifier/inverter system, based on a trickle charged 144V battery; or
- where computers are used, a UPS with a battery of about 30 minutes capacity as standby, supplemented by a generator set.

In all cases the internal AC output is galvanically isolated from the primary power supply and supplies smaller installation earth free with 150V or where UPS is required with 220V single phase. Larger installations are fed with 220/380V, three phase. The DC for the interlocking is supplied by rectifiers. Signals are AC fed with normal voltage day and night. Track circuits are primarily fed by AC. Point machines are DC or AC single phase operated, points operation is simultaneous. An earth leakage detector continuously checks for insulation faults.

At large stations, the power supply system is designed individually.

CFL

The primary power source is the public power supply. The DC for the interlocking is supplied from a trickle charged 48 or 60V battery. Signals are AC fed with normal voltage during daytime and reduced voltage at night. Track circuits are primarily fed by AC, impulse TCs by DC. Point machines are DC, AC or three phase AC operated, points operation is staggered. The power supply is galvanically isolated from the primary power supply and earth free. For testing purposes and functional reasons, battery negative pole, AC supply neutral and star point of the three phase points machine supply network are connected. An earth leakage detector continuously checks for insulation faults.

In case of primary power supply failure the battery takes over for a limited time, also feeding motor generator sets or static invertors. In case of prolonged primary power supply failure a diesel generator takes over.

NSB

The primary power source is the public power supply. The DC for the interlocking is supplied directly from a rectifier. Signals are AC 95–105Hz fed from AC/AC converters with normal voltage during daytime and reduced voltage at night. Track circuits are primarily fed by 95–105Hz AC. These frequencies are chosen to prevent interference from harmonics of the traction current. For convenience the signals are fed from the same network. Point machines are AC operated 220V 50Hz single or three phase, or by $16\frac{2}{3}$Hz single phase. In more recent computer interlockings 220/380V 50Hz three phase is used. Indication lamps are generally DC fed from the interlocking supply. In PLC controlled interlockings 24V AC or DC is used for indication lamps. The power supply is earth free. An earth leakage detector continuously checks for insulation faults.

In case of primary power supply failure the traction network on electrified lines, UPS for 2 to 12 hours, or a diesel generator takes over.

NS

The primary power source is the public power supply. A synchronous rotating AC/AC converter generates 3kV 75Hz for signalling purposed fed parallel on both sides of the track to all consumers along the open line, with changeover facilities in case of disturbances. 75Hz is chosen to prevent interference from parallel overhead high tension lines. The 12V DC for the interlocking cabin and for level crossings on the open line is supplied from individual trickle charged batteries. Similarly, 28V batteries supply the non-safety related relays. On the open line, in relay houses and relay cases 110V 75Hz, derived from the

3kV supply, feeds track circuits and block signals with constant voltage directly. Rectifiers feed the 12V relays. Signals inside stations are fed with normal voltage during daytime and reduced voltage at night. Point machines are DC fed, supplied from a trickle charged 136V battery. Points operation is staggered. The DC supply for interlocking and points is earth free and continuously detected. Regular checks once a year are made to detect insulation faults in the AC network.

In case of primary power supply failure the batteries, where provided, bridge the changeover gap, until the second power supply takes over. Interruption of the AC supply affecting signals and track relays during this time is tolerated.

BV

The primary power source is the BV owned power supply. The DC for the interlocking is supplied directly from a rectifier. The voltage depends on the type of interlocking. Signals are AC fed with normal voltage during daytime and reduced voltage at night. DC track circuits are primarily fed by AC. Point machines are DC operated, supplied from a rectifier. In recent computer interlockings, three phase point supply is used. Indication lamps are generally AC fed, however, where circumstances demand, DC feeding is also used. The internal power supply is earthed, external circuits are floating. An earth leakage detector continuously checks main signals and points detection circuits for insulation faults.

Constantly charged batteries are used at remote level crossing installations, both for relay circuits and for signals. Level crossing track circuits are directly battery fed for availability.

In case of primary power supply failure a second power supply or a diesel generator takes over.

To ensure uninterrupted power supply to vital computers, a battery backup is used.

5.6 Standards and legal obligations

ÖBB

The '*Österreichisches Eisenbahngesetz*' (Austrian Railway Law) governs the railway safety system. The Ministry of Traffic or a nominated expert checks a manufacturer's documents, development steps and proof of safety. Tests are conducted in the factory and on site. The Ministry finally issues a type approval for equipment or a system, which applies for all further installations. For changes, the same procedure is followed. ÖBB assures function initially in the factory and on site. The main directives for railway safety systems and checks are laid down in regulation 'ÖVE-T3 *Elektrische Eisenbahnsicherungsanlagen und - geräte*'. International (UIC), national (ÖVE, ÖNORM), and VDE and DIN standards are applied.

NMBS/SNCB

It is at the discretion of the railways to determine the scope of tests and proofs of safety for the approval and acceptance of railway signalling systems.

SBB

For Switzerland the signalling principles are laid down in federal regulation 'Articles on Design and Operation of Railways' 1983, articles 38 to 45. The Federal Office of Transport has delegated to SBB the authority for approval and acceptance of new signalling equipment and for delivering the operating licence within their own network. UIC, ORE and VDE standards are applied. Quality assurance follows EN29001.

DB AG

The '*Eisenbahn- Bau- und Betriebsordnung*' (EBO) in Germany governs construction, equipping and operation of standard gauge public railways. This regulation, supplemented by the '*Eisenbahn Signalordnung*' (ESO), prescribes certain signals, signal aspects, safety conditions (for example flank protection), block systems and train protection. Interlocking systems conform to the relevant UIC Recommendations and VDE– (German electrical engineers) 0831 and other standards. Interlockings meet DB AG's specification Mü 8004.

A '*Bundesbahngesetz*' (BbG) Federal Railway Law delegates responsibility for safe and orderly testing, operation and maintenance of movable and fixed installations exclusively to DB AG.

RENFE

In Spain the 'Standard about Electric Interlockings Operating and Safety' applies, incorporating national and international standards. The Ministry of Public Works and Transport specifies the regulations governing level crossing installations.

RENFE has standardized all devices and elements used in interlockings. National and international standards have been used in the specifications.

SNCF
All signalling products have to satisfy the SNCF specifications and are approved after tests in the SNCF laboratories.

BR
In Great Britain vital parts of the interlocking comply with British Standards, for example BR930A for relays.

FS
Standards for all-relay interlockings are defined in detail in a number of technical documents. The most comprehensive are the technical rules '*Apparati centrali elettrici con comando ad itinerari del tipo a pulsanti*' and a set of schematic circuit diagrams for each type of interlocking. For computer interlockings FS standards essentially comply with IEC 65A and the drafts CENELEC EN50126 WG5B, EN50128 WGA1 and EN50129 WGA2.

CFL
CFL accept standards applied in the home countries of its European suppliers.

NSB
NSB themselves are responsible for setting standards and for verifying internally that these standards are met. The standards are usually based on international standards, but stronger codes of conduct are used when necessary.

NS
In the Netherlands, NS themselves have responsibility for setting the rules and regulations.

BV
In Sweden, the railway company itself has responsibility for setting the rules and regulations. The Swedish Railway Inspectorate is the supervising authority which checks that the rules are adequate for the purpose and that the railway fully accepts its responsibilities. The Railway Safety Law demands that the railway should be safer than any other means of transportation. Safety should be continuously increased, but nothing is said about the construction of railway signal installations. Interlocking systems conform to the relevant UIC recommendations and NUTEK, the Swedish electrical board, standards.

6 Block Systems

F. DE VILDER

Contents

6.1 General	222
6.2 Operating conditions	222
6.2.1 Single track lines	222
6.2.2 Double track lines	222
6.2.3 Criteria for the choice of a block system	222
6.3 Survey of block systems	222
6.3.1 Definitions	222
6.3.2 Characteristics – features	223
6.4 Description of block systems	226
6.4.1 Token working block	226
6.4.2 Telephone block	227
6.4.3 Manual block	229
6.4.4 Relay block	231
6.5 Automatic block systems	231
6.5.1 Non-centralized automatic block	231
6.5.2 Centralized automatic block	232
6.5.3 Coded current automatic block	232
6.6 Procedures in case of disturbances	233

Illustrations

Fig. 6.1	Block systems application chart	224
Fig. 6.2	Radio electronic token block (RETB), in use in Scotland	226
Fig. 6.3	Zugleitbetrieb (DB AG)	228
Fig. 6.4	Signalisierter Zugleitbetrieb (DB AG)	228
Fig. 6.5	Tokenless working block (NS)	229
Fig. 6.6	Manual block with electromechanical interlocking, in use on NMBS/SNCB	230
Fig. 6.7	Intermediate block signal, combined with a distant signal on the right (DB AG)	232

6.1 General

The basic function of a block system is to control the spacing of trains. Most block systems provide facilities to activate other systems and devices. So, in an extended way, it can be said that a block system is used to:

- avoid a head-on collision;
- avoid an end-on collision;
- activate the appropriate announcement of a level crossing or fixed staff warning systems, according to the direction of traffic;
- activate other signalling systems, according to the direction of traffic such as crocodiles equipping the signals, access to private siding key operated points, automatic route setting and automatic train stop.

6.2 Operating conditions

6.2.1 Single track lines

Single track lines are installed for low traffic density. Relevant economic evaluations are affected by the geography of the regions to be crossed. In mountainous regions, for instance, the level of investment on a substructure for a double track line is prohibitively high.

For the same reason (ie saving costs), single track lines are normally equipped with a simple signalling system.

Avoiding collisions becomes the predominant requirement. Often only a small number of the safety requirements are implemented, with the remainder being replaced by standard instructions and rules.

6.2.2 Double track lines

For higher traffic density, double or multiple track lines are necessary.

The need to avoid a head-on collision is solved by allocating a direction to each track. Thus the predominant requirement for a double track system is to avoid an end-on collision.

Because the traffic density is higher, a more sophisticated block system is necessary.

6.2.3 Criteria for the choice of a block system

A railway company utilizes more or less complex systems according to both the risk and the cost level.

Commonly used criteria are:

- the type of traffic:passenger trains, freight trains, industrial tracks;
- the line speed;
- the traffic density;
- the fact that if a railway network is electrified, the cost of the power supply for Automatic Block Systems is reduced.

The development of block instruments also reflects the development of mechanical, electromechanical and electrical interlockings.

Increased traffic conditions and increased automation requires additional facilities to increase possibilities of line protection.

Similar additional facilities are also required for increased protection of staff working along the line and of trains running over the line in case of signal failures or in case of bypassing the block conditions during other disturbances.

6.3 Survey of block systems

6.3.1 Definitions

The spacing of trains is achieved by dividing the line into one or several block sections. The number of sections and the length of such a section depends on the traffic density.

A block section is protected by a signal:a starting and home signal at each end of the line and possibly a block signal at the entry of each block section.

A block system is said to be:

- a system with *closed track* if the block signals are normally closed. The block signal is only cleared to permit a train in a non-occupied section;
- a system with *open track* if the block signals are normally showing proceed. The block signal returns only to danger when a train occupies the section.

The block system is *permissive* when entry into a block section with closed signal is authorized on the initiative of the driver only, given a number of precautions commonly used on automatic block type lines.

The block system is *absolute* when entry into an occupied block section is forbidden on the initiative of the driver only. A block system with closed track is always absolute.

Manual block systems are mostly absolute.

A block system can be *automatic* when the detection of a train in a block section is achieved automatically by track circuits or axle counters.

A block system is said to be *manual* when actioning of signals and observation of the presence of a train requires the presence of a local signalman.

Block systems are applied to both unidirectional and bidirectional lines.

Co-operation between signalmen is always needed when a manual block system is used. When an automatic block system is used, however, co-operation between signalmen depends on whether or not facilities for the automatic change of the direction of traffic are provided.

The logic for this operation can be performed by either relay circuitry, solid state interlocking, all kinds of mechanical, electromechanical and electrical interlockings, or simply by documented telephone calls.

BR uses token working for single lines: the signalman gives the driver a visible token of authority to travel over the single line. Both following and opposing moves are prevented whilst the token is in the possession of one train.

Block systems are also used for the announcement of trains at the receiving station, and for automatic setting of entrance routes to the stations.

6.3.2 Characteristics – features

6.3.2.1 Single lines
The change of direction of traffic takes place upon request and if conditions are met, such as block in normal position.

6.3.2.2 Double or multiple lines
There are several different cases for double track lines

Occasional single track operation
In this case the block is unidirectional. If, for example, a track has to be closed for repair work, a change to reverse working can be carried out by reconfiguring the equipment.

The telephone block procedures are required. The throughput of trains is reduced by the lack of block signals in the wrong direction.

Signalled wrong line working
In this case the block is normally unidirectional, on most of the networks. This block is used when limited bidirectional working is occasionally needed, for example in the case of track repair work.

The driver can see that he will be running on the wrong track by a special subsidiary aspect on the starting signal (except FS).

The entry into the arrival station is controlled by a shunt signal.

On BR a main aspect with junction indicator is used at both the entrance and exit to the wrong line.

The throughput of trains is reduced by the lack of block signals in the opposing direction. Simultaneous running in the same direction on both tracks is not permitted.

On NS the 'wrong track' applies to a track of a double track line which is not equipped for bidirectional working, and therefore unsignalled.

Bidirectional working
This type of operation is used on lines with high traffic density to increase the throughput of the line. Trains may run simultaneously in the same direction on both tracks of a double track line or in both directions on the third track of a triple line.

The equipment is as for single track lines with the exception of the FS.

The number of block signals is not necessarily the same for both directions. It depends on the local requirements and traffic density.

The block signal is located before the end of the block section (25m on ÖBB, 20m on FS, 9m on NMBS/SNCB and 9m on NS), in order to prevent the signal changing too early to the stop aspect.

Therefore the distance between two opposite signals corresponds to this neutral section, from 18m on NMBS/SNCB up to 200m on CFL. On RENFE such disposal does not exist.

The layout of the signals for wrong direction working is governed by the same principles as applied for those of the normal direction. Mostly, normal and wrong signals are placed back to back for economic reasons (common lineside boxes) taking into account the rules for the neutral section.

In some administrations, for example on NMBS/SNCB, the wrong direction signals are flashing.

On some railways the driver can see that he will be running on the wrong track by a special subsidiary, eventually flashing aspect on the exit signal (see Chapter 1). The entrance signal to the next station is also completed by a subsidiary aspect announcing the end of wrong direction traffic working.

As the normal direction of travel is on the right or

Block system Railway	Token Working Block	Telephone Block	Manual Block Electromechanical Interlocking	Relay Block	Non-centralized Automatic Block	Centralized Automatic Block	Code Current Automatic Block	'Normal' Traffic Direction L=Left R=Right	Routes Single and Double Track
	6.4.1	6.4.2	6.4.3	6.4.4	6.5.1	6.5.2	6.5.3		
ÖBB	–	1730km *Zugleitbetrieb* 520km *Zugleitbetrieb*	*Felderblock* 176km	ZG62 1823km	*Selbstblock* 924km	*Zentralblock* 450km	–	L/R	5623km
NMBS/SNCB	–	858km	324km	–	2250km	–	–	L	3432km
SBB	–	–	*Felderblock* 73km	DC coded block 975km	DC coded block 1800km	250km	–	L	3098km
DB AG	–	300km *Zugleitbetrieb* 190km *Signalisierter Zugleitbetrieb*	*Felderblock* 5350km Carrier frequency block TF–block 71 1050km	*Relaisblock* 1320km	*Selbstblock* 15 260km	*Zentralblock* Zb S65 Zb S600 3600km	–	R	27 070km
RENFE	–	6113km	BEM 1200km	–	2944km single track 1970km double track 610km double track two-way working	–	–	R	12 837km
SNCF	–	CAPI 3258km (computer aided telephone block) ECLAIR 34km (computer and radio assistance to driver)	BMU (unified manual block) 2903km BMVU (single track manual block) 3949km	–	BAL (luminous automatic block) 10 027km BAPR and BAVB (automatic block with 'carre' protections and long length block: 2692km TVM 1037km	–	–	L	23 900km

BR	KT(key token) 500km (falling) RETB (radio electronic block) 830km (static)	–	Absolute block Scottish Region tokenless block BRB tokenless block 5260km (falling)	–	TCB (track circuit block) 9950km (rising)	–	–	L	16 540km
FS	–	3150km	5094km	–	Axle counter block: 3306km 'Fixed current' block: 704km	–	3812km	L	16 066km
CFL	–	21km	–	Semi-automatic 45km	175km	25km	–	R/L	266km
NSB	–	Manual dispatching 1255km	–	–	single line: 2588km double line: 107km	–	–	R	3950km
NS	–	*Centraal telecom blokstelsel* 75km	–		single line: 931km double line: 1774km	–	–	R	2780km
BV	–	5360km decreasing <5% of all traffic	–	–	single line 4470km double line 1200km	–	–	L	11 030km

Fig. 6.1 Block systems application chart

left-hand track, see **Fig. 6.1**, the associated signals are placed on the right or left side accordingly.

The signals for the opposite direction are placed on the opposite trackside, possibly above the track to avoid positioning them between the tracks.

6.4 Description of block systems

6.4.1 Token working block

The 'one train working' method of operation is usually applied to branch lines that only require one train to be on the complete line at any one time. Hence no passing places are provided. Historically a wooden train staff, inscribed with the limits of the branch, was used as authority to proceed when handed to the driver. This block system is in use on BR and is called *key token* (KT).

6.4.1.1 Electric token block

Token working relies upon the driver being given a visible token of authority to travel over the single line. A number of particular systems exist, the most common being the *electric token*. Instruments are provided at each end of the single line section which are electrically connected. After withdrawal of a token from one machine, a second token cannot be obtained from either instrument until the one already drawn is placed in the other instrument at the far end of the section or returned to the instrument from which it was taken. Both following and opposing moves are thus prevented while a token is in the possession of one train.

Each token is inscribed with the geographical limits of the movement it authorizes.

In-section sidings are catered for with the provision of intermediate instruments.

This system requires signalmen to be in attendance at each instrument. In order to reduce the number of signalmen, train crew operated instruments are often provided. With this arrangement train crew operate the instruments under the supervision of a remote signalmen. The system is known as *no signalman remote key token working*.

Token working is usually performed over dedicated lineside circuits but these are often too expensive to provide or maintain in the rural areas to which this system lends itself. As an alternative the public telephone network can be used in conjunction with fail-safe reed frequency transmission providing a level of security.

6.4.1.2 Radio electronic token block

Radio transmission is used as the control centre to train communications media for the signalling system known as *radio electronic token block* (RETB). The system provides for the exchange of electronic tokens, which give authority to enter single line sections of railway, between a central interlocking and the train. The driver receives his token as a message displayed on his cab unit, for example 'Block Entrance Station WELSHPOOL – Block Exit Station NEWTOWN'.

Fig. 6.2 Radio electronic token block (RETB), in use in Scotland

A number of sections of line and passing loops can be controlled from one central point. The only trackside equipment required are marker boards, trailable point mechanisms and associated indicators. Local staff are not required and routes may be operated at night and weekends without having to pay overtime to a large number of staff.

The system has helped to secure the future of a number of rural lines where costs exceed revenue.

The control centre comprises a solid state interlocking (SSI), VDU, keyboard and a radio base station. Line coverage is provided by a number of FM fixed radio stations and repeaters, each working at different frequencies. Transmissions from the control centre are sent to the first fixed station which re-broadcasts the message to trains within the area covered by that station. The repeater transmits the message onto the next base station where the process is again carried out of transmitting to trains in that area and repeating the message on.

The system can cope with six repeaters which allow for seven areas before the signal to noise ratio and group delay become unacceptable. To notify train crews of frequency change between fixed station areas, trackside channel change boards are erected. The system works as an open network with all transmissions heard by all those connected. The transmissions are simplex and 100% radio coverage is not a requirement. Complete coverage is only necessary at those locations where token exchange takes place.

On board the train is a mobile radio and RETB cab box. The cab box contains a microprocessor system which checks incoming data messages and their content and which drives the token text display. All trains running within the system are identified by their cab box identity. This is unique and held in fusible link PROM. All messages contain this identity and the SSI keeps track of each train by use of it.

Token exchange relies upon signalman and driver establishing voice contact over the radio and following predefined message exchanges. Following this the tokens are actually transmitted to or returned from the train by data telegrams.

Handshake exchanging takes place at 1200 bauds FFSK (fast frequency shift keying) to establish the link between control centre and mobile. The tokens are transferred in ASCII form in a number of packets. These are sent three times and must be received so that a complete token is constructed – no error correction is employed.

Portable units are available for locomotives not normally fitted and for engineering teams. Special engineering tokens can be issued to an engineer's portable unit, allowing work gangs to take possession of a section of line in between train movements, directly from the work site. This feature has greatly improved the on track work time of engineering gangs.

As time progressed, tokenless working block was invented. As its name suggests, tokenless block does not require the driver to be in possession of some tangible evidence as authority to enter the line. *Permission to proceed is given purely by lineside signals.* These are interlocked through block instruments at each end of the line, controlled by signalmen and the sequential operation of track circuits and treadles by the train as it enters and departs each section. A number of variations of the system exist for historical reasons.

6.4.2 Telephone block

Telephone block is used mainly on low traffic lines with small or no passenger traffic. Signalmen communicate by means of telephone between neighbouring stations following a standard procedure in prescribed repeated wording:

- signalman X rings up Y, to announce a train and to ask if the line is clear
- Y confirms that the section XY is free, and accepts the train from X
- signalman X clears the start signal. The starting signal is set at danger when the train has passed this signal
- X confirms to Y that the train has entered the section XY
- the arrival of the train is announced by Y to X when the train has arrived inside the station complete and the home signal is at danger [This step, however, is not carried out on FS.]

These procedural steps are recorded for each train in log books in both stations.

In case of level crossings, the telephone line is interrupted at each level crossing. On CFL and FS the telephone line is not interrupted. The CFL and DB AG gate keeper can listen in with a parallel telephone. Gate keepers at the level crossings listen in on the conversations between X and Y to establish train arrival times. Telephone block is normally used for lightly used and low speed lines.

Telephone block procedure is also frequently used on lines when block equipment is inoperative. It relies entirely on the integrity of the people involved and on strict adherence to the rules.

On SNCF single track lines that are not equipped with a manual block system, the telephone controlled block is reinforced by an computer system which manages the inter-station block sections and the block operations. This is known as the computer aided block system (CAPI). These lines may or may not have signals. If signals exist, they are not connected to the computer system.

This system, though not forming a safety system itself, makes a contribution to improving the safety and operation of single track lines with light traffic.

ÖBB, SNCF and NS have developed a version with a centrally located dispatcher who regulates the operation of trains by communication with the drivers using telephone or radio. The drivers are responsible for requesting permission for departure and for announcing departure and line clear.

- ÖBB : *Zugleitbetrieb*
- SNCF : ECLAIR
- NS : *CentraalTelecomblokstelsel*

This system, though not forming a safety system itself, makes a contribution to improving safety and also to lowering costs on lines with light traffic, by the reduction of permanent staff at stations.

On DB AG's network, there is no telephone block as described above. Instead there are only about 300km lines with 'Zugleitbetrieb' – see **Fig. 6.3**.

Here there is only on 'Zugleiter' in a central station who communicates with the train driver. There are no further signalmen along the line. This system, based on communication, is only allowed for freight train lines. Because this kind of service is not supported by signal safety, it will be replaced eventually.

It is a different matter with the '*Signalisierter Zugleitbetrieb*' (SZB) (see **Fig. 6.4**). In this case there

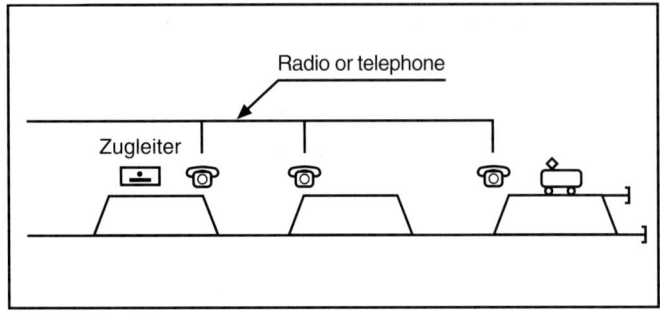

Fig. 6.3 Zugleitbetrieb (DB AG)

is only one 'Zugleiter' in a central station; no further signalmen are along the line.

This system, however, provides *full safety*.

At each station there is a signalbox with complete track free detection with loops and tail detector or axle counters, simple light signals and electric point machines or spring points.

This system has been recently developed and is known as SIG L90 Type or Stw vB depending of the trademark of their respective manufacturers.

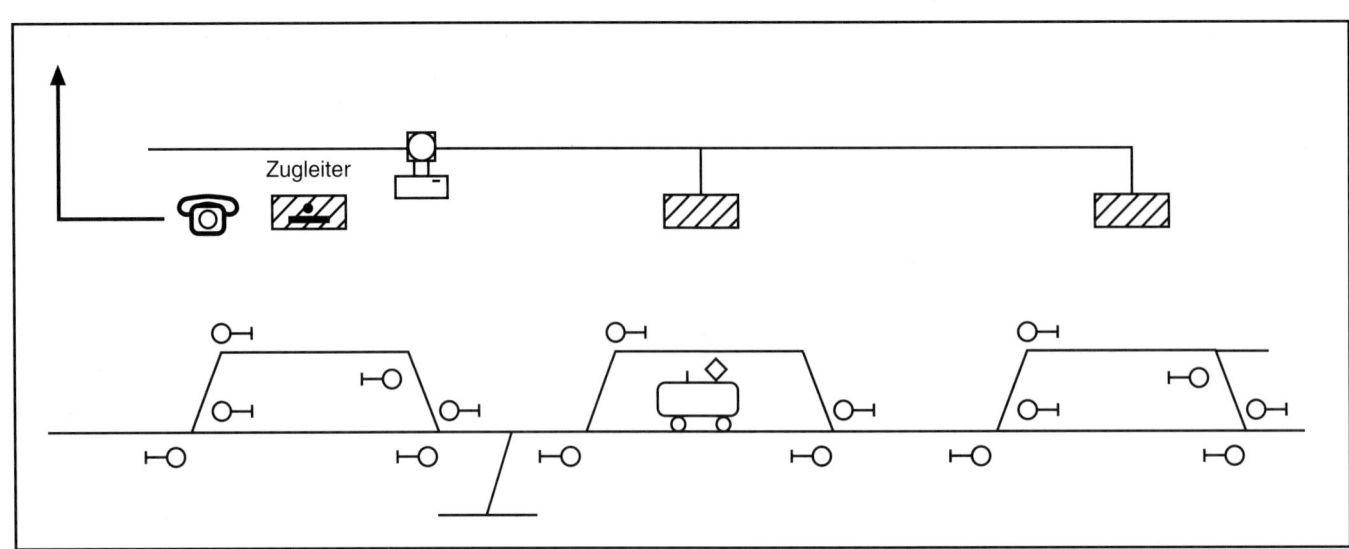

Fig. 6.4 Signalisierter Zugleitbetrieb (DB AG)

DB AG will equip about 50 lines with these systems in the near future.

- On SIG L90 during regular operation, the clearance of the signal at a distant station is managed by a personal computer (PC) supported by a colour monitor.
- For Stw vB a simple version without steering connection between Zugleiter and distant station – though with full safety – is effected; the train driver clearing the signal in a distant station by means of a remote control unit working with infra red techniques (similar to those of a TV).

All these systems have the following in common:

- they form a full safety system;
- there is no staff at stations along the line;
- they are equipped with radio communication for substitute operation in case of disturbance or for requesting permission for departure.

On NS an equivalent system is used on some low density single lines with small simplified stations equipped with relay interlockings. The typical layout for these stations is shown in **Fig. 6.5**. The route from the entrance signal into the arrival platform track is normally set and locked, this signal showing a yellow aspect preceded by a green aspect shown in the warning signal. The stations' exit route is locally controlled by the train driver. When pushing the plunger, provided that no opposite exit route is set from the next station on the line and both line and opposite entry route are not occupied, the entry route is cancelled and the exit route set which controls the point in the reverse position.

After locking the exit route the signal will clear, showing a green aspect.

The departing train will release this route, unlocking the points, as soon as the train is fully in rear of the station's entrance signal. Then the points are automatically controlled to normal and the entry route is set and locked, enabling the entrance signal to clear again. The train can now either reverse into the other platform track or move on to the next station. In both cases a new exit route to the line can be set from either side of the line as soon as the train has cleared the line. If the exit route is set from the side where the first train is leaving the line this train must be clear from the points as well.

In the case that through train operation is required a simplified remote control can be applied in order to control the exit signal centrally.

6.4.3 Manual block
(Manual block with electromechanical interlocking)
With the invention of the dynamo electrical principle it was only a matter of time until it was applied to provide electromechanically enforced interlocking of signals while using the same operation sequence as the telephone block.

At each end of the line a block apparatus is installed. This apparatus is electrically linked.

The operation is such that blocking of signals is effected locally by hand, and unblocking is effected electrically from the remote end.

Adherance to the operation sequence is electromechanically enforced, including signal operation and train presence in advance of home signals. No train detection equipment other than rail treadle and overlaid insulated rail is required.

To each block line with unidirectional traffic the direction of travel is allocated permanently. On bidirectional lines the block maintains its set direction until it is changed manually.

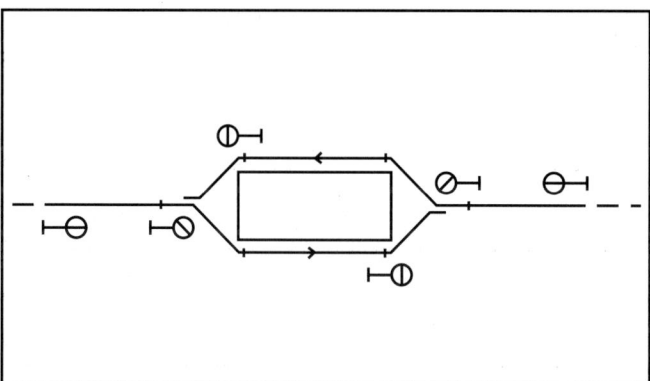

Fig. 6.5 Tokenless working block (NS)

Operation sequence

Normal status: the previous train has traversed the section XY and step (d) below was executed.

The block apparatus in Y had been operated earlier to relinquish '*permission*' to X to dispatch one or more trains, until requested again.

Fig. 6.6 Manual block with electromechanical interlocking, in use on NMBS/SNCB

This procedure blocks all of Y's starting signals and transmits to X for unblocking of the starting signals.

The direction of the block direction of travel is indicated in the block apparatus.

The procedures are as follows:

(a) signalman X rings up Y to announce a train and to ask if the line is clear.
(b) Y confirms that the section XY is free and accepts the train from X.
(c) in X an exit route is set and the respective starting signal cleared.

When the train has departed and the starting signal proved at danger again, the block apparatus is operated by X to block all X's starting signals. Simultaneously, '*train on line*' is transmitted to Y. This is the precondition for the next step (d).

Note: On ÖBB the block apparatus is operated when the signal is cleared.

(d) after the complete train has been *observed* by the signalman the block apparatus at Y is operated to transmit '*line clear*' to X. This operation unblocks X's signals again. This operation is only possible once 'train on line' was received, the train had passed the clear home signal (detected by both on insulated rail and a treadle), and the home signal was subsequently normalized.

On the FS network the system is slightly different:

- when a train has traversed the section XY and the operation cycles of both apparatuses have been completed, no direction is defined in the system any longer.
- line clear is not transmitted to X after the complete train has been observed at Y.
  Line clear will be implicit in the permission which will be relinquished for a subsequent train.
- permission is relinquished to X to dispatch only one train.
- transmission is by means of direct current pulses, which are also used to transmit messages from one signalman to another, for instance: type of the train to be dispatched.

Transmission is by means of:

- low frequency AC:
  – on ÖBB, SBB and DB AG: *Felderblock*
  – on NMBS/SNCB: *Blokstelsel met gekoppelde toestellen Block enclenché*
  – on RENFE: BEM
  – on BR: Scottish Region tokenless block and BRB tokenless block
- frequency division multiplex superimposed onto the telephone line:
  – on DB AG: TF–Block 71
- pulses of direct current:
  – on RENFE: BEM
- encoded current pulses or frequencies:
  – on SNCF: BMU on double track lines; BMVU on single track lines

6.4.4 Relay block

Relay block is used in relay interlockings linked to interlockings equipped with electromechanical block apparatus. The operation of the block is simulated with relay circuits to provide an acceptable interface.

Transmission is by means of:

- low frequency AC:
  - on DB AG: *Relaisblock*
  - on CFL: semi-automatic block
- encoded pulses:
  - on ÖBB: ZG62
  - on SBB: DC coded block

By installing train detection equipment, the block system may be made to operate semi-automatically (automatization of step (d) above).

6.5 Automatic block systems

The automatic block system is widely used by the European networks.

6.5.1 Non-centralized automatic block

Non-centralized automatic block is used for single and double track lines.

The block is locked by a set route into the block section or by a track occupation. The starting signal is unblocked automatically after the train has released the route, and has left the block section and the next block signal has changed to stop (the home signal of the next station if no block signals).

The normal status of a block signal for normal direction is free. It changes its aspect to danger only when a train has entered the block section behind the signal.

On BV first block signal (at station limit) is normally closed for safety reasons. Proceed aspect is displayed when the next block section is cleared, direction locked and departure route set with the signal.

The aspects of the intermediate signals are transmitted back to the interlocking cabin on ÖBB, DB AG, RENFE, or are transmitted from signal to signal on NMBS/SNCB, FS and NS.

The block signal displays a proceed aspect again as soon as the block section is free and the next signal covers the previous train.

The whole line must be equipped with track circuits, or axle counters.

When a departure route is set up, the block system is checked as regards correct direction and block in normal condition (no other route set up and no disturbances), including no occupation of the block section. The block is now locked: any other departure route as well as a changing of the direction is inhibited.

When the exit route is set, the starting signal will then display a proceed aspect. The next exit route may be set for a subsequent train when the previous train has passed the first block signal.

Conditions for reversal of train traffic are:

- reversal must be initiated from that side, where there is outgoing traffic, the station having the direction of traffic.
- there cannot be a locking for the reversal.
- all track circuits between the two stations must be free, or no train between the two stations in the case of axle counters.
- the block must be in normal condition, including all integrated components, no exit route lined up, no disturbances.

All European railway administrations here invented different names for their non-centralized automatic block system:

- on ÖBB and DB AG, it is called '*Selbsblock*', and is based on axle counters or track circuits.
- the NMBS/SNCB call it merely 'Automatic block'.
- on SBB, they call it 'DC coded block'.
- on SNCF, the automatic block is divided into:

(i) *luminous automatic block* (French acronym BAL), for short block sections on lines using track circuits;
(ii) *automatic block with restricted permissivity* (French acronym BAPR) with long block sections, using either track circuits or axle counters. The term 'restricted permissivity' comes from the constraints imposed on the driver when approaching an occupied section;
(iii) *automatic block in use on the high speed lines* (French acronym TVM).

- on BR, it is called TCB (track circuit block).
- on FS, they use two systems:

(i) *axle counter block*, based on the comparison of axle counting at the entrance and the exit of a block section;
(ii) *fixed current automatic block*, based on track circuits.

- on BV all double track lines have bidirectional automatic block and parallel running at normal speed is allowed.

Fig. 6.7 Intermediate block signal, combined with a distant signal on the right (DB AG)

6.5.2 Centralized automatic block

Centralized block equipment is installed in adjacent stations interlocking cabins for easier maintenance and better protection. It controls intermediate block signals.

A major advantage of centralized block equipment is being able to have the train detection information of the line available in the signalbox. This may be used, for example, for setting a substitute signal. If the distances allow it all the block signals of both directions are controlled from one station.

Train detection is mostly carried out by axle counters.

Generally the same centralized automatic block equipment can be used both on single and double track lines.

As the block signals are controlled from the equipment room of the signalbox the block functions between starting and block signals in the one direction and between the block and home signal of the opposite direction are provided by a centralized block system. There is no transmission over line cable required between the signal controls.

On ÖBB and DB AG this centralized automatic block is called '*Zentralblock*'.

6.5.3 Coded current automatic block

The system installed on the FS network is based on coded track circuits covering the whole line. Every code is associated with a given situation of the line ahead.

At every boundary between block sections a coded current equipment associated with relay logic is provided to receive and identify the incoming coded current, to control the block signal accordingly and to generate and transmit the coded current to the adjacent block system.

The basic information about the situation of the line ahead, such as the number of free sections and allowable speed value, is conveyed on the rails by the coded currents, which could also be received by trains through receiving coils and car borne decoding equipment, thus providing a cab signal function on board.

Lineside wires are used when additional functions are to be supplied, such as additional speed value information or change of traffic direction on the line.

Coded current equipment is relay based for the oldest installations and solid state for the most recent ones. It is allocated in lineside cabins near the line signals or in the interlocking cabins in the stations.

The operation is fully automatic and allows for the continuous updating of signal aspects according to the situation of the line ahead.

Registration is provided for the basic steps of operation of the system.

On double track lines, when required, the double sense of traffic on each track can be allowed, introducing additional lineside signals and switching the coded current equipment. Particular relay logic provides the necessary safety conditions to allow the change of traffic direction.

6.6 Procedures in case of disturbances

In case of failure to interlocked manual blocks, emergency unblocking is provided by means of sealed auxiliary equipment – a counted and documented act.

Generally, for automatic block systems, the telephone block procedures will be required in case of failure.

If a signal or block equipment is inoperative, a paper order is necessary.

An alternative to a paper order is the use of a subsidiary signal controlling restricted speed within the station (ÖBB, DB AG, BR). The subsidiary signal may be set, on condition that the train is approaching the signal. It may be necessary for the signalman to have the line clear of the block section confirmed by his colleague in the next station.

Clearing of starting or block signals and bypassing the required block conditions is provided under certain failure conditions and after completion of a determined procedure according to operational rules. This applies only for electronic interlockings on ÖBB.

Typical reasons for disturbances are:

- track circuit failure;
- axle counter failure;
- block cannot be released because for example the red lamp on the home signal has failed;
- direction of traffic cannot be changed.

This is not a complete list of all possible failures that are considered but it provides a variety of conditions where the operation will still have to be maintained.

On the block sections with axle counters it is possible that after a discrepancy of counting axles in and out of the section the axle counter can be manually reset. This action is preceded by a formal procedure.

Typical applications

- *SNCF* introduced the term 'restricted permissivity'. The term comes from the constraints imposed on the driver when approaching an occupied section. If he does not have verbal permission from the controller or the station regulator, a wait of 15 minutes is compulsory. This avoids the consequences of having to proceed at sight alone over a long distance, when it is more convenient to wait until the traffic has cleared the preceding section.
- *ÖBB, SBB*: The general aim is to clear the signals as long as the failures are proved to originate from technical reasons and are feasible at reasonable cost.

  In the interlocking a special command was introduced to override a disturbance and to be able to clear a departure signal to a disturbed block or to clear an integrated block signal. Before the operator gives this command he has to fulfil the relevant regulations – he has to check that no conflict with other trains can happen. The command is documented.

  The purpose of this method is to go on the line with the maximum speed. If the block is disrupted it makes no sense to go slowly where the route is safe: it only causes a unforeseen delay.
- *SBB* developed the line prohibition mode of the block system.

  The line prohibition is a technical means of preventing trains from entering a block section; thus route setting from either station is not possible anymore without special procedures. Even the direction of traffic cannot be changed under the prohibited mode.

The prohibition can be switched on for a particular line by the signalman having the right of way. It can only be switched off by mutual operation of the signalmen of both stations.

Line prohibition is required for:
(a) securing the line according to the operation rules;
(b) switching the adjacent line to single line working;
(c) clearing the starting signal of either station using bypass facility of block as protection against unintentionally setting of routes in the opposite direction by either manual or automatic operation.

The line prohibition is indicated at either station on the mimic diagram by a white bar.

Block signals are normally in automatic operation and cleared according to the passing by of trains. Each block signal is usually provided with a prohibition lock to prevent it from clearing if automatically required but not allowed.

The block signals at level crossings take the function of starting and home signals respectively according to the direction of traffic. They are provided with the automatic facility as station signals.

Change of direction of traffic is possible if the other line is in prohibited position. This mode of operation requires a reduced speed of 80km/h for the train running on the opposite track: the signal that is normally used for the left-hand track is switched for use for the right-hand track.

If the line and/or block signal prohibition cannot be switched on, the operators have to ensure by themselves that no route has been preselected and no signal in this line is in automatic mode. Only then and after the formal procedure has been completed may the driver be authorized to pass the signal at danger. For this he will be given a formal train order.

The dispatcher will set the route as far as possible, for example individual point operation, individual point locking, shunt route etc.

- *FS* have also developed a line prohibition mode, which is quite similar to that of SBB.

  It applies to double track lines equipped with automatic block, prevents setting of routes towards the prohibited track and releases level crossings. On the lines where simultaneous running of trains in the same direction on both tracks is not allowed, the change of direction of traffic is possible only if the other line is in prohibited position. This mode of operation requires speed not exceeding 180km/h. The line prohibition mode is indicated at both stations by a red light on the illuminated diagram.

- *ÖBB* use a track change equipment if a track disturbance continues for a long time. For example, if a track cannot be used because of repair or maintenance, they install a mobile track change equipment located in a container. This usually consists of four points and four signals, and is installed in the interrupted track and adapted to neighbouring equipment.

7 Radio Systems in Signalling

A. FISHER

Contents

7.1 General	236	
7.2 Definitions of radio systems	236	
7.3 Application of radio systems	236	
7.4 Current usage of radio systems	236	
7.4.1 Voice communications and non-safety data transmission	238	
7.4.2 Driver only operation	238	
7.4.3 Shunting operation	238	
7.4.4 Central control operation	238	

- 7.4.5 Remote control by radio — 238
- 7.4.6 Track to train and other communications — 238
- 7.4.7 Signalling — 239
- 7.4.8 Warning systems — 239

- 7.5 Network requirements — 239
- 7.6 The future — 239
- 7.7 Conclusion — 239

Illustrations

- Fig. 7.1 Tabular presentation of the use of radio systems — 237
- Fig. 7.2 Signal and telecommunications equipment on board a locomotive. Radio equipment on left-hand side — 239

7.1 General

The application of radio systems to signalling has developed by varying degrees within railway administrations over the past two decades. The limitations of application are dependent on the safety criteria placed on the system, the level of infrastructure investment and the overall need to adopt other methods of communications. Within this chapter the diversity of all these constraints is clearly described and a summary of actual use is also shown to identify the extent to which each listed railway administration makes use of radio systems within its operating architecture.

7.2 Definitions of radio systems

Radio is defined as the transmission and reception of messages by electromagnetic waves of radio frequency without connecting wires. The normal frequency ranges available are defined below, but system requirements, national licensing and international agreements preselect the most suitable bands for use.

Very high frequency (VHF70–88MHz)

This is suitable for wide area coverage applications but is restricted by its susceptibility to interference both from electric traction systems and due to weather conditions creating adverse propagation from other similar sources.

Very high frequency (VHF155–220MHz)

This is more suited for wide area coverage applications as the interference susceptibility level is significantly reduced for both electric traction and propagating conditions. The propagation capabilities are also reduced.

Ultra high frequency (UHF420–470MHz)

This is very well suited to local area communications and where a secure high integrity link is required because of its high immunity to electrical interference. A set of channels in this band have been allocated for international railway use, although in some countries the national restrictions prevent their use.

Some investigations into the use of 900MHz carriers have also taken place and some advantages can be identified but to date no widespread use of this range is in operation.

7.3 Application of radio systems

Radio systems have traditionally been used to enhance static telephone communications to permit 'mobile' links to be made. A cost effective signalling solution to long thin railways today would undoubtedly include a radio backbone system which would cater for all signalling and telecommunication needs. The acceptance of something new in signalling has traditionally been veiled in safety decisions and yet, where radio systems are concerned, the safety of the system is an integral part of the communications protocol.

The use of a radio system may affect the signalling systems availability which in itself may be seen as safety critical. Obviously, the need for high integrity radio systems goes hand in hand with a high capacity operating railway where the signalling is radio based.

This link between safety and availability and the relationship of 'fail-safe' communications in the radio environment has limited the development of such systems to be used as enhancements or contributory systems.

7.4 Current usage of radio systems

A brief summary of usage by railway administrations is listed together with a diagrammatic representation of this information.

ÖBB

Radio systems are used for shunting and train radio for *central control operation* within ÖBB. Communication takes place between the Control Centre and trains, and from the Centre it is possible to call trains selectively or collectively.

NMBS/SNCB

NMBS/SNCB use radio systems mainly for communication purposes. A limited use is made of radio systems to exchange short data messages.

SBB

With SBB radio is not used in any safety system, although it could be used for tokenless block, which is currently based on coded voice frequencies, which would allow the use of radio.

When a new ATP System was evaluated, the use of radio transmission was also studied. However, it was found that the topography of Switzerland with its mountains and valleys made it very difficult to achieve sufficient radio coverage in an economic way.

	ÖBB	NMBS/ SNCB	SBB	DB AG	RENFE	SNCF	BR	FS	CFL	NSB	NS	BV
Signals	X			X		X	X		X			
Level crossing			X#			X						X
Train routes			X#									X
Token						X	X					
Shunting	X		X	X		X		X		X	X	X
Voice communications	X	X		X	X	X	X	X	X	X	X	X
Track warning systems				X		X						
ATP										X		X

\# Application on private railways only

Fig. 7.1 Tabular presentation of the use of radio systems

DB AG
With DB AG, addressable radio systems with high transmission security are used for signalling to monitor signal aspects and voice communication (also for the purpose of shunting).

General radio communication systems are used for shunting purposes with special operator action for potentially dangerous movements. Radio is also used for the remote control of unmanned shunting engines.

RENFE
Radio is not currently in use for signalling applications.

SNCF
Currently SNCF does not use radio directly in its safety system, although it considers that radio systems make a contribution to the safety of train movement. A variety of projects for control and command systems, based on the 900MHz radio, are currently being studied.

At present SNCF identifies systems as follows:

- the track to train radio link;
- the establishment radio which encompasses all radio;
- systems other than the track to train link.

BR
With BR radio transmission is used as the control centre to train communications medium for the signalling system known as radio electronic token block (RETB). The system provides for the exchange of electronic tokens, which give authority to enter single line sections of railway, between a central interlocking and trains. Voice communications also use radio systems.

FS
Radio is not at present used for signalling applications. Radio systems are used to support operations in marshalling yards and stations and maintenance activities along the lines. Systems for track to train and onboard voice communications are being installed in areas not covered by the existing FS system.

Projects are under way to take advantage of public 900MHz mobile radio systems to allow voice data communications for service and passenger applications as well as the realization of a special wayside network for emergency communications.

CFL
CFL use radio systems for communication purposes only, in shunting (non-fail-safe) and for ground to train radio links (95% coverage of the network).

NSB

Radio systems are used to support operations in marshalling yards, stations and maintenance activities along the lines. Radio communication with position control related to signal numbers is presently being installed.

NS

NS use radio systems for communication purposes and the remote control of shunting movements within marshalling yards.

BV

With BV radio is used for low speed control of level crossing and shunting activities. Train radio communication allows links between train drivers, guards and traffic control centres. It is also used for providing ATP data which complements track based beacons to provide a radio block system.

7.4.1 Voice communications and non-safety data transmission

General voice communications systems over radio are used for various applications. A description of these systems and applications follows. These are not linked to country application, and may vary slightly to accommodate local administration changes and rules of operation.

7.4.2 Driver only operation

The running of passenger trains in driver only operation (DOO) has led to the requirement for secure communications directly between the train and signalbox. It is crucial that the signalman is in discrete contact with the correct driver and that the location of the train is correctly identified, as the radio system is used to exchange instructions regarding rules and regulations. Besides call security, high availability and continuous coverage are two other system requirements. As an aid to safety it is seen as important that signalmen and drivers are able to communicate at any point on the railway.

7.4.3 Shunting operation

Radio can be used to assist in the operation of shunting, but is used in a fail-safe way since there is obviously potential danger in these manoeuvres. In general if the link fails then the shunting procedure must be ended.

An example of this process is now described.

A special radio set is used which provides a continuous carrier the presence of which is monitored continuously. This signal, modulated by the shunting supervisor's radio is also monitored continuously. As long as this signal is received in the driver's cab the driver can continue through his manoeuvre. However, if the signal should stop, he is obliged to halt immediately. There is also local monitoring of the generation of the signal on the shunting supervisor's radio set, and speech has priority over the signal.

7.4.4 Central control operation

On secondary lines with low traffic and low speeds the interlocking system may have reduced functionality. A line consisting of several small stations can be controlled from a signalman in the centre. The signalman is connected to all train drivers by radio and gives them commands to start or stop.

7.4.5 Remote control by radio

Radio systems for remote control purposes, are also used as follows:

- remote control of shunting engines which allows a single operator to act as both driver and shunting supervisor. The transmission is fail-safe, such that if the carrier is interrupted the engine has its brakes applied.
- remote control of level crossings, used where train traffic is slow and light. This allows the treadle normally used to trigger barrier closure to be dispensed with, and involves the use of a portable transceiver.
- remote control of the pushing engine by the pulling engine in the case of super heavy trains, over a digital radio link.

7.4.6 Track to train and other communications

Track to train radio links are used to provide radio communications in speech mode between the line regulator, the station staff, the drivers, the train crews and workers along the length of the track.

Apart from its obvious effect on the regulation of traffic movement, the track to train radio forms a useful addition to signalling systems for the safety of traffic, even if it was never intended that it should do so in the strictest sense of signalling.

In the form of a compatible extension to the traditional system, a data transmission track to train radio link which allows the simultaneous transmission of speech and data messages is now available.

This type of link is ideal for the transmission of a radio alarm when the signalling is activated automatically with the detection by the signalling system of an entry into an already occupied block section, for example, thus constituting a very useful recovery message.

Additional radio connections within a train are possible, for example, a connection between two train drivers, necessary for co-ordination in the case of double traction, or connection to the guard of the train, or connection for an announcement from the control centre to the passengers.

Local links are used for the preparation of trains for departure, for brake tests, for splitting, unhooking and reformation of trains, for control of special train composition procedures, for the maintenance of installations, work parties on the track and train operated staff alarm warning systems.

7.4.7 Signalling

The use of radio for signalling purposes has been limited, but some applications exist. RETB is an example of such a system which relies on radio transmission. The system provides for the exchange of electronic tokens which give authority to trains to enter single line sections of railway.

As already mentioned the use of radio for remote control of track based equipment is of course possible, but this is usually protected with some form of direct locking when the train arrives at the site of the equipment.

7.4.8 Warning systems

Radio is used to transmit the absence and presence of trains to trackside working groups. Different tones are sent to indicate approaching trains on adjacent tracks and the track upon which work is being carried out.

7.5 Network requirements

The application of radio communications within an administration network varies according to functional needs, but an integrated network of communication is required, radio forming a constituent part of the network.

For local communications where the operating range is short and the operating site is not static, single frequency handportable and mobile radios are usually employed.

For sites where a wider coverage is required and where a static area is concerned such as a control centre, station or depot, fixed station radio with an aerial system is usually employed.

For wide area network requirements which encompass more operational systems such as train radio and other functional needs radio control areas covering 200–300km^2 are established. This requires anything up to 20 fixed station sites, positioned according to topology and operational need, to be established.

In general the radio system in today's operating railway complements the communication network to provide links between mobile units, road vehicles, locomotives, control centres and the railway and national telephone networks. At present, when used for signalling purposes, radio links tend to be isolated from the main communications network.

7.6 The future

As signalling systems is general become more complex and the need for more safety on the trackside increases, reducing trackside equipment and the placing of signalling into the cab is becoming more attractive. The most convenient method of achieving this is with a radio based system. Work is already being done to provide ATP type data within a train using radio links, and this is just the beginning with projects such as ASTREE on SNCF, DIBMOF on DB AG (discussed in Chapter 14).

7.7 Conclusion

When radio was first introduced to improve local communications its acceptance and growth over the last decade has provided facilities for many aspects of the signalling operation. The potential for further growth is very high and the European standards which apply will allow for integration between administrations on a large scale.

Fig. 7.2 Signal and telecommunications equipment on board a locomotive. Radio equipment on left-hand side

8 Internal and External Safety Conditions

J. PIZARRO

Contents

8.1	General	241
8.2	System design	242
	8.2.1 Safety and dependability	242
	8.2.2 System lifecycle	242
8.3	Safety	243
	8.3.1 General	243
	8.3.2 The safety organization	243
	8.3.3 Safety integrity (WG5B)	243
	8.3.4 The technical safety assurance	248
	8.3.5 Software case: general	248
8.4	Traditional relay circuitry	248
	8.4.1 Design of circuits	250
	8.4.2 Tests	251
8.5	Reliability	252
8.6	Duplication of elements	252
	8.6.1 Transmission lines	252
	8.6.2 Duplication of control	252
	8.6.3 Duplication of the power supply	252
	8.6.4 Duplication of hardware	253
8.7	Maintenance	253
8.8	Homologation	253
8.9	Quality control and reception of materials	254
8.10	Other measures	254
8.11	Standardization	254
8.12	Relays	255
	8.12.1 Introduction	255
	8.12.2 Classification of relays	255
	8.12.3 Basic conditions	255
	8.12.4 Characteristics of safety related to reliability	256
	8.12.5 Characteristics of materials used in the manufacture of relays for signalling	256
	8.12.6 Characteristics of contacts	257
	8.12.7 Main tests in relays	257
8.13	Signalling cables	258
	8.13.1 Materials	258
	8.13.2 Construction	258
	8.13.3 Electrical tests	258
	8.13.4 Mechanical characteristics	258
	8.13.5 Delivery	258
8.14	Cable laying systems	258
	8.14.1 General	258
	8.14.2 Suspended cable systems	259
	8.14.3 Underground cable systems	259
8.15	Staff licensing	262
8.16	Analysis of the different administrations	262
	8.16.1 Design of the interlocking circuits	262
	8.16.2 Cable laying	262
	8.16.3 Cables used	263
	8.16.4 Installation of local control panels and alternative lines (CTC systems)	263
	8.16.5 Installation of equipment and systems	264
	8.16.6 Specifications, norms and standardization	264

Illustrations

Fig. 8.1	Factors affecting dependability of transport systems (WG5B)	242
Fig. 8.2	System lifecycle for a guided transport system (WG5B)	244
Fig. 8.3	Safety lifecycle	245
Fig. 8.4	Risk analysis	246
Fig. 8.5	Safety organization	247
Fig. 8.6	Software development lifecycle (WGA1)	249
Fig. 8.7	Software safety route map (WGA1)	250
Fig. 8.8	AVE Madrid-Seville line	251
Fig. 8.9	N–type relay: fail-safe type	255
Fig. 8.10	C–type relay	255

8.1 General

One of the main objectives of railway administrations is the safe transportation of economic quantities of passengers and goods in the shortest possible time. At the same time major efforts are made in the prevention of possible accidents, whilst endeavouring to increase the carrying capacity of the lines and the commercial running speed of the trains. It is in this respect that railway signal engineering and systems play an important role in the attainment of the above objectives.

On one hand safety designed railway signalling systems make it possible for trains to move between stations and inside stations, preventing the human factor from committing mistakes that could turn into accidents. They also provide train drivers with the appropriate information regarding the permitted speed for each movement, in relation to the position of other trains in the vicinity. In this chapter the important measures that guarantee that good working signalling installations are safe in accordance with safety standards are described.

On the other hand, the grade of service adopted by the signalling systems will allow an increase in the carrying capacity of the lines when provided with interlockings, automatic blocks, centralized traffic control, two-way working tracks, driving assistance systems and automatic train control systems. All railway signal engineers must pay special attention to mandatory safety requirements in the design and development of signalling systems, in order to achieve the highest level of safety.

To achieve this goal it is necessary to describe the characteristics of the fail-safe working of a signalling system and then apply criteria to demonstrate that all appropriate care has been taken and the objectives fulfilled.

Under the authority of CENELEC, European norms have been propounded by working groups such as:

- *WG5B*: dependability for guided transport system (reliability, availability, maintainability, safety).
- *WGA2*: railway applications: safety related electronic railway control and protection systems.
- *WGA1*: railway applications: software for railway control and protection systems.

These projects, co-ordinated jointly between railway administrations and signalling equipment manufacturers, are complementary to the long tradition of safety considerations applied to the technical solutions of signalling problems.

As the subject is important it is worthwhile to give some definitions:

- *availability*: the ability of an item to be in a state to perform a required function under given conditions at a given instant of time or over a given time interval, assuming the required external resources are provided.
- *dependability*: the collective term giving a qualitative description of *availability* and *safety* and their influencing factors: *reliability*, *maintainability* and *maintenance support performance*.
- *maintainability*: the probability that a given active maintenance action, for an item under given conditions of use can be carried out within a stated time interval, when the maintenance is performed under stated conditions using stated procedures and resources.
- *reliability*: the probability that an item can perform a required function under given conditions for a given time interval.
- *fail-safe*: a signalling system installation is said to be intrinsically safe or fail-safe if, when a failure affecting safety takes place, the system will then pass to a situation known and accepted as safe, that means freedom from unacceptable levels of risk.

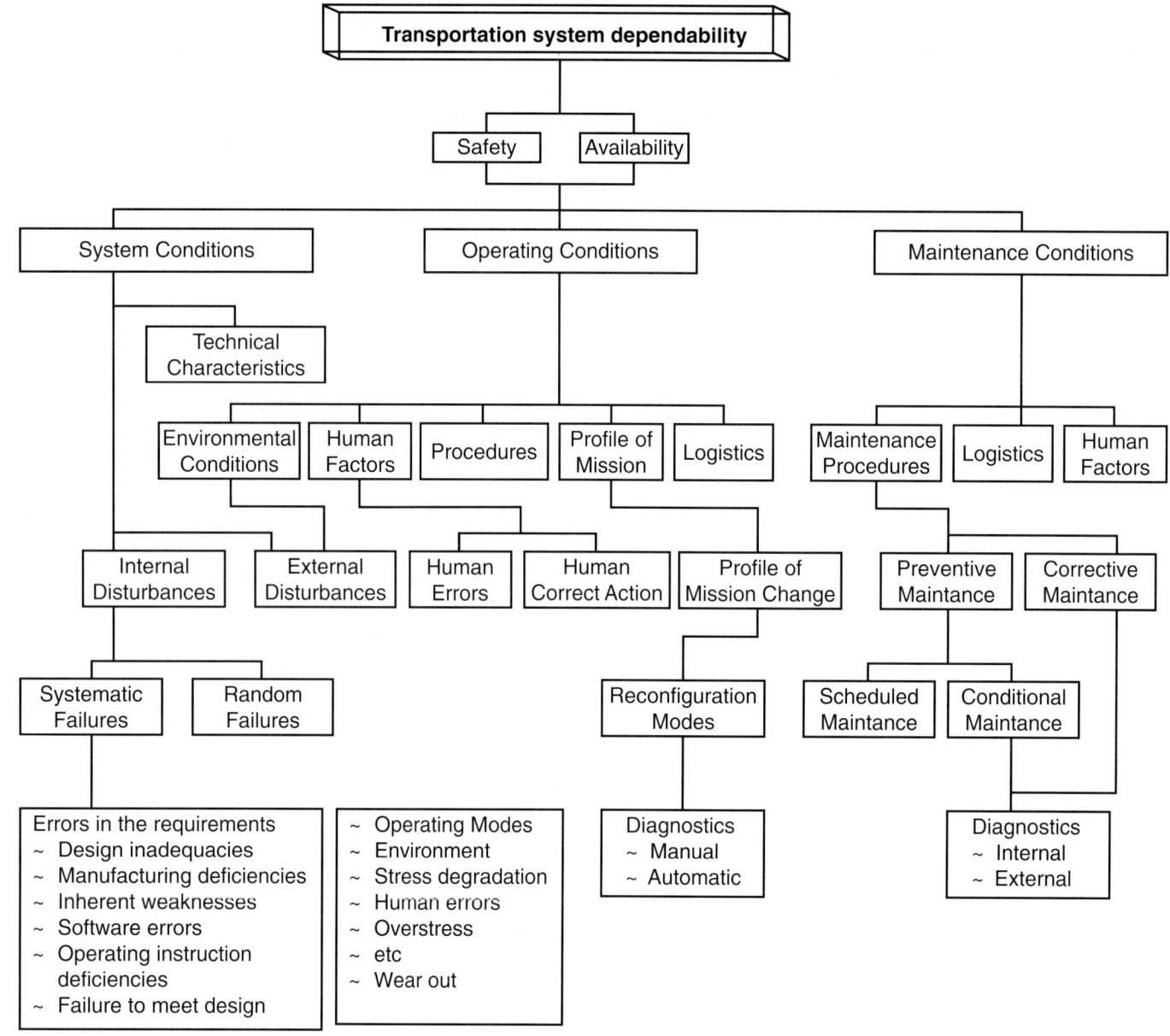

Fig. 8.1 Factors affecting dependability of transport systems (WG5B)

8.2 System design

8.2.1 Safety and dependability

Fig. 8.1 clearly identifies the close relationship between safety and dependability and this relationship is under conditions that are:

- those incorporated in the system itself;
- those coming from the proper use of the system;
- those related with maintenance operations.

Operations and maintenance of the signalling systems introduce human factors which, nowadays, are the more unsafe and risky conditions.

8.2.2 System lifecycle

The process for the specification and demonstration for the dependability elements is based on the system lifecycle and can be summarized as:

- a clear definition of requirements which are traceable to measurable targets.

- the assessment of failures and their impact on dependability and where necessary the reduction of this impact.
- the assessment of risks and their impact on safety and, where necessary, the mitigation of these risks.
- the assessment of human factors and due allowance for these factors.
- provision of evidence that the requirements have been met.

Lifecycle

The system lifecycle describes the various stages of the system from conception to its final removal from service, and it is fundamental to the understanding and implementation of this European norm.

A system lifecycle which is appropriate to a guided transport system is shown in **Fig. 8.2**.

This lifecycle is adopted as it emphasizes the need to consider the application requirements during the specification phase. It also shows that the activities of verification and validation are tasks that require planning and action during the specification phase and design.

The definition phase is where the scope of the project is identified and a team of suitably qualified personnel assembled in order to assess the functional needs before moving on to the next phase.

It is recognized in this standard that the definition phase is equally applicable to the guided transport system authority or to the guided transport system supply industry when embarking on a speculative development.

The specification phase will, after a process of analysis, application considerations and apportionment of RAM targets from some overall global target, result in a specification that gives quantifiable and achievable targets.

It is recognized that the specification may be raised either by the guided transport system authority or by a guided transport system supply industry undertaking a speculative development.

The phase of design and implementation may be undertaken either by a group within the guided transport system authority or in an organization external to the guided transport system.

For the purposes of this standard, the group undertaking this work are known as the design authority.

Under this phase the need to plan ahead for verification and validation activities is essential in order to ensure comprehensive checking at stages throughout this phase to demonstrate that requirements are being met.

In some instances, the process of validation may be undertaken by the guided transport system authority or their nominated representative in order fully to examine the design and implementation phase, its documentary evidence, analyses and test results in order to achieve an independent confirmation of its suitability.

Note that the design phase given is equally applicable to the design of new equipment, new systems or to the use of proven equipment in a particular application.

The achievement of safety targets includes the need to demonstrate functional correctness (systematic integrity).

The operation phase encompasses all the activities associated with operating a system in the environment of a guided transport system. A particular point to note is the need to monitor the achieved performance in order to take remedial action if the original expectations are not met and also to aid the compilation of a random access memory database.

8.3 Safety

8.3.1 General

The fail-safe level of a guided transport system is evaluated or, at least, approved by the official authorities (governments, etc) of the railway administrations concerned.

To achieve such evaluation or approval it is mandatory to:

- identify a given organization;
- make a risk analysis in order to clarify them as described **Figs. 8.3** and **8.4**.

8.3.2 The safety organization

The safety organization shown in **Fig. 8.5**, or a similar one, has to be settled as a part of the safety process (WG5B).

8.3.3 Safety integrity (WG5B)

Safety integrity is defined as 'The likelihood of a system complying with the specified safety requirements under all stated conditions within a stated period of time'.

244 INTERNAL AND EXTERNAL SAFETY CONDITIONS

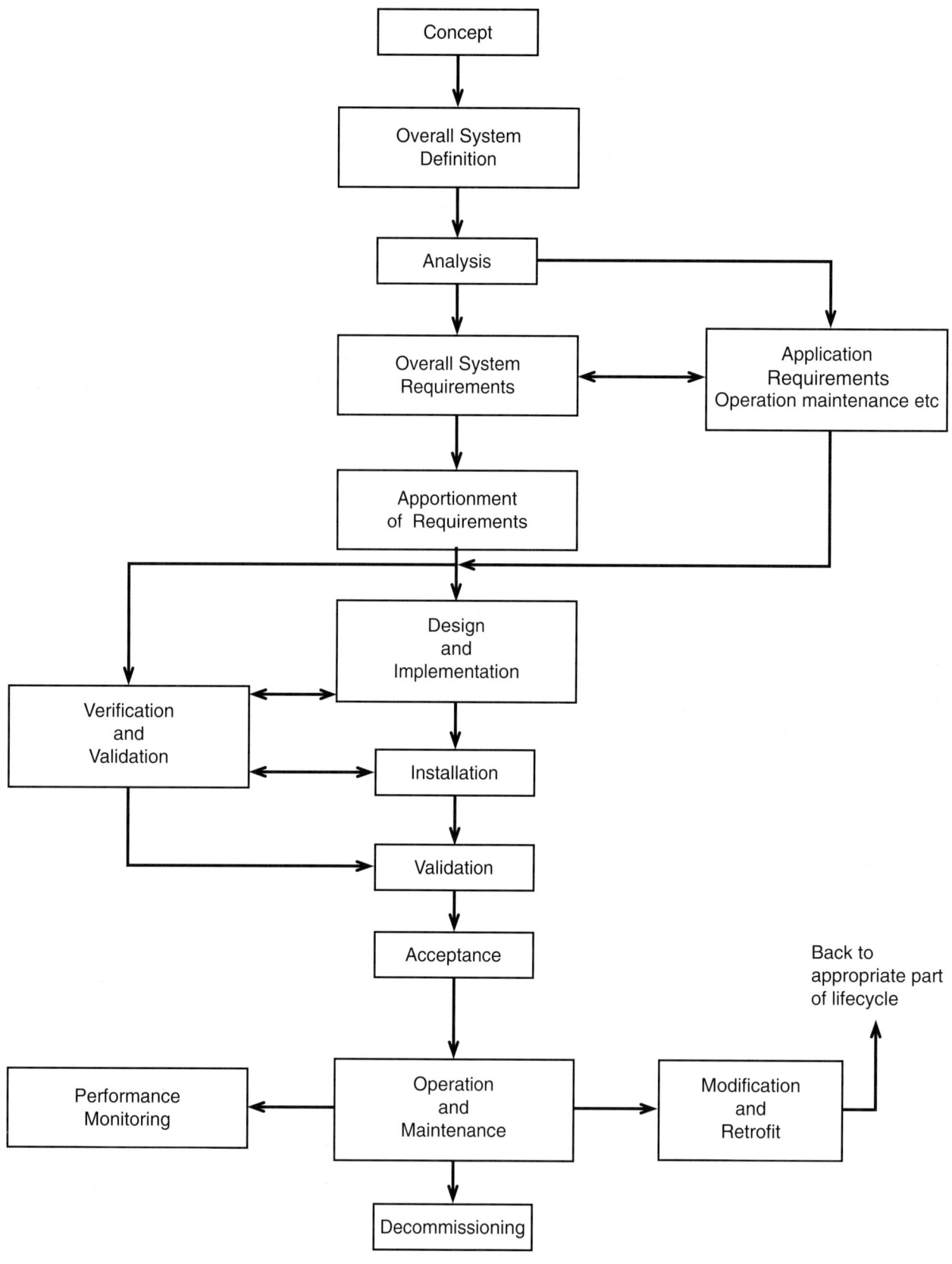

Fig. 8.2 System lifecycle for a guided transport system (WG5B)

INTERNAL AND EXTERNAL SAFETY CONDITIONS 245

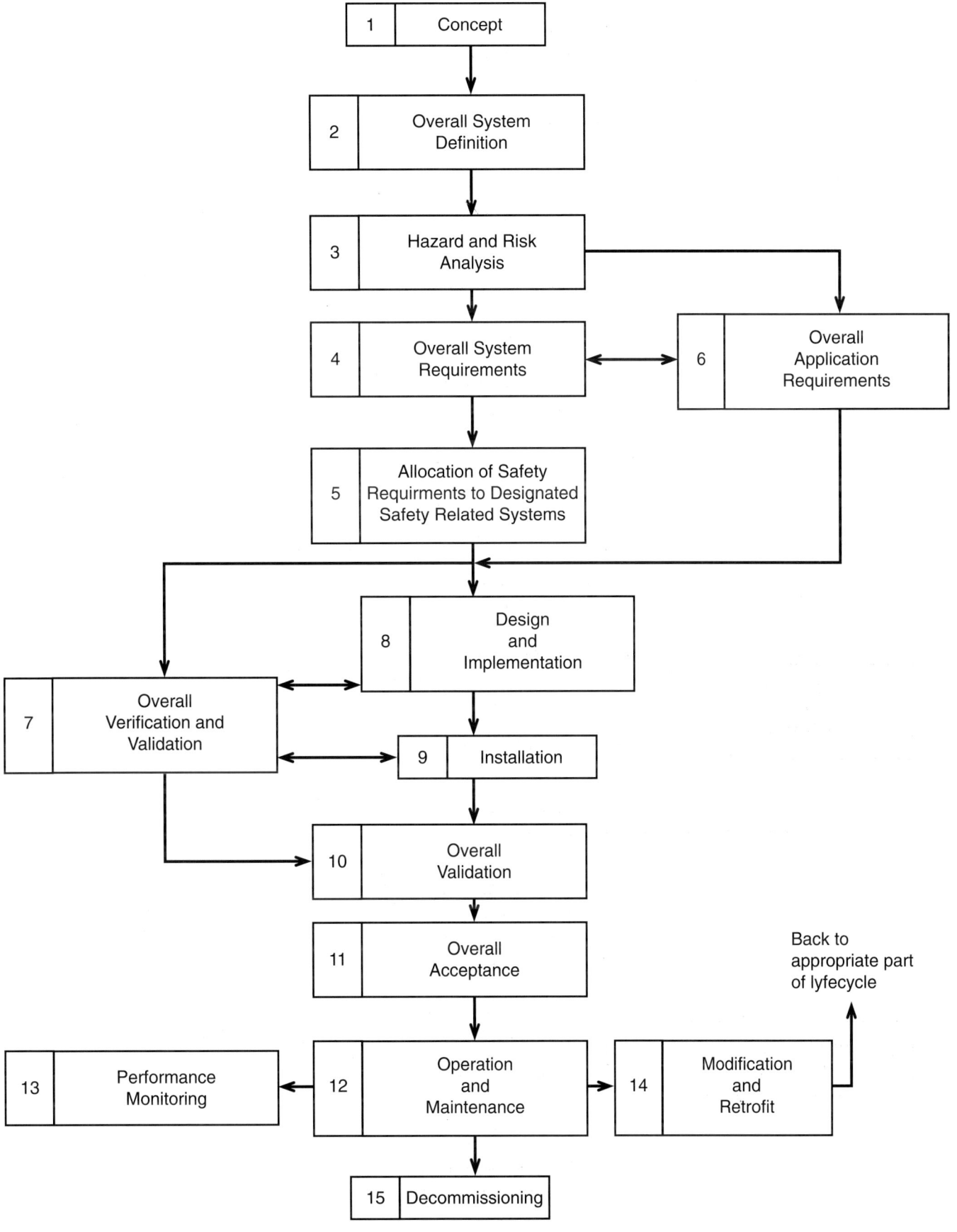

Fig. 8.3 Safety lifecycle

Category	Description	Definition	
		Consequence to personnel	Consequence to service
4	Catastrophic	Fatalities and/or multiple severe injuries	
3	Critical	Single fatality or severe injury	Loss of a major system
2	Marginal	Minor injury	Severe system(s) damage
1	Negligible	Possible single minor injury	System damage

Fig. 8.4 Risk analysis

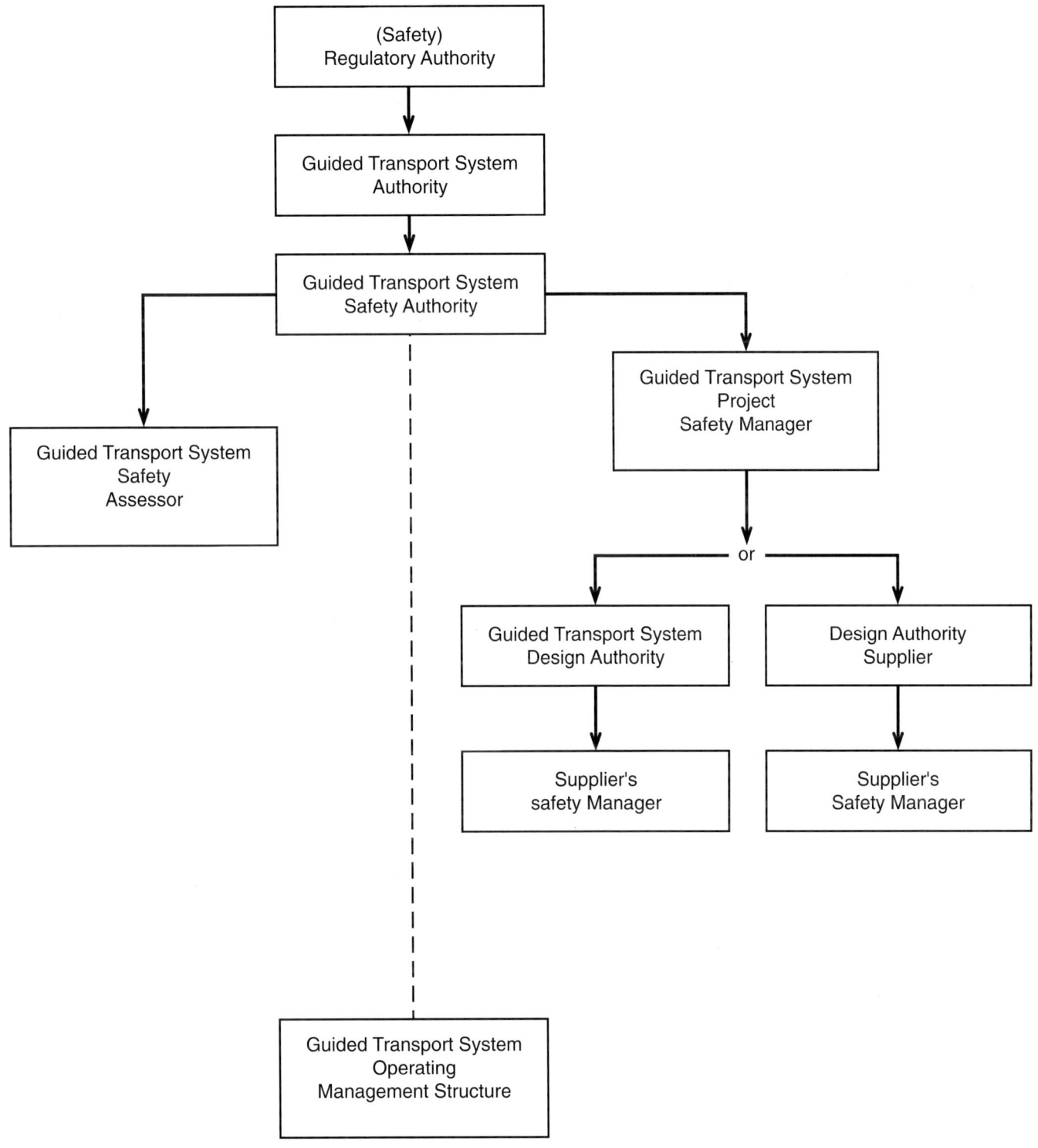

Fig. 8.5 Safety organization

There are four levels of safety integrity: safety integrity Level 4 is the highest level of safety while safety integrity Level 1 is the lowest – the levels are defined later in this section.

The safety integrity level is a measure of the tolerable level of risk, and as such the guided transport system safety authority must set or agree the target safety integrity level for a system.

Safety integrity levels provide a consistent means of expressing the tolerable level of risk throughout a guided transport system and between this one and the given support industry.

At the guided transport system and major system level, numerical targets are in general not appropriate, and for such systems safety integrity levels are identified as follows:

- safety integrity levels for major systems are identified from the risk classification and the criteria for tolerable levels of risk.
- target safety integrity levels are apportioned to the subsystems taking account of the safety integrity level identified for the major system and from second and third stage risk assessments.

8.3.4 The technical safety assurance (WGA2)

In addition to the evidence of quality and safety management, a third condition has to be satisfied before a system, subsystem or item of equipment can be accepted as having met its safety requirements. This consists of technical evidence for the safety of the design, which should be demonstrated in a document called the 'Technical Safety Assurance Report'. This forms part of the safety case or proof of safety for the system/subsystem/equipment.

The Technical Safety Assurance Report should include all the technical facts that are relevant to the safety of the design, including (or giving reference to) all supporting evidence; for example, design calculations, test results and safety analyses. The document should be arranged under the following chapter headings:

- assurance of correct operation;
- operation under fault conditions;
- operation with external influences;
- safety related application conditions;
- test specification and results;
- quantitative safety analysis.

8.3.5 Software case: general

Software has a reliability of its own – ie the probability of getting the intended result after submitting the input. For software safety it is necessary to look at the whole system – the software must be seen in the context of the particular application. It is meaningless to speak about a safe program for sorting, if there is no hazard possible due to an erroneous sorting algorithm.

The current state of the art means that neither the application of quality assurance methods (the so-called preventive measures) nor the application of software fault tolerant approaches can guarantee the absolute safety of the system. There is no known way to prove the absence of faults in reasonably complex safety related software, especially the absence of specification and design faults.

The principles applied in developing high integrity software include, but are not restricted to:

- top down design methods;
- modularity;
- verification of each phase of the development lifecycle;
- verified modules and modules' libraries;
- clear documentation;
- auditable documents; and
- validation testing.

The different software integrity levels in this standard require different levels of assurance that these and related principles have been correctly applied; in order to clarify them as described in **Figs. 8.6** and **8.7**.

8.4 Traditional relay circuitry

Before developing the fail-safe relay circuitry the engineer responsible for its design has to know which kind of relays are available, safety type or not, since each one will necessitate designs of different characteristics.

By their very nature, safety relays satisfy the conditions of safety without any special verification about their correct working.

However, fail-safe conditions must be assured through supervision and redundancy circuits – ie safety supervised security.

Both designs have been used for many years without any apparent ascendancy of one over the other. Railways operating intrinsic relay philosophy are BR, SNCF, FS, NS (with supervised relay philosophy), ÖBB, DB AG, CFL, BV, NSB; the following use both on their networks: CFL, NMBS/SNCB, RENFE.

INTERNAL AND EXTERNAL SAFETY CONDITIONS 249

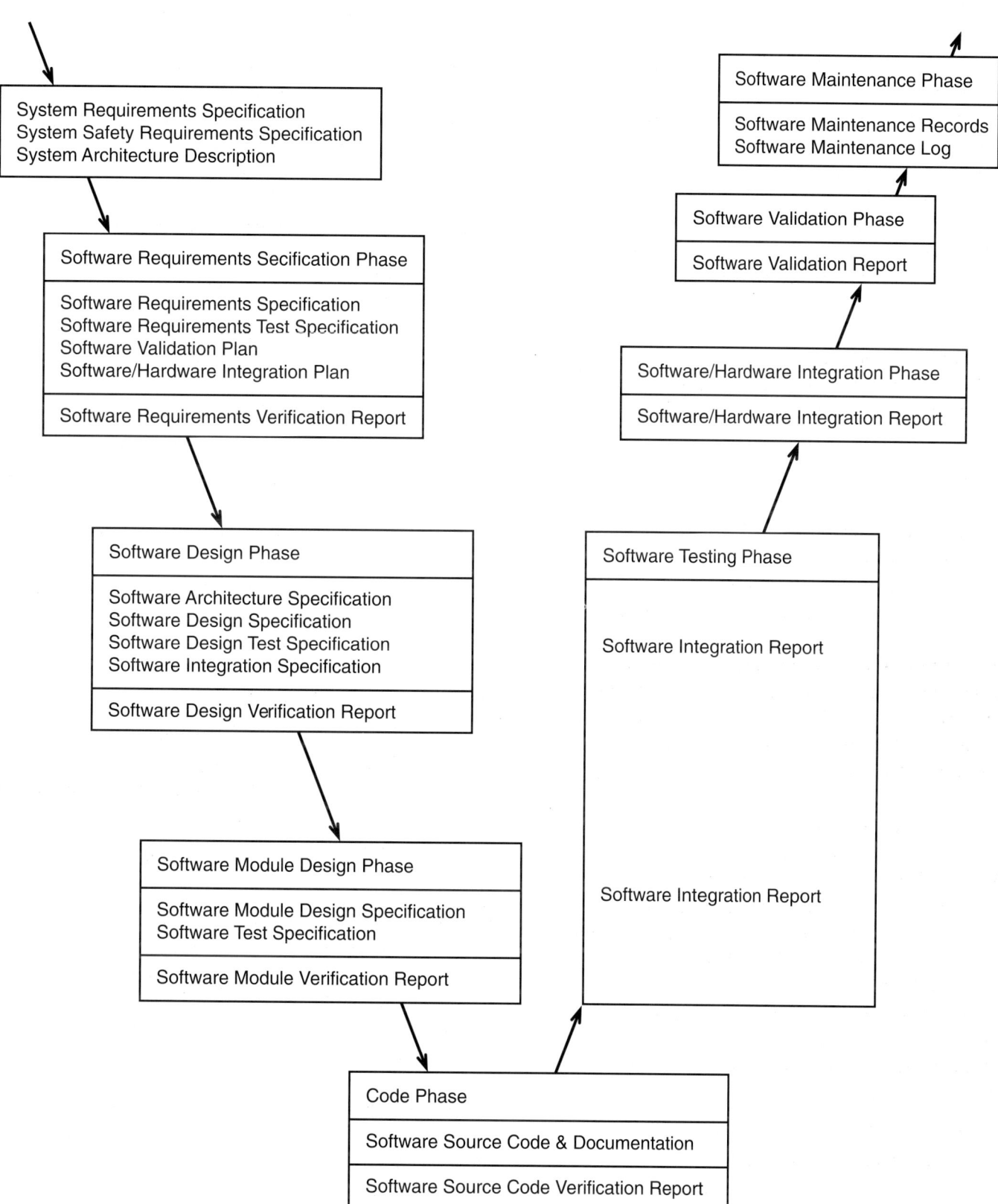

Fig. 8.6 Software development lifecycle (WGA1)

The following are examples of normal conditions for the application of relays for interlocking circuits on RENFE:

- all safety contacts to be positioned on one side of the relay coil with the other side of the relay coil connected to zero potential.
- auxiliary contacts to be positioned at the beginning of the circuit, or on the side connected to the ground coil.
- in circuits relating the relay coil with signals, driver, etc, two wire circuits must be used, with both preferably cut off by relay contacts.
- each signal light unit must be provided with an isolation transformer.
- special care must be taken with the use of the repeater or band relay contacts, picking out properly the place where all said relay contacts must or must not be checked out.
- the design of electronic equipment for application in railway signalling systems, such as:
  - (i) flashers
  - (ii) time delays;
  - (iii) converters;
  - (iv) track circuits.

Consequently, special steps should be taken, such as the duplication of the hardware or the software and for comparison to be made of the results, accepting them as correct only when they are alike.

More detailed criteria and of other steps to be taken in connection with the signalboxes are discussed further in the chapter on interlocking (Chapter 5).

8.4.1 Design of circuits

According to the topography of the stations and tracks, the commercial speed of the trains and their braking characteristics, and their future exploitation needs, the railway administration should first determine:

- both normal working and shunting routes in the stations;
- compatibility or incompatibility of each route with respect to other routes;
- protection of flanges;
- signal indications and spacing.

Once the engineer has the station and route exploitation programme together with an approved signalling plan, he can then develop the

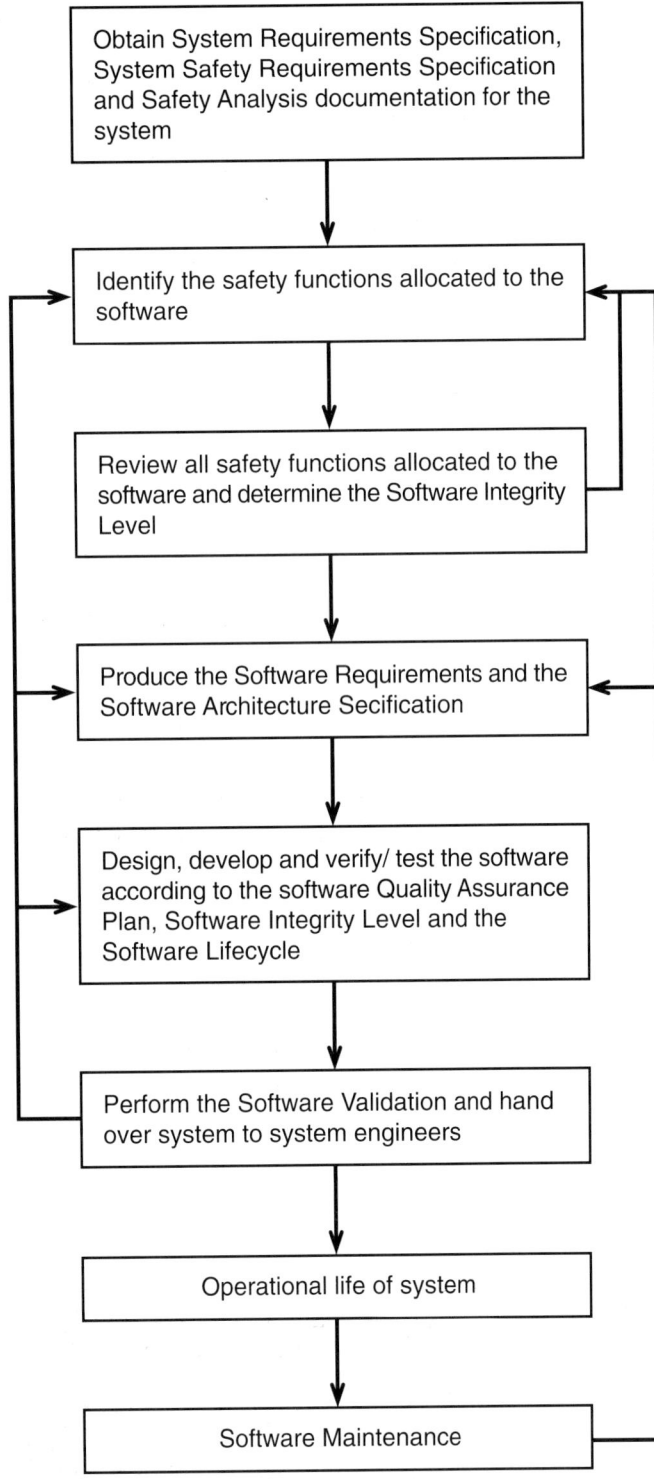

Fig. 8.7 Software safety route map (WGA1)

corresponding circuitry or software, checking that in each executed route:

- the position of switches is correctly placed and interlocked;
- the train detection system, if any, protecting the traffic of trains is free;
- incompatible routes cannot be made;
- signalling indications in each route are the correct ones;
- the manually controlled level crossing is closed and interlocked, if it is necessary for this route. With an automatic control operated by approach lock, the signal is cleared afterwards.

In order to complete the signalling circuit design with as few mistakes as possible, it is necessary:

- to have a general exploitation normalized for the whole of the railway network;
- for the exploitation programmes to be developed with the aid of computers, especially in the cases of the larger installations;
- for circuits to be standard as far as possible;
- for computer aided design (CAD) systems to have been employed.

8.4.2 Tests

Once the design of the signalling system circuits has been completed and manufactured, the installation of all works can be undertaken. This also marks the beginning of the system testing phase, during which a thorough controlled testing programme has to be carried out to ensure that everything functions properly. In this respect tests of the signalbox must be carried out jointly between the manufacturers and the railway administration to confirm that each and every condition of safety demanded for the installation has been met in accordance with the signal system specification.

Some of the most important security measures include the following:

8.4.2.1 Track circuits
- Adjustment of the working values of track circuits in accordance with their specifications.
- Exact position of the insulating or electric joints.
- Checking of phase opposition in those cases where it is required.
- Checking the train shunt is within the permissible limits.
- Concordance of the track circuits on the ground with their representation on the control panel.

8.4.2.2 Signals
- Focusing and positioning of signal heads to obtain the appropriate visibility.
- Adjustment of the lamp voltages and checking that the relays operate within the current specified.
- Concordance of the signalling indications on the ground with their representation on the control panel.

8.4.2.3 Points and crossings
- Checking that in those switches fitted with drives or bolts, there is a loss of proof of the switch tongue coupled with its own opening, to the dimensions specified.
- Checking in those switches to force open the point.
- Checking of movement with the driving handle.
- Checking that the local control of switches is fitted with electrical drives.
- Checking the discordance and that the switch does not reach its final position within a permissible time.
- Concordance of positions and indications of switches on the ground with their representation on the control panel.

8.4.2.4 Other measures
- Checking the correct insulation of cables.
- Checking that the installation of signals, cabinets, boxes, has been done according to plans in such a way they are not foul of the train gauge.

Fig. 8.8 AVE Madrid–Seville line

8.5 Reliability

Below are some of the most important measures to be taken in order to obtain a higher rate of reliability in the signalling system:

- duplication of elements
- tests
- maintenance
- homologation
- quality control and reception of materials
- other measures

8.6 Duplication of elements

One of the main measures to be taken into consideration to ensure the continuous functioning of installations, is to duplicate some parts of the system.

These measures make installations much more expensive. But if cost analyses are made, and compared to the variety of problems created in railway traffic operation when certain kinds of failures occur it will be seen that it is necessary, in certain cases, to duplicate some elements. The following paragraphs detail some of these measures.

8.6.1 Transmission lines

On installations with centralized traffic control (CTC) it is convenient to have an alternative transmission line. The alternative line can be given in different ways:

- by the use of a national telephone line circuit.
- by the use of another line circuit in a secondary cable.

In both cases, the changeover of lines must be able to be made from the CTC district control office and/or locally, in stations.

It is also useful if both lines – the principal and the alternative – are closed in a loop.

8.6.2 Duplication of control

It is very important, and relatively cheap, to provide the different stations on a CTC line with a local control panel.

This solution reduces any confusion and allows the operating staff to take local control of the affected stations. In order to limit maintenance work on the local panels, operating staff will only be provided with the indispensable facilities and drives, which will be out of service when the station's local control is not taken through them.

The operation of the local control can be made:

- in a normal way, by an expressed concession from the CTC's operator.
- in case of emergency, without the authorization of the district control office, through the use of certain command devices which will be normally sealed.

8.6.3 Duplication of the power supply

It is particularly important to have one totally reliable power supply system.

Incidents due to failures in the power supply to station interlockings, as well as routes on automatic block, are those with the greatest effect on the regulation of traffic, causing installations to stay completely out of service and perhaps disrupting very long routes.

This is why special precautions must be taken to ensure against power supply outages.

At present, with railway networks almost completely electrified, the primary power source is obtained from the overhead traction power supply. Whilst this supply is considered reliable, there are occasions when breakdowns do occur for one reason or another. These could have serious repercussions on the signalling systems if they relied upon a single supply. However, most modern signalling systems are provided with a duplicated power supply, resulting in an uninterrupted 24-hour in service availability for most of the time.

The secondary power supply may be produced by an emergency electricity generator, equipped with additional equipment which ensures a smooth automatic changeover from main power supply to the standby supply under failure conditions and without disruption in the signalling indications supply.

To provide for any possible incident that might occur on the main line because of trouble at the intermediate power supply transformers, it is good practice to provide the intermediate posts with a bypass disconnecting switch system, telecommanded from the CTC district control office, or from substations and overhead electrical control offices.

Input and output disconnecting switches and their corresponding bypass switches should also be remotely controlled from the transformers of traction substations, in order to supply the line from various locations, if required.

8.6.4 Duplication of hardware

It is necessary to duplicate computers in those installations provided with CTC equipped with electronic driving and checking systems, in the lower and upper levels.

Breakdowns in the computer must be determined through a routine procedure.

There must be a constant self-control of reserve computers with the aid of a verification on line program. Computers' input, for example the connection of a keyboard, does not have to be commuted – its parallel operation will be sufficient.

8.7 Maintenance

System maintenance is of vital importance and should it be reduced, will almost certainly result in breakdowns, both of equipment and systems.

If breakdowns occur frequently trains will be delayed and the travelling public inconvenienced. In addition, railway traffic security will be considerably reduced, as many timed trains will have to run using a telephone block system, ignoring red or extinguished signals, making wrong track lines in non-two way working track blocks.

Remedial action can be made in two ways:

- before the breakdown
- after the breakdown

The first has great advantages over the second, since potential breakdowns can be detected before they occur. This is the service called preventive maintenance. All equipment and installations are checked out regularly in anticipation of a possible breakdown.

Generally, all maintenance work means great expense; this explains why the 'before' action plan must be carried out on a planned and regular basis.

If maintenance is correctly planned, a high percentage of the work of the staff engaged can be accurately programmed. As always there can be unforeseeable breakdowns and it is very difficult to achieve a 100% success rate.

Initially there should be a programme for maintenance works, listing all maintenance activities together with the corresponding dates for their execution.

These activities will be directed from the information given by the suppliers of equipment and systems, and from ratios fixed by the railway administration concerned, compiled from historic data.

A list of materials for each maintenance project to be carried out, as well as spare parts or components to be renewed, together with the labour required, can be calculated from previous data.

Amongst the most important benefits achievable with preventive maintenance are:

- improved security;
- improved reliability;
- longer working life for equipment and installations;
- less interruption to train services due to signalling equipment failure;
- better utilization of human resources;
- reduced need for spare parts.

At present, due to the current technological progress of signalling systems employed when introducing electronic interlockings, integrated driving and control systems, and automatic operation systems, it is very important to ensure adequate vocational training for maintenance staff.

There are also contents lists available which include forms that could be used to control preventive maintenance works on signals, track circuits, and point operating apparatus.

8.8 Homologation

In order that both equipment and systems required for use in railway signalling comply with the specifications of safety and reliability demanded, standards should be established so that neither equipment nor systems will have to pass the corresponding on line tests to verify the execution of the reliability conditions.

Once the established tests have been passed, they will be homologated by the corresponding railway administration.

Homologation is necessary in the following circumstances:

- when the supplier introduces modifications in equipment or systems which have already been tested.
- when it is related to new equipment or systems correctly developed by the initiative of suppliers, or because the railway administration has asked for it.

8.9 Quality control and reception of materials

Normally, signalling equipment and systems are not manufactured by the railway administrations. Instead, the administrations rely on experienced suppliers, with manufacturing and development guarantees of equipment and systems.

Despite this, the administration must carry out the necessary tests in order to control the security and reliability of all equipment to be made during each manufacturing process.

Previously, according to the type of materials to be examined, the necessary groups, the kind of inspection they must pass, the kind of sampling, the maximum number of failures accepted in each group, must all be determined.

Once suitable groups have been determined, all tests specified for each material will be carried out. For example, control of:

- raw materials;
- finishing and external appearance;
- verification of geometric measures and tolerance;
- electrical tests;
- electromagnetic tests;
- mechanical tests;
- working tests under varying conditions of temperature and humidity.

Since these quality controls cannot always be reproduced in the laboratories of the railway administration, it may be necessary to employ companies specializing in quality control or laboratories that specialize in certain tests.

8.10 Other measures

In addition to the measures already considered other precautions must be taken, which though difficult to specify, are undoubtedly going to contribute to improved reliability in the functioning of the security installations:

- protection against atmospheric storms by using appropriate dischargers for transmission lines and track circuits. Use of surge diverters in the transformer boxes of the power transmission lines.
- use of optical fibre cables for transmission on CTC lines.
- use of screened cables offering a reduction factor in those sections running parallel to power lines.
- installation of air conditioning equipment and filtering of dust in premises housing signalling equipment.
- use of protections and top quality paints in order to avoid corrosion in outside equipment.
- use of the appropriate type of packing capable of protecting equipment against clashes and frictions that can harm their structure.
- use of the appropriate transportation means, so equipment can arrive at its destination in the best condition.

8.11 Standardization

Standardization is considered in three different categories:

- production of equipment according to established standards, so their parts can be interchangeable with others and continue working correctly, even if they are purchased from a different supplier.
- as the application of elements which have been designed to be used in different signalling fields so that, under certain circumstances, they could be adopted for use as part of alternative equipment or systems of signalling equipment.
- the modular design of certain parts of the signalling equipment.

Standardization nearly always improves safety, since it considerably reduces the possibilities of a manufacturing error. Improved quality can also be obtained and therefore reliability should be increased.

However, there are some disadvantages:

- with standardization, designs might be retained even if they have imperfections.
- high expenses when changing a design increase resistance to the introduction of innovations and improvements.

With regard to the use of components, computers, etc, not specially manufactured for railways, these will only prove to be advantageous if there is an economic profit, with security and a maintained reliability.

The main considerations regarding the use of components not specifically manufactured for railways, would be related to the possibilities of:

- using the latest technology;
- use of modern and well proven software methods;
- use of compilers of reputable make.

8.12 Relays

8.12.1 Introduction

In order for a signalling installation to function with the levels of safety and reliability demanded, many requirements must be satisfied; success depends both on the components' features and the circuits related to them.

Without doubt, present day relays are the most important elements of electric interlockings making up the installation. That is why this chapter pays particular attention to everything related to their characteristics and the materials used.

In general terms it may be said that when the performance of relays of certain specific characteristics changes, then the complexity of the circuits to be used will increase. These circuits, besides achieving their primary function, must also monitor correct relay operation, regarding those characteristics that differ from the ideal ones.

Reference to characteristics hereafter is intended as a framework of the main features of signalling relays employed on the RENFE network.

8.12.2 Classification of relays

Two types of relays are described below.

8.12.2.1 N–type relays (non-supervised relays) (see Fig. 8.9)

These relays must, by themselves, satisfy the safety conditions, without receiving help from other relays and without any special verification of the circuits in their working regarding shunts between contacts, and safety drop. The type of circuitry acceptable with its use is known as N–type.

Fig. 8.9 N-type relay: fail-safe type

Fig. 8.10 C-type relay

8.12.2.2 C–type relays (supervised relays) (see Fig. 8.10)

With their use, these relays must guarantee conditions of safety through an additional checking of their behaviour, from certain points of view (relay drop, shunt), through the appropriate supervision circuits, and the doubling of, at least, certain vital circuits (redundancy). The correct installation of circuits for these relays is called C–type.

8.12.3 Basic conditions

There are two basic conditions that satisfy relays, with maximum guarantees:

- condition of non-dislocation of contacts.
- condition of the safety drop away time of relay.

8.12.3.1 Characteristics of non-dislocation of contacts

Non-dislocation of contacts means that under no circumstances can two or more contacts of the opposite character (front–back) close simultaneously.

The above condition is necessary for both N and C–type relays. Guarantees of achievement cannot be substituted by special provisions in the circuitry.

As for the condition of non-dislocation, it must be borne in mind that:

- the operation of contacts must be originated only by the action of the mobile armature. Indirect command by means of other springs is not permissible.
- the inclusion of relay construction restraints are such as not to allow any front contact to close when a de-energized contact keeps faulty closed and vice versa, even in the case of applying energizing or landslide forces higher than

nominals. In the case of N–type relays (unwelding front contacts) interest must be centred in the first of both conditions.

- the stroke of armature is limited by front and back butt ends.
- the opening of front and back contacts, and strokes of armature with both contacts opened, should be higher than the values given.

8.12.3.2 *Characteristic of the safety drop away time of relay*

In contrast, the condition of safety landslide (that which must accurately guarantee N–type conditions), is not considered as achieved with the exact level by C–type relays which, notwithstanding, can be used in many functions if special supervision circuits controlling landslide are used, thus preventing dangerous situations.

For reasons that extend further than the safety behaviour of relays (for example, shunts between cable conductors), it is necessary to supervise the landslide of at least one relay representative of each function, even when using N–type relays, if these functions are vital for safety circulation.

Regarding the condition of safety drop away, the following features should be considered:

- in proportion, front contacts must have at least one of its elements of carbon or carbon silver that becomes unwelded.
- there should be adequate strength and reliability in the armature's revolving system.
- the armature's drop away and, therefore, the safety opening of front contacts, must be made by the action of gravity regardless of other forces.
- the minimum distance of air gap must be guaranteed by safety butt ends.
- relays must have individual gangway covers.
- electric and mechanical design must be effected with high quality materials, so that these materials can always work below their limit values.

8.12.4 Characteristics of safety related to reliability

8.12.4.1 *For N and C–type relays*

- Useful life of relays must be greater than 10 million operations when energized with a current equal to the nominal current. One operation consists of one excitation and one drop. The average number of operations must be 30 per minute.
- Useful life of contacts with a nominal load must be two million operations.
- One relay is considered to have a possible three-year period in (reasonable) storage before being put into service. It will still be considered reliable without being checked or examined during the time, and again later, once in service.

8.12.4.2 *For N–type relays*

A plug in execution with mechanical code excludes the possibility of insertion in an inadequate base.

8.12.4.3 *For C–type relays*

The condition for execution with plug in mechanical code is not necessary for this kind of relay since, normally, they are installed in modules of geographic circuits. Nevertheless, terminals must accomplish the condition of good welding.

8.12.5 Characteristics of materials used in the manufacture of relays for signalling

All non-metallic parts used in relays and bases must be non-hygroscopic, auto-extinguishable, and must not suffer deformations and damage because of torsions or cracks; when they are transparent, they must be permanently transparent. This must be true even when exterior temperatures range from $-10°$ celsius to $+70°$ celsius and the relay is energized up to its nominal tension.

Materials that may become fungal or deteriorate if exposed to sunlight, should not be used.

Insulating materials used to manufacture relays must satisfy the test indicated in the paragraph on isolation resistance (see **8.12.7.1**).

All parts, either separately or jointly, should be corrosion resistant, or should be treated to become corrosion resistant.

Different metals, which might come into contact with each other, should be selected and protected so they can reduce the effect of electrochemical action to a minimum.

All materials used in construction should be physically and chemically inert for exterior temperatures between -10 and $+70°$ celsius.

Special care must be given to weldings. A non-corrosive building up material should be used, and no excess should be removed afterwards.

All screws must be firmly fixed to prevent them moving or coming loose while in service.

The electric power of materials which form the part of the connector's base that makes contact with terminals of relay, should be between −0.55V and +0.10V.

8.12.6 Characteristics of contacts

Contacts must be able to support 3A continuously for N–type relays, and 2A for C–type relays, when they are closed, without being damaged or becoming overheated. They should be able to commute tensions of at least 250V.

Commutation of power and currents indicated in each relay's own characteristic must be accomplished during the useful life of the relay.

Simple contact elements should be used, but in all contacts there must be one element with a convex surface.

Working contacts must be carbon/silver unwelding or carbon silver/silver blend, and silver/silver for the back contacts.

Contact elements should not be off centre in respect to one another. Opening and closing of all contacts must be done simultaneously.

When contacts are established they must make a minimum stroke of 0.2mm.

When energizing or drop is complete, there must be at least one contact force of 25g for unwelding contacts and 20g for delayed action contacts.

When the relay is in the attraction position or in complete drop, the maximum resistance of unwelding carbon/silver contacts will be of 0.2 ohms, and 0.05 ohms for back contacts. This resistance will be measured by making a continuous current of 100mA pass through the contact. These resistance values are related to new relays, admitting a maximum 20% sliding from the initial values after the relay's useful life.

8.12.7 Main tests in relays

8.12.7.1 Isolation test

This test is carried out to determine the isolation between the conducting parts isolated from the relay.

The resistance of isolation is measured under a continuous tension of 500V, between the following parts:

- between the different coil windings;
- between each winding and mass;
- between one contact's joined outputs and joined outputs of the remaining contacts;
- between each contact's output in its opening position;
- between mass and all the joined outputs of contacts.

The isolation resistance measured must be higher than 2 megohms.

8.12.7.2 Test of dielectric strength

This test is carried out immediately after the previous test of isolation resistance.

Isolation between each winding and the other parts of the relay, as well as contacts between contacts and the relay body, should be such that it will resist one testing tension of 2000V rms/50Hz for one minute.

The testing machine must supply one current of between 10 and 20mA in short circuit. The test begins by putting into effect a tension with a value equal to half of the test. The tension grows progressively until reaching the prescribed value.

Contour outlinings, perforations or superficial crepitations must not be carried out during the test.

8.12.7.3 Contact's unwelding test

As indicated above, front contacts of N–type relays must be unwelding and must satisfy the conditions indicated below, when the relay is at an ambient temperature and is being fed with its nominal tension.

Short circuit test

The two constituent elements of a front contact close one circuit protected by a 6A fuse of retarding feature, fed by a controllable DC supply so it can supply a current of up to 200A. The contact will stay with no sign of welding, and will be established until the fusion of the fuse.

Test with discharge of a condenser

The two constituent elements of a contact should be joined to a dry paper 200 microfarads condenser +/− 5%

The condenser must be charged up to a continuous tension of 330V, and then be discharged on the contact by the proximity of its elements. *No sign of welding should be noticed.*

In contacts of C–type relays, the possibility of a low percentage of weldings is taken into consideration, as well as in the metallic contacts of N–type relays. These contacts must pass the unwelding tests previously described, observing a maximum percentage of weldings below 5% of the number of contacts tested.

8.12.7.4 Vibration test

The correct functioning of relays should be checked when conducting the vibration test described below, and when closed contacts will not open and opened contacts will not close.

Relays must be fixed to the platform of the vibrating machine in their normal working position, by means of their normal fixing device. The bobbin connection cables and contacts of the relay must be arranged so that they will not create obstacles other than they would create when in service.

The relay must be working, becoming energized or de-energized, at least once in each vibration direction.

The frequency rate should be within 5 and 22Hz, and with a maximum amplitude of 1mm, or a frequency within 22 and 50Hz and a maximum acceleration of 2g.

Abnormal electrical functioning should not occur during the test. When the relay is withdrawn from the vibrating machine it will go through a test to check that connections and fixing elements have not suffered any metallic deterioration or loosening.

8.13 Signalling cables

Railway signalling cables play an important role as the bearer of signalling circuits between the control centre/signalbox/relay room and externally located equipment.

An engineering specification containing details of cable making materials, cable composition and required physical and electrical characteristics is produced by the signal engineering department within a railway administration. The specification is then employed to direct cable manufacturers in the production of a standard and quality controlled cable product complying with the limits laid down in the specification.

A rough guide to the contents of a typical cable specification is listed below.

8.13.1 Materials

- Conductor
- Conductor insulant
- Insulation
- Sheath
- Outer sheath, serving and additional overall protection as specified
- Cable filling

8.13.2 Construction

- Conductor insulation
- Stranding
- Colour coding or other identification
- Cable filling, if required
- Interior sheathing
- Outer sheathing

8.13.3 Electrical tests

- Conductor resistance
- Insulation resistance
- Dielectric strength
- Further electrical tests as specified by the engineer

8.13.4 Mechanical characteristics

- Mechanical resistance
- Insulation shrinkage
- Bending radius

8.13.5 Delivery

- Cable drums, normally provided to separate specification which provides for any specific drum dimensions suitable for trackside cable laying operations

8.14 Cable laying systems

8.14.1 General

For the safety and reliability of one of the external installation components – ie the cables that join the interlocking cabin with the different external elements – special caution must be taken with the route system selection and the performance of cable laying.

The right selection of the cable route system will prevent cable failure, make easier cable laying and thus reduce the chance of interference and failure, contributing to a safer and more reliable installation.

On the other hand, the correct route selection will also allow future variations in cable numbers to be carried out whilst guaranteeing damage limitation to existing cables and permitting new cable laying to be carried out at minimum cost and shortest time.

In respect of their varying characteristics and different results two methods of cable laying are discussed:

- suspended cable systems
- underground cable systems

8.14.2 Suspended cable systems

In considering the option of providing a line plant system along the railway line, the use of a suspended cable fastened, perhaps, to an existing overhead pole route provides the more economical alternative.

This method is particularly attractive when providing a cable run for communication purposes on a station to station basis.

Two types of cable have been in use since the 1960s:

(1) a cable with an in built catenary, which supports the cables and is tied off on existing overhead line poles or some appropriate alternative fixing.
(2) a separate catenary wire is suspended along the line of railway and the cable fixed to the catenary by the spinning action of a second wire. This is sometimes referred to as a spun cable.

It should be pointed out, however, that this type of cable should not be considered for high security circuits because of the likelihood of cable damage from a variety of sources, such as:

- wind and storms;
- lightning;
- corrosion of the catenary wire;
- maladjustment of cable height at level crossings;
- vandalism by shot gun.

The two types of cable mentioned above are *not* suitable for an installation in AC electrified areas.

8.14.3 Underground cable systems

8.14.3.1 Service tunnels

This system is normally employed in very important stations with complex railway cable systems, such as central workshops and marshalling yards. The use of service tunnels is determined by the installation of a large number of cables, in addition to electrical conduits.

Their construction is part of the station or railway complex itself, which is why their specific features and dimensions are determined in the corresponding project.

Service tunnels must have watertight directional illumination from entrance and exit doors, and they should incorporate drainage, natural or forced ventilation, and gas and fire warning systems.

In tunnels, cables and electrical conduits are installed on opposite sides of the tunnel.

Cables should be set on horizontal trays, with each separate service kept independent in its own tray. Generally, communication cables will be set on the upper trays, safety installation cables on the intermediate trays, and those for power supply on the lower trays.

8.14.3.2 Layout of conduits

Underground PVC conduits are normally used for general cable house connections, substations, crossing of tracks, platform crossings and roads.

The construction of underground conduits includes the necessary control boxes and manhole chambers.

The following types of conduit layouts can be used:

- normal installation of conduits (in normal ground, rocky ground and ballast areas);
- crossing of tracks;
- crossing of other conduits;
- special layouts of conduits;
- trenches;
- switch flangeways.

Installation of conduits in normal ground, rocky ground and ballast areas

The normal provision of conduits are constructed underground by means of a trench, with PVC tubes in the inner part set in concrete and then buried, so the upper part of concrete remains 60cm deep in the ground.

Whenever it is possible, the layout of conduits must go on the outside part of the electrification poles and away from the track cess. If the latter is impossible, the largest track spacing must always be the one to be used.

Layouts of conduits must not be allowed to go under the track, except at the crossing points.

To avoid damage in possible derailments or by lateral displacements, siting of the conduits should try to separate them from the track as much as possible.

The minimum allowable distance for layouts of conduits, from the edge of the trench to the nearest track, should be made in accordance with a safety gauge.

Normally, layouts of conduits must be straight in each section.

To determine the number of conduits, the following quantities should be allowed:

- one conduit for power.
- at least two conduits will be required for communications.
- for safety installations, a number of conduits are required for each installation; bearing in mind each conduit must be occupied by a maximum of two-thirds of its surface.
- in the platform area there must be one conduit for lighting.
- layouts of conduits will have at least one conduit in reserve.

Layout of conduits in normal ground
This means a layout of conduits whose trench will be excavated in normal ground. Normal ground is one that results from the sedimentation of remains from natural erosion and debasement of mineral and organic remains mixed with alluvial stones.

Layout of conduits in rocky ground
This is a layout of conduits where the trenches in which it will be constructed have to be excavated in rocky ground.

Rocky ground is the one formed mostly by earth mixed with clayed silicates, where the presence of rock will be obvious all along the line, is more than 20cm thick, and where its construction requires other means for excavation.

Layout of conduits in ballast areas
This is a layout of conduits where the trench for its construction must be excavated in a ballast area. A ballast area is that area of the platform that, due to its proximity to the rail, is affected by the ballast forming part of its slope, and is 30cm thicker.

Crossing of railway tracks
These crossings should always be carried out at right angles to the track, by means of the corresponding layout of conduits.

In the route, in a crossing of railway tracks corresponding to the set of main cables, control boxes and manhole chambers must be built on both ends of the crossing. At minimum, there should be four conduits.

The crossing of railway tracks for secondary cables, both in route and in stations, should be made with a minimum of two conduits, and without building control boxes or manhole chambers at the ends. Conduits crossing the railway must be long enough to extend beyond each rail by 1.8m.

Because of the importance of these crossings special attention is required to ensure their installation is carried out correctly.

Crossings with other conduits
Crossings of conduits with other cables not in keeping with the installation must be avoided whenever possible, if they are situated in trenches or other channels, as well as water lines, conduction of gas or other fluid.

Special layouts of conduits
As a general rule, normal layouts of conduits should be used in stations between entry signals and trenches in the middle of the track.

Notwithstanding this, there are occasions when ground cannot be crossed over by trenches, so alternative methods must be provided. For example, installation of galvanized steel tubes to cross over drainages and, generally, all those areas in which the ground doesn't allow the construction of layouts of conduits already described.

Trenches
Usually, this kind of channel will be constructed along routes between stations.

When making the trench, spoil with sharp edges should not be left around the ends, as it may damage the insulation of the cables. The excavation must be made with sufficient slope so as to assist the drainage of water.

A layer of sand or clean soil of about 10cm thick must be laid at the bottom of the trench, so cables can stay perfectly settled. The cable must be placed on top, with another layer of sand or clean soil over that. Its height will depend on the kind of trench used.

When the conduction's outlining is made through areas with permanent humidity, or in areas where it is possible to have water filtration, it is better to use fine gravel.

At a later date, there may be a need to carry out repairs or new cabling, so, in order to warn about the presence of the cable, a continuous row of tiles should be laid in its longitudinal dimension, in the transversal direction to the axis of the trench. A plastic grid may be used as a continuous visual warning instead of the tiles; in that case the *colour* will be the distinguishing mark for the type or types of cable in the trench.

The grid's transversal dimensions must be 20 or 40cm, and 50 to 100cm long, to allow the installation to be continuous.

Colours to be used, according to the type of cables, must be:

- power cables: *yellow – bright orange*
- telephone and control remote cables: *green – strong yellow*

The plastic grid must allow an enlargement of up to 30% without getting broken, so it will not suffer any damage and can be a warning when excavation works are carried out in the cabling area.

In any ground, the distance from the grid to the upper part of cables must be at least 25cm, in order to prevent cables from being touched when a tool is driven into the ground.

The upper part of the trench should be filled in with natural soil from the excavation.

According to its depth, to house the different cables, trenches can be classified as:

- 80cm trenches
- 110cm trenches

Trenches of 30cm are used for cables with several functions – either signalling or telecommunications cables, which can be buried at the same level.

The feed cable should be housed in this type of trench, and at the same level of the other cables when it is impossible to make a 110cm trench. In this case, the high tension cable will be protected from the rest, separating it with a row of tiles along all the layout of conduits.

Trenches of 110cm must be used for multifunction cables, one of which will be the power cable, which must be buried at a different level from the rest.

Electric cable conduits
Generally, electric cable conduits are used in those cases when ground conditions do not permit the specified trench depths.

Electric cable conduits are singles, doubles or triples, according to the different functions of cables (power, signalling and remote control).

Electric cable conduits must be fitted with a concrete lid to prevent spoil or rubbish from getting into the cable housing.

An excavation 25cm deep should be made, as described, so the switch flangeway can be buried.

A layer of clean soil (about 10cm deep) or concrete plate must be set on the bottom of the excavation, allowing the setting and levelling of the switch flangeway.

Coupling of two switch flangeways should be done with a fibre cement slump plate. This ensures that the joining of the cement lids does not coincide with the coupling of two switch flangeways, which gives more firmness to the whole.

The upper part of the flangeways' lid should be covered by a layer of soil about 5cm deep, which prevents unauthorized tampering. This is unnecessary in those cases in which they go along the tracks and stations where the flangeway's lid is at the same level as the public walkway.

The minimum free dimension of the switch flangeway is given by the diameter of the cable or cables housed inside it, so they can leave the same free cable space.

8.14.3.3 Safety measures

The following precautions must be taken when preparing trenches and layouts of conduits:

- trenches must be dug with a minimum of advance time in relation to the rest of the construction.
- during the works, and in order to prevent any possible accidents, trenches must be protected and signposted in the normal way; for instance, using plastic and coloured tapes, ropes with red hangings, fences, pointing out the places where people can pass through safely. The trench must be covered with steel plates and wooden panels.
- where the ground is paved, only the exact and necessary surface should be opened, by mechanical or manual means that can make the cut as clean as possible.
- trenches in platforms must not stay opened at night. Only the stretch in which works are going to be made should be opened each day. On the same day, the platform should be closed and sealed.
- it is important to pay special attention to the tamping of the stones that are going to fill in the trench, especially in the crossing of tracks where a perfect packing of ballast will also be required.

8.15 Staff licensing

The professional signal engineer, through his training and experience, acquires an intuitive sense of fail-safe situations, which prevails in all aspects of his work. This highly developed fail-safe culture, which pervades the signalling profession, has contributed greatly to the excellent safety record of the industry.

Today the complexity of modern signalling schemes and the commercial pressures for fast implementation demand greater discipline and a higher degree of formality in working methods and procedures. Within this environment it is apparent that the competence of the staff performing signalling tasks must be assured.

To this end a licensing scheme, administered by the Institution of Railway Signal Engineers, exists with the following aim: to ensure that work on railway signalling and telecommunications assets, the malfunction of which could compromise railway safety, may be undertaken only by staff who have been properly trained and who have demonstrated their knowledge, ability and skills to perform specific tasks by satisfying the requirements of a licensing agent accredited by the Institution.

The scheme is based on the principle of accreditation whereby the Institution lays down the rules and standards which are applied by railway and contracting organizations acting as Accredited Licensing Agents. It caters for all UK railway and contracting organizations and is designed to be adaptable to overseas organizations, complying as it does with EC regulations for such bodies.

Licences detail the activities the holder is competent to undertake, such as specifying, design, installation, testing, commissioning and maintenance. Examination, testing and record of practice are essential elements of obtaining and renewing a licence.

8.16 Analysis of the different administrations

8.16.1 Design of the interlocking circuits

Generally, the development of interlockings of electric circuits and the software of the electronic interlockings is made by the companies supplying equipment and systems; always based on the specification and the exploitation criteria prepared by the railway administration.

The railway administrations have engineers prepared to control and approve the circuitry plans made by the suppliers, and they can even make modifications to the existing installations.

8.16.2 Cable laying

The aerial cable laying systems have generally been superseded by all railway administrations, and replaced by the following methods:

ÖBB
The cable laying is done in the trenches and, in new installations, in cable routes.

NMBS/SNCB
In the stations, the cable laying is carried out by a system of conduits laid in a straight line in the trenches.

SBB
Cable laying is carried out mainly in conduits.

DB AG
According to the environmental conditions, the system chosen for the cable laying is one or the other.

RENFE
Usually by cable trenches along the open trackside and by conduits in station areas. For very high speed lines concrete cableways are employed for both open line and station areas.

SNCF
Most cables for recent installations are installed in trenches or, more precisely, covered cable-beds, laid alongside the track path.

BR
Cables are normally laid alongside the line in concrete troughing, although various designs of plastic have been used.

FS
The cable laying is mainly carried out in conduits of concrete laid in trenches. Lids, with a minimum width of 50cm, can constitute the service walkway.

In tunnels plastic troughs are used. They have very good characteristics in case of fire (excluding PVC or other materials, which can develop toxic or corrosive gas), and are fastened on the walls by means of flat iron shaped stirrups.

CFL
Ways of concrete (beton) troughs are used in all new installations.

NS
In general, trenches in station areas, concrete conduits are used for better accessibility.

The depth of trenches varies, from a minimum of 50cm to a maximum of 80cm.

Normally, simultaneous laying of signalling cables with power cables with tensions above 1000V is not made.

8.16.3 Cables used

ÖBB
Cables are normally armoured. Cables with a reduction factor are only used if absolutely necessary, for example, for lengths of 6.5km.

NMBS/SNCB
Armoured. Cables with reduction factor, not for signalling purposes.

SBB
Armoured star quad cable and with reduction factor for lengths above 4km.

DB AG
Not armoured and with reduction factor in certain cases.

RENFE
Armoured and with reduction factor in certain cases (EMI).

SNCF
Not armoured.

BR
Not armoured.

FS
Only the automatic block power supply cable (1000V) is armoured. Cables with a reduction factor are used only when necessary. In this case, the maximum allowed longitudinal induced voltage must not exceed 60V in normal operation and 430V under fault conditions. In tunnels it is mandatory to use cables with appropriate characteristics (no fire propagation, low toxicity index, absence of halogens and corrosive gas, low gas opacity).

CFL
Armoured, if necessary, cables with reduction factor are used.

NS
Armoured.

8.16.4 Installation of local control panels and alternative lines (CTC systems)

ÖBB
At present, transmission lines are not duplicated. Stations have local control panels. In the future, with the concept of the operation control system, whether there shall be local control or not, will be selected according to the importance of the station. The number of transmission lines, singled or doubled, is closely connected with this item and can also be selected individually.

NMBS/SNCB
At present, the procedure is the same as on ÖBB. In more recent installations, however, the transmission chain is completely doubled, suppressing the need for a local control.

SBB
Two lines are used; one is for remote control systems and the other for automatic control of the system.

DB AG
The transmission line is not duplicated and local control panels are set up. Transmission lines have been duplicated on the very high speed line.

RENFE
Local control panels and alternative transmission lines are always used.

SNCF
Local control panels (mimic or VDUs) are commonly used, even on high speed lines, though usually simplified versions.

BR
Emergency local control panels have been provided with some installations. Present technology using SSI does not allow the use of ELCPs.

CFL
The transmission line is not duplicated, and local control panels are always installed.

NS
Duplicated transmission lines by independent cables. Local control panels are not being installed.

8.16.5 Installation of equipment and systems

- in some countries, the installations of external equipment is always carried out by the companies supplying the equipment and systems (NMBS/SNCB, DB AG, RENFE, FS, NSB).
- in other countries it is undertaken either way, by the supplying companies or by the railway administration (ÖBB, CFL, SBB, SNCF, BR).
- in NS, the construction, assembling and wiring of signals cases is done by suppliers and a subsidiary company of NS. The installation is done by an NS subsidiary.

8.16.6 Specifications, norms and standardization

ÖBB
There are only operational specifications. The signalling lamps, the switch motor and power supply of field units are normalized. A study is being made for the standardization of level crossings.

NMBS/SNCB
It has a complete set of the specifications, techniques and norms. There is a low level of standardization.

SBB
Technical specifications and regulations have been developed. There are two signalling suppliers, which are not interchangeable.

DB AG
Security, reliability and availability norms have been developed, as well as specifications of obligatory observance. The point machine and all types of track circuits are interchangeable.

RENFE
Technical specifications and performance documentation have been developed for all signalling projects. This technique allows for the standardization of equipment and materials, and for compatibility of supply. It further ensures improved accuracy in installation testing.

SNCF
Specifications, norms, methods, etc, have been developed. The different equipment to be supplied by companies to carry out the same functions must be completely compatible, both electrically and mechanically.

BR
Specifications and norms have been developed. There is a high degree of standardization, although different suppliers' designs remain their own intellectual property.

FS
Technical specifications and rules have been established by FS either for installations and for outdoor equipment. Most important equipment is fully standardized at level of shop drawing. FS has property of shop drawings or, at least, the right to use them for calls for tenders.

CFL
Technical specifications and norms have not been developed. Standardization has a low level and there is no interchange between the elements of different manufacturers.

NS
Technique specifications and norms have been developed. Standardization only reaches the level of components.

9 Automatic Train Protection and Control

J. CATRAIN

Contents

9.1	General	266	9.4 European train control system (ETCS)	274
9.2	Intermittent systems	268	9.5 Automatic train protection specification (ATPS) – BR	275
	9.2.1 Signal repetition on board	268		
	9.2.2 Speed supervision system	270	9.6 Present systems in Europe	275
9.3	Continuous systems	272		
	9.3.1 Centralized systems	272		
	9.3.2 Decentralized systems	273		

Illustrations

Fig. 9.1	*Crocodile*	268	*Fig. 9.5*	*Summary of automatic train protection and control systems in current use by European railway administrations*	276
Fig. 9.2	*INDUSI – track transponder*	269			
Fig. 9.3	*KVB – cab display equipment*	271			
Fig. 9.4	*TVM – on board equipment*	273			

9.1 General

From the beginning train driving was left as the responsibility of the train driver. Initially without alternative means other than sightseeing driving, it quickly became apparent that higher levels of speeds and traffic increased the necessity to provide visual signals. But in fact major accidents were registered by driver failure whatever the reasons, together with a requirement for a stricter mental and physical selection of drivers. It became obvious at the beginning of this century that it was necessary to force trains to stop automatically when they were in danger. This concept came in conjunction with the emergence of the automatic braking system.
Automatic train protection (ATP) was born, together with cab signalling at the beginning of the twentieth century.

In the years that followed, it was still difficult to imagine a driverless system, and efforts were made to develop assistance to the drivers who were and still are, highly skilled, professional people. It was mainly a system whose main objective was to check the driver's vigilance.

When, in the 1960s, there was such great progress in electronic developments offering computers, data transmission, and, of course, equipment cost reductions, it was more evident that a driverless system could become a reality.

The first railways to be considered were the Mass Transit Systems, because they operate in a protected and closed environment and the total investment with regard to traffic remains cost effective. Driverless systems are now widely employed on Mass Transit railways with little doubt as to their efficiency and effectiveness.

For trunk railways, the problems are different: the environment in which the trains move is much more immediate, with level crossings, animals, etc; also the investment cost is much higher due to the variety of rolling stock, passenger coaches, wagons, and the length of the network with respect to the larger number of drivers.

One of the main reasons leading to the introduction of ATP and control was the need to increase train speeds to above 200km/h. At such speeds human beings cannot react to conventional signalling in a safe manner. But the higher speeds do allow an increase in line capacity by employing a lower quantity of lineside signals.

The prevention of accidents and single driver operation were also good reasons to install ATP and control on all railway networks.

A more flexible system becomes a necessity in terms of efficiency, availability and safety on railway transportation systems, which have the reputation of being the safest of all public transportation modes.

What does 'driverless' mean? Not necessarily the total absence of human beings in the cab.

In fact the present situation may be summarized as follows:

- *Total driverless system* implemented mainly on mass transit systems, sometimes with the opportunity of a simplified driver's operation in a degraded mode. In this mode it has to be clearly understood that rolling stock has been specially designed for this type of operation.
- *Automatic train control* with driver acting as supervisor on an equipped line section and becoming a fully responsible active driver on a non-equipped line section.

  The system may be able to operate locomotive traction and brake equipment (automatic train operation: ATO) under the control of speed levels elaborated in real time (automatic train protection: ATP). The ATO system is not required to be fail-safe by itself, but it is mandatory for the ATP system.

  The ATC mode requires accurate subsystems as all the operating actions are under the control of the system. That is true not only for the system itself but also for the outside data that will make it safe. This includes the speedometer system, the localization, the dynamic and static characteristics of the train, and line data. The system availability is also a dominant factor, so there will be no dramatic change in fail-safe level with degraded mode.
- *Automatic train protection.* Under this system the driver remains the operator with all his actions under the scrutiny of the ATP system, which permanently determines if the train running is following instructions the driver has to respect completely.

  At this stage, either the driver is supplied with speed information on his desk or he is not. In the first instance, since it has to follow this instruction the ATP system must be fail-safe (in accordance with lineside signals) and highly reliable in order not to confuse the driver with two different types of information. In the second case, he has to follow instruction from lineside signals, the ATP system being the 'parachute' if the driver makes a driving error.

These systems may be divided into two main categories:

- continuous systems (system with continuous transmission).
- intermittent systems (system with intermittent transmission).

A continuous system means that data is transmitted permanently between lineside equipment and trains. This is the more significant system in terms of performance but is very costly, especially the track equipment. Continuous systems are used for large sections of lines where an increase of capacity is or should be a must.

An intermittent system means that data is transmitted from place to place, but this is less effective since data is not refreshed in real time. Intermittent systems are cheaper and allow for a quicker and larger protection of risky areas with the same amount of investment.

A good alternative is the in between system – the semi-continuous system. As a cost effective solution it enables data to be continuously refreshed in small zones in order not to slow trains without due cause, such as a case of signal opening after having been registered by train as closed.

It is important to stress that ATP, ATO and ATC have all been defined in varying ways. Some definitions are given below.

UIC – Collection of Signalling Terms 1972
- ATO: System covering the acceleration, running and braking of trains automatically on information supplied.
- ATC System: The system that effects an emergency brake application, if the permissible speed is exceeded.
- No definition of ATP.

UIC – ETCS Project 1991
- ATP: Discontinuous or continuous speed monitoring with or without cab signalling.
- ATC: ATP plus continuous display in the cab of variables aiding the driver to drive as closely as possible to the monitored speed limit and/or supervision curve when braking.

IRSE – Railway Control Systems 1991
- ATP: A system that enforces obedience to signals and speed restrictions by speed supervision is known as an automatic train protection (ATP) system.

Intermittent systems

During the steam age, with the existing technology, mechanical solutions favoured and developed all along the networks of Great Britain, France, Germany, Switzerland, Belgium and Spain.

They consisted of a discrete repetition on board the train of warning information either by visual, acoustic alarm or both. These systems used contact or contactless equipment between the track, as the source of information, and the train.

They included an acknowledgement button operated by the driver, the omission of which triggered the automatic braking. Such a procedure implied that the driver would not forget the information given and would operate braking in due time – a controversial issue.

Some of these systems have now been improved, drawing from actual experience and from developments in technology.

More sophisticated functions were therefore incorporated. These included:

- supervision of braking after acknowledgement by the driver;
- speed supervision by continuous check of the authorized speed in accordance with data memorized from place to place (Ebicab, KVB, ZUB 121, TBL);
- automatic stopping of trains passing signal at danger.

Continuous systems

Continuous systems emerged later and integrated more and more sophisticated functions. In this category various solutions may be classified as follows:

- cab signalling with overspeed detection (BACC/FS);
- cab signalling with fail-safe speed monitoring (ATB/NS);
- cab signalling with speed monitoring (TVM/SNCF, Eurotunnel);
- speed control which must be safe (LZB/ÖBB, DB AG, RENFE) and used either in connection or not with cab signalling.

9.2 Intermittent systems

Intermittent systems include two categories:

- signal repetition on board;
- signal supervision.

9.2.1 Signal repetition on board

This type of intermittent system is installed in order to increase the driver's vigilance.

Its main function is to check that the driver has observed a warning signal. If so, he has to press a button prior to passing the warning signal; if not, emergency braking is automatically applied. The procedure is recorded on a paper tape, together with the train speed. This system has been achieved with different technology:

- electrical contact between train (brush) and metallic piece in the track (Crocodile);
- magnet or electromagnet influence (ISR 72/ZSI 90/ZST 90 – AWS);
- inductive coupling (INDUSI and AFSA).

The disadvantage of these systems is their low transmission capacity and the fact that the driver must not 'forget' this information.

9.2.1.1 Crocodile (NMBS/SNCB, CFL, SNCF)

This system supervises the driver's reaction to the restrictive aspects of distant signals. The driver has to acknowledge the information prior to passing that signal – if not the brakes are automatically applied.

The system delivers audible and visual information to the driver.

- Signal clear:
  – Brief action on the bell – no visual information – no action from the driver.
- Restrictive signal:
  – Flashing visual light is displayed (does not exist on CFL network).
  Passing a restrictive signal, the bell gives a short 900Hz tone and the driver has 5 seconds (SNCF, CFL) or 7 seconds (NMBS/SNCB) to react, by pushing the vigilance button.
  The flashing yellow light then switches to steady, up to the next clear signal.

Lines under 40km/h maximum speed are not equipped with this system.

In the track, a metallic piece includes slopes for activating the locomotive brush. This metallic piece is

Fig. 9.1 Crocodile

connected either to positive or negative polarity of a battery, the inverse polarity being connected to the rail.

On board, a mobile brush is connected to sensitive relays which select the polarity, with the return being made by the wheels.

This equipment gave good results, but with increasing speed and higher traffic levels it was necessary to have more available transmission options.

Thus when the maximum speed was increased from 160km/h to 200km/h, in addition to Crocodile equipment, SNCF installed on those line sections a pre-warning equipment based on induction coupling (60–100kHz).

The pre-warning signalling system includes all signal information transmitted by the Crocodile system plus the signal information related to speeds over 160km/h – ie the pre-warning signal and flashing green aspect.

The switching from the Crocodile system to the pre-warning system is automatically operated when the train runs over an activated pre-warning beacon. On board, a display showing 'B' illuminated indicates to the driver that he is entering a 200km/h line section. The train limited to 160km/h is only concerned with the Crocodile system and not equipped with the additional pre-warning system.

The pre-warning signal on the track consists of the letter 'P' for the speed indicator board signalling system, or the flashing green for the automatic block signalling system.

When the pre-warning signal is closed, drivers are informed by a 'P' flashing display together with a bell sound. The 'P' symbol is illuminated for 12 seconds and the sound is stopped by pushing the acknowledgement button.

Pre-warning signalling is associated with speed supervision. On board, this equipment permanently checks that the speed from 220km/h to 160km/h is achieved below a predetermined curb in the memory of the on board computer.

If the predetermined conditions are not fulfilled, the complete emergency braking is applied and a red display 'URG' is illuminated in the cab.

9.2.1.2 Automatic warning system AWS – (BR)

There is little functional difference between AWS and Crocodile equipment except for the technology used. On the operational side it will be noticed that the driver may release the brake application as soon as he operates the vigilance button. This is not the case with the Crocodile system where emergency braking is applied instead of service braking. The system operates by intermittent magnetic induction between the track and the engine. The track equipment consists of two magnets:

- permanent magnet type, south pole oriented;
- electromagnet from a 12 or 24V supply, and energized only when the signal displays the green aspect.

The role of the first magnet is to activate the on board equipment and of the second to give the information 'North field' for green aspect; then the bell will sound for 1 second and the visual indicator will show a full black disc. If there is no 'North field' (ie a signal not showing a green aspect) then a horn will sound continuously and, without a driver's reaction, will automatically stop the train. If the driver reacts by pressing a button then the horn will stop and the on board indicator disc will display a black–yellow aspect in order to remind the driver of the caution aspect displayed by the side line signal. At the next signal, the first magnet restores the indicator to the all black position.

AWS equipment is fitted on approximately 98% of the BR network where its use is appropriate. It excludes RETB routes, minor secondary lines and freight branches. All main lines and rolling stock are AWS equipped.

9.2.1.3 Signum – (SBB)

This system belongs to the same family as before, except it is a contactless system and has no transmission for a clear signal. Since 1976 the system has been extended to an absolute stop aspect.

Therefore, the aspects are: Stop, Warning, Clear. When the driver passes the stop aspect the emergency brake will be applied without pre-warning since the driver normally would have been warned at the distant signal that he may expect a signal at danger.

9.2.1.4 INDUSI – Induktive Zugsicherung – (ÖBB, DB AG)

This contactless type system supervises the driver's vigilance as well as providing a 'limited' speed control (punctual as for the Signum one) based upon a fixed distance between distance and main signals and time.

A passive track transponder in the track may be either short circuited (no signal–clear aspect) or tuned at 2000Hz (signal at danger) or tuned at 1000Hz (warning signal). A transponder at 500Hz is used for speed supervision in the approach to a main signal at danger (40, 50 or 65km/h) depending on the category of trains. Emergency braking is applied if speed is higher than a predetermined speed at this place depending of the train category.

Fig. 9.2 INDUSI – track transponder

As for other systems, the driver has to acknowledge the distant signal displaying a warning aspect. If not, 4 seconds later the brake is activated. Depending on brake capability of train (three categories can be selected on board by switch), the speed reduction is checked after 20, 26 or 34 seconds for 95, 75 or 60km/h respectively. A main signal overrun at danger will bring an immediate automatic train stop. A special override button allows the driver to pass the signal at danger (2000Hz influence) with a supervised speed limited at 40km/h as long as the override button is operated.

A new version, I60R (INDUSI 60 Rechner), introduced in 1992 provides the same functions together with added facilities such as extended data entry features, improved braking supervision and serial interface to an electronic data logger. It executes semi-continuous speed supervision in the approach to a danger signal. This system monitors train speed of up to 160km/h. All DB AG main lines and rolling stock engines are equipped with INDUSI. With ÖBB all lines at maximum speed greater than 120km/h are also equipped with this system.

9.2.1.5 ASFA – (RENFE)

Contactless and inductive with five frequencies between 60kHz and 111kHz, this system has the same functions as the INDUSI ones. Time reaction given to the driver for pushing the acknowledgement button is limited to 3 seconds – before emergency automatic braking. It also includes the transmission of 'clear' signal and can be used up to maximum speed of 220km/h.

For all of these systems it can be seen that there is no redundancy at all and that they may be switched off when not operating without any restriction regarding the train running – ie no speed limitation.

In addition, since they are very simple they have the added advantage of robustness. Transmission presents a problem area because of mechanical contact and reliability of relays. Some improvements have been carried out in order to dispose of the problems. They no longer meet the needs of modern railways.

9.2.2 Speed supervision system

In this category, much more information is transmitted from the track to the train. This information includes the distance to the next signal, the gradient of track between two signals and then with signal indications, and on board information such as train length, braking characteristics and maximum speed allowed. All this information allows the system to calculate the theoretical curve of the train speed and in real time to supervise the accurate speed of the train. If the speed is in excess, alarms are given to the drivers, and service and/or emergency brakings are then applied dependent upon the speed excess.

This final function is the same for all existing systems in Europe; differences occur in technology, the transmission path, the drivers' operating rules and the adaption to the different locomotives type and lineside signalling equipment.

9.2.2.1 Transmission beacon locomotive TBL – (NMBS/SNCB)

This system transmits from place to place by means of beacons. Data to the train is provided by the lineside signals, as they are observed by the driver, so long as the data does not change in the meantime.

Currently the NMBS/SNCB network is partially equipped with a TBL basic program called TBL 1 which consists of three main functions:

Stop
Whenever a red showing signal is passed by, an emergency braking follows automatically.

Memor
Indicates on board the train whether the last passed signal was a stop signal (for trains) or a shunting signal, and intervenes whenever necessary.

Supervision
Automatically applies the brakes whenever the driver does not acknowledge the passing of a restrictive signal, a similar sequence to that of the Crocodile.

The Crocodile exists together with TBL 1. When a train is equipped with a TBL antennae and a Crocodile brush, the TBL has priority. Whenever the TBL 1 information is missing, the Crocodile applies.

Another improved system under consideration is called TBL 2. This differs from TBL 1 in three main ways:

- increase of transmission capacity;
- complete calculation of braking curves;
- 'real' cab signal.

The main functions of TBL 2 are:

- braking curve monitoring for track occupation, permanent or temporary speed restriction;
- maximum line or train speed supervision;
- train stop when passing signal at danger.

The beacon, not located in the axis of the track for automatic detection of the running direction, permanently emits a message of five data codes, one code being selected out of ten, ie 100 000 combinations (TBL 1). The following generation (TBL 2) beacon permanently emits a message with 119 useful bits.

Redundant coding is used. Data is effective if one half of the message transmitted in one way is similar to the second half transmitted in a different way.

Messages are stored up to the next beacon – ie to the next signal or intermediate beacon.

The speed supervision is computed on board from parameters such as distance between the present speed and the restricted speed to observe, average gradient (positive or negative) for the section, train braking characteristics (deceleration, time lag), train length, train category (maximum speed allowed), and speed tachometers.

The most restrictive speed supervision is always taken into account.

The TBL system uses an active power supplied beacon in the range of 100kHz; the on board equipment is triplicated, two out of three systems, whilst the trackside equipment is only duplicated, two out of two systems. The choice being a better availability for the rolling stock during its mission. For the trackside equipment maintenance monitoring can easily reduce the working interruption time.

9.2.2.2 Contrôle de Vitesse à Balises KVB – (SNCF) – EBICAB–L 10 000 – (NSB, BV)

The transmission link between the track equipment and the train is made through beacons located at trackside signal locations where the data is collected.

The main functions achieved by the system are:

- to collect data from the signalling equipment and line description, then to transmit them through the beacon transmission unit;
- to collect, on board, the speed value from tachometers;
- to memorize all train description data;
- permanently to compute the speed supervision of the train in real time;
- to warn the train driver of overspeed;
- to activate emergency braking in an abnormal situation;
- to register additional data for operating purposes;
- to display data on the driver's desk.

At a maximum speed of 300km/h, eight messages can be transmitted from the track to the train. Four of them have to be read identically by the train for message validation. The message itself includes four codes of eight bits each. Amongst the eight bits, four are used for coding and the remaining four for validation of the code. Three codes are used for data and one for synchronization purposes.

Data received from the track equipment confirms:

- speed limit at the given signal;
- speed limit at the next signal;

Fig. 9.3 KVB – cab display equipment

- distance to go between the given signal and the next signal;
- equivalent profile (gradient) for the distance to go.

Data entry on board:

- speed limit authorized for the given train by 10km/h;
- train length by 100m;
- maximal deceleration on flat profile in m/sec^2;
- train category;
- real speed given by tachometer or inductive captors.

Regarding the operating rules, SNCF does not display speed information on the driver's desk, only stop or proceed. The driver relies on lineside signals only.

On the NSB and BV networks, the driver receives optical information from lineside signals as well as from the system on board.

The ATC system is considered to be fail-safe. The driver may accept the more elaborate speed information presented to him on the ATC panel.

The optical lineside signals give only limited speed information and are used as a fall back should the ATC system fail.

The service braking operation is under the driver's control on the SNCF network, with assistance being given by alarm when the speed exceeds the maximum allowed speed + 5km/h. So the driver has not to 'back' his driving on the system, this will only react in dangerous situations with emergency braking applied to stop the train.

The philosophy is different on BV network, where the driver may let the system operate the braking application.

In fact, the different rules are due to the physical differences between the two networks and also SNCF's desire to obtain quick benefits from the system by installing it in the most sensitive locations (temporary speed restriction, important junctions in terms of daily train number, high speed line – above 160km/h) with pre-warning equipment being removed.

The beacon is of the non-powered type, energy being received from the 27mHz signal permanently emitted by the locomotive. This is particularly interesting for temporary speed restrictions, since no power supply is necessary.

On board, the equipment consists of one set of computer hardware and two different sets of software with the output being compared before execution.

Datalogger and maintenance facilities are available.

On the BV network, when the on board equipment is out of order for any reason, the train speed is, in accordance with traffic rules, limited to 80km/h. On SNCF, if on board equipment is not in service, the driver has to limit the train speed to 160km/h even if the train is allowed to reach a higher speed.

9.3 Continuous systems

The installation of the continuous system is more expensive than the intermittent system due generally to the fact that they give an improved traffic capacity for the lines and offer such operational facilities as high speed or very high speed, regulation, energy savings as well as no lineside signals or reduced quantity of signals. They also offer the chance of full automatic train control.

The main functions remain the same, ie continuous trains speed supervision.

There are two categories:

- centralized systems: LZB (DB AG, ÖBB, RENFE)
- decentralized systems: TVM (SNCF), ATB (NS), BACC (FS)

9.3.1 Centralized systems

9.3.1.1 LinienZugBeeinflussung LZB – (ÖBB, DB AG, RENFE)
The main function of this vital system is to monitor trains continuously, routinely as well as in emergencies.

By a full duplex link with a line centre through an umbilical cable laid into the track, each train receives continuous train control commands.

The centralized line centre may monitor up to 16 loops of a maximum 12.7km length, ie about a maximum 100km of double track section line. This centre is connected to all interlockings, as well as to adjacent centres for in and out relationships between centres. Both links are permanent.

The entrance and exit of LZB section are fully automatic.

When entering a LZB section, trains transmit to the centre their specific characteristics such as braking performance, train length and category once and thereafter continuously repeat their localization, actual speed, braking characteristics and internal operational data.

The computer using available data, sends to every train, at 1 second intervals, a control command message which gives the length of the available braking distance and the localizations of braking initiation in respect of the braking capacity of the train. It also generates the target data for the driver – ie the distance to the nearest target and speed reduction allowed there. Operating information is also transmitted.

The on board computer, using the data, permanently calculates the maximum permissible speed, continuously controls the driver's indicators and monitors the train speed. This permits fully automatic train operation.

Another important feature is the ease with which line centre staff, through the operator terminals, can access the system to obtain temporary speed restriction, to query the system status and to receive information from trains.

For safety and availability reasons, the on board equipment includes three computers on the two out of three principle with duplication of tacho generators and receiver antennae. The line centre is also fitted with three computers on the two out of three principle.

This system is presently applied for speeds up to 300km/h, with mixed traffic automatic braking without driver's assessment, and reduction of lineside signals, with the acceptance of reduced capacity of the line for non-equipped trains. The reduction of lineside signals is only applied to new lines.

9.3.2 Decentralized systems

9.3.2.1 Transmission voie machine: track to train transmission TVM – (SNCF)

Although this system is connected to a central control centre, it is based on a decentralized architecture as far as the train control is concerned.

Used on very high speed lines, it is linked to the signalling status of the line, ie the track circuit layout and associated equipment. Transmission is achieved in one way – track to trains.

In other words, the train localization is given by the occupancy of a definite track circuit with a flat length of 1500m, the origin for which is the protected zone.

Fig. 9.4 TVM – on board equipment

Two versions of TVM are in service; TVM 300 on the Paris–Lyon line and TVM 430 on the Paris–Lille line. These are described in Chapter 10.

From centres, spaced every 12km, dual computers produce messages of 27 bits transmitted by the frequency modulated track circuit acting as the carrier of very low frequency signals of 0.88 up to 17.52Hz, bandwidth by steps of 0.64Hz.

The messages include speed limit at the end of the section, lower speed if any for the following section, distance of the block section, gradient and operational instructions.

The on board message is decoded several times per second and permanently supervises the actual speed of the train in respect of the braking curve calculated by the computer from data given by the message.

The odometer drift, if any, is corrected each time an electrical track circuit joint is crossed.

With the train being under the driver's control the system will actuate emergency braking if the speed overrides the permitted speed limit.

Ground and train borne equipment is duplicated for availability reasons. The fail-safe process is given by the coded monoprocessor method, which ensures a very high level of data coding whatever the structure of the computer hardware.

This system allows speeds of up to 350km/h on lines without lineside signals.

9.3.2.2 Automatische Trein Beïnvloeding ATB – (NS)

The ATB system is a fail-safe system which invokes an emergency braking up to a standstill in cases where overspeed is detected without braking activation from the driver within a few seconds. It is based on coded track circuit transmission to the engine of one out of five speed limits (40, 60, 80, 130 and 140km/h).

Since the system cannot distinguish between 40km/h and stop, ATB is mainly a protection system for speeds over 40km/h; up to this speed the responsibility remains with the driver who has to follow the instructions given by the lineside signals. On the other hand the 40km/h speed limit, which applies when no code is received, facilitates the running of trains on the basis of driving on sight in the event of signalling failure.

The driver is allowed to increase the speed before having seen the less restrictive aspect in the next lineside signal if the cab signal speed information changes to a higher value. In that case, however, the driver has to take into account that because of the limitation in available speed levels the cab signal may display a higher speed than the local maximum speed as shown by fixed lineside speed signs and signals. For example, a cab signal – 130km/h; lineside signalling – 90km/h. Therefore the lineside signalling is decisive whilst the cab signalling is there for support.

The driver must of course apply the brakes in order to decrease the speed if the cab signal speed information changes to a value lower than the actual train speed before having seen a more restrictive aspect in the next lineside signal, due perhaps to route cancelling.

Audible information is given when the speed information changes (gong) and when an overspeed is detected without braking activation from the driver (bell). A buzzer will sound every 20 seconds if the speed limit 40km/h is displayed and no brakes are applied.

In any event the driver has either to activate the brakes or to push the acknowledgement button in order to avoid an emergency braking.

Due to electromagnetic interference from powerful chopper locomotives and higher traffic levels, NS has decided to improve this system and to implement the improvements in three stages

- *Step 1*: More speed levels and braking curve supervision for functions, implementation of intermittent transmission (beacon) plus semi-continuous transmission (inductive loop) in order to eliminate track circuits later.
- *Step 2*: Advisable speed in terms of operational management instead of giving only the maximum allowed speed.
- *Step 3*: Transmission from the train to the track in order to eliminate track circuits, under a moving block system principle.

9.3.2.3 Coded track circuit automatic block BACC – (FS)

Designed as a headway system, perhaps more as cab signalling, this coded track circuit system is mainly a four code system on the network, extended up to nine codes for high speed. A speed control device has been added on locomotives able to achieve speeds higher than 160km/h.

Information relative to gradient and distance to go is not available.

The speed control system, as a separate item, ensures the comparison between actual speed and limit speed. When actual speed exceeds the limit speed, emergency braking is applied until the speed comes below the authorized speed, resetting of brake is under the control of the driver.

9.4 European train control system ETCS

As shown in the accompanying table (**Fig. 9.5**), existing train control systems on each of the 12 railways concerned have very different techniques and technologies, as well as very different driver operating rules. So it is hopeless, in a short term, to consider a unique European system. However, Europe is now a fact and rail travel across borders will increase more and more. As a result, it is imperative to study this problem on a European basis, starting from the existing infrastructure and evolution towards common standards. The European Union has already launched such programmes as Eurocab, Eurobalise and Euroradio leading to greater European understanding.

ETCS will be the result of multi-functional railway requirements and technical specifications by the industry.

The structure of ETCS should be open to a modular approach which presents the opportunity to obtain equipment at a lower price than that for separate systems. Space reduction and maintenance cost savings will be to the benefit of all railways.

The scope of the ETCS project is described as follows.

From the specifications, the different manufacturers concerned will be in a position to develop innovative solutions under the normal competition, synonymous with progress.

The main problems to overcome are:

- To develop a very powerful transmission system, full duplex, between trains and the operation centre. The transmission will not be limited to the existing needs but will be capable of extension beyond existing signalling problems.

  UIC has already decided to introduce the 900mHz global system for mobile communication (GSM) as a European railway system. This new system will also cover the needs of ETCS.

  Data transmission by beacon should remain a necessity in order to solve local problems, localization, or back up problems. A technical group called Eurobalise has to produce a specification.
- To develop an accurate and safe odometry system not only based on the train wheels with their slip-slide problems, but also able to give a very precise localization of the train. This is a very important matter as far as the introduction of the moving block concept is concerned.
- To specify how to determine braking characteristics and train length integrity safely.
- To determine what kind of interface driver system has to be specified taking into account the existing operating rules.
- To plan precisely the different steps to be followed in order to achieve a smooth upgrading from the existing situation to a full integrated European system.

This chapter represents a very short summary of what has to be done during the coming decades. It is an exciting programme and needs the real will of all the concerned participants, both railways and industry.

9.5 Automatic train protection specification ATPS – (BR)

BR currently rely on the AWS system and skilled train drivers. Like many European railways they face major problems which now render the introduction of more sophisticated equipment for safer trains mandatory.

BR has also recently undertaken two pilot schemes to study the application of ATP.

BR general aims are therefore conditioned by the following expectations and restraints:

- the system should supplement rather than replace existing fixed block signalling.
- the system should be employed on all BR's lines and BR's traction fleet.
- the system can be used selectively (sprinkling for risky areas).
- no modification of operational rules either for the fixed or on board equipment.
- the system should not reduce the existing train running performance.

In this respect, the specification should include:

- on board speed calculating system – ie no full duplex transmission required.
- intermittent system for cost effectiveness as the basis of the system with options, place to place, to be continuously linked with track to train communication, mainly in the vicinity of signals.

Basic requirements are:

- continuous supervision of train speed;
- monitoring of train braking with full service brake application and release when warning conditions do not apply, standstill not required;
- train running direction detection;
- train stops when passing a signal at danger;
- rolling away protection.

Driver's interface does not provide cab signalling information. The cab display indicates the most restrictive supervised speed constraint except when warning or intervention sequences are applied.

9.6 Present systems in Europe

Fig. 9.5 represents a summary of automatic train protection and control systems in current use on the 12 railway administrations covered by this book.

ÖBB
Austria – INDUSI – LZB

NMBS/SNCB
Belgium – Crocodile – TBL 1 – TBL 2

SBB
Switzerland – Signum – ZUB 121 SBB/BLS

DB AG
Germany – INDUSI – LZB

RENFE
Spain – ASFA – LZB

SNCF
France – Crocodile KVB – TVM 300 – TVM 430

BR
Great Britain – AWS

FS
Italy – BACC 1 – BACC 2

CFL
Luxembourg – Crocodile

NSB
Norway – EBICAB 2

NS
Netherlands – ATB

BV
Sweden – EBICAB 2 and L 10 000

*NB: This list does not tally in all cases with the table in **Fig. 9.5**.*

Fig. 9.5 Summary of automatic train protection and control systems in current use by European railway administrations

Railway	1 ÖBB	2 ÖBB	3 NMBS/SNCB	4 NMBS/SNCB	5 SBB	6 DB AG	7 DB AG
TYPE OF EQUIPMENT							
Manufacturer	Siemens	Sel Alcatel Siemens	–	ACEC	Integra	Sel Alcatel Siemens	Sel Alcatel Siemens
Trademark	INDUSI	LZB	Crocodile	TBL 1	Signum	INDUSI	LZB
Year of introduction	1963	1993	1930	1987	1934	1934	1965
BASIC FUNCTIONS							
Transmission wayside to train	Inductive Intermittent	Inductive Continuous	Brush contact Intermittent	Inductive Intermittent	Inductive Intermittent	Inductive Intermittent	Inductive Continuous
Info displayed to driver	Yes	Yes	Yes	Yes	Lamp	Yes	Yes
Acoustic warning to driver	Yes	Yes	Yes	Yes	Buzzer	Yes	Yes
Speed check	With tacho-generated actual speed	Comparison between actual speed and received info	No	No	Optional	Comparison with tacho-generated actual speed	Comparison between actual speed and received info
Acknowledgement by driver	Yes	Yes	Yes	Yes	Yes	Yes	Not required
Automatic braking	Yes	Yes	Yes	Yes	No	Yes	Yes
Data recording	Yes	Yes	Yes	Yes	Yes	Yes	Yes
Other information	–	Compatible with INDUSI	–	–	–	–	Compatible with INDUSI
PHYSICAL CHARACTERISTICS							
Physical layer	Tuned resonant circuit Carborne magnetic coil (2)	Loops of 300m long twisted every 100m 4 antennas	Iron bar 20V DC fed	Beacon	Track magnet	Tuned resonant circuit	Loops of 300m long twisted every 100m
Track to train transmission	500Hz–1000Hz –2000Hz	36kHz carrier 400Hz modulation 1200bds 83 bits telegram	(+) and (–) polarities	Inductive 100kHz FSK	Magnetic inductive	500–1000–2000Hz	36kHz carrier 400Hz modulation 1200bds 83 bits telegram
Train to track transmission	No	56kHz carrier 200Hz modulation 600bds 41 bits telegram	No	No	No	No 200Hz modulation	56kHz carrier 200Hz modulation 600bds 41 bits telegram
Power supply for wayside eqt	Not required	220 or 750V AC	Battery	AC + battery	Not required	Not required	220V or 750V AC
Power supply for carborne eqt	24V DC	24V DC	–	–	36V DC	24V–110V DC	24V DC
Wayside equipment data with dimensions (mm)	– 1258×198×170	1 box for 6 loops 754×225×855	– 5.2m long	–	– 990×270×220	– 1258×198×170	1 box for 6 loops 754×225×855
• position	35mm over rail level	–	90mm above rail level	–	45–50mm above rail level	35mm above rail level	–
• installation	–	–	1.31m long 15cm between track and beacon axles metallic brush	–	One magnet mid rails, one magnet 280mm out of rails	–	–

Railway	1 ÖBB	2 ÖBB	3 NMBS/SNCB	4 NMBS/SNCB	5 SBB	6 DB AG	7 DB AG
Carborne equipment data with dimensions	Cabinet	Rack 19"	–	–	Rack 19"	Cabinet	Rack 19"
Typical logic equipment	Relay–electronic	Microprocessor	Relay	Microprocessor	Analog–CMOS–relays	Relay–electronic and microprocessor	Microprocessor
Speed measurement	No	Train equipment	No	No	No	No	Train equipment
OPERATIONAL FEATURES							
Info transmission capacity	3 bits Stop warning Speed check	67 useful bits track to train 28 useful bits train to track	1 bit	5 decades	Stop warning	3 bits	Track to train: 67 useful bits Train to track: 28 useful bits
Max. allowed train speed km/h	160	300	–	160	160	160	300
Info from track to train	Stop warning Speed check	Braking curve Allowed speed Target distance Other info	Warning clear	Stop warning Clear shunting	Polarities	Stop warning Speed check	Data for braking curve, target distance and speed Add. info
Info from train to track	3 freq. cont. transmitted	Train location Brake characteristics Max. deceleration Actual speed	None	No	–	3 frequencies continuously transmitted	Train location, Brake characteristics Max. deceleration Actual speed
Info displayed to driver	Yellow lamp (1000Hz signal)	Allowed speed Target distance and speed Add. info	Lamps and sound	Lamps and sound	Red lamp: stop Yellow flashing: warning	Various lamps	Allowed speed Target distance and speed Add. info
Interface between wayside and signalling equipment	Control contacts	Serial/parallel interface via cable	Relay	–	Control of magnet in lamp proving circuitry	Control contacts	Serial/parallel interface via cable
SAFETY RELIABILITY							
Protection of information	None	Checked code hamming distance: 4	None	Checked code 2 out of 3 computers	Decoding only if info available in antivalent form	None	Checked code hamming distance: 4
Protection against wayside eqt failure	Regular check	2 out of 3 computers	None	Periodic test	Duplication of magnets	Regular check	2 out of 3 computers
Protection against carborne eqt failure	Speed restriction to 120km/h	2 out of 3 computers	None	Periodic test	Circuitry design and robust equipment	Fail-safe eqt – speed restriction to 100km/h	2 out of 3 computers
Protection against wayside removal	None	Absence of telegram: automatic braking	None	In lineside signal	–	None	No telegram received: automatic braking
EXISTING APPLICATIONS							
Extent	Main lines, some secondary lines	Western Austria	All network	Partly	SBB network and private railways	All traction units	–
Quantity of carborne equipment	1200	25	1000	150	1500	18 000km main lines	370
Quantity of wayside equipment	9400	50km double track	4755	2000	15 000	4000km branch lines	1200km double track

AUTOMATIC TRAIN PROTECTION AND CONTROL

Railway	8 RENFE	9 SNCF	10 SNCF/ EUROTUNNEL	11 SNCF	12 SNCF	13 BR	14 FS
TYPE OF EQUIPMENT							
Manufacturer	Dimetronic	–	CSEE	CSEE	GEC Alsthom	–	SASIB
Trademark	ASFA	Crocodile	TVM 430	TVM 300	KVB	AWS	–
Year of introduction	1972	–	1993	1981	1989	1956	1985
BASIC FUNCTIONS							
Transmission wayside to train	Inductive Intermittent	Brush contact Intermittent	Inductive Continuous	Inductive Continuous	Inductive Intermittent	Inductive Intermittent	Inductive Continuous
Info displayed to driver	Lamp	Lamp	Yes	Yes	Lamp and digital display	Warning dial	Target speed lamp
Acoustic warning to driver	Bell	Yes	Yes	Yes	Buzzer	Short tone bell Continuous horn	Buzzer
Speed check	Yes Previous beacon with red aspect	No	Comparison between actual speed and authorized speed	Comparison between actual speed and received information	Yes	No	Yes
Acknowledgement by driver	Yes	Yes	Yes	Yes	Yes	Yes	Yes
Automatic braking	(1)	Yes	Yes	Yes	Yes	Yes	Yes
Data recording	Yes	Yes	Equipment of train	Equipment of train	Yes	No	Yes
Other information	–	–	Compatible with TVM 300	–	–	–	–
PHYSICAL CHARACTERISTICS							
Physical layer	Beacon	Trackside iron bar fed by 20V DC	Track circuit	Track circuit	Beacon	Track magnet	Track circuit
Track to train transmission	Inductive frequency between 60 and 111kHz	(+) and (−) polarities	1700–2000–2300–2600Hz (27 different simultaneous modulations)	1700–2000–2300–2600Hz (18 different modulations)	Inductive	Magnetic induction	2 frequency carriers 50 and 178Hz modulated
Train to track transmission	No	No	No	No	No	No	No
Power supply for wayside eqt	Not required	–	24V DC	24V DC	Not required	12–24V DC	150V AC
Power supply for carborne eqt	24V DC	–	72V DC	72V DC	12V DC	–	24V or 110V DC 77/150V DC
Wayside equipment data with dimensions (mm)	LC resonant circuit 600×250×104	5.2m long	–	–	–	1570mm long	ZI bonds 370×615×355
• position	40mm over rail level	axis of track 90mm above rail level	–	–	–	axis of track	–
• installation	254mm between track and beacon axles	–	–	–	–	165mm from track magnet	–

(1) In case of no acknowledgement from the driver if the overspeed overcomes a fixed value

AUTOMATIC TRAIN PROTECTION AND CONTROL 279

Railway	8 RENFE	9 SNCF	10 SNCF/ EUROTUNNEL	11 SNCF	12 SNCF	13 BR	14 FS
Carborne equipment data with dimensions (mm)	–	–	Racks	Racks	–	–	525×440×286
Typical logic equipment	Electronic	Relay	Microprocessor	Electronic	Microprocessor	Relay magnets	Microprocessor
Speed measurement	No	No	Train equipment	Train equipment	–	No	Tacho generator, toothed wheel and magnetic transducer
OPERATIONAL FEATURES							
Info transmission capacity	1 bit	1 bit	27 useful bits	18 signalling information	4 × 8 bits telegram	2 bits	9 active codes
Max. allowed train speed km/h	220	220	350	300	300	300	255
Info from track to train	Signal aspects and speed check	Clear warning	Data for braking curve and display (distance, gradient, speed)	18 speed values	Speed allowed Target distance and speed Gradient	Clear warning	9 speed values
Info from train to track	None	None	None	None	None	No	No
Info displayed to driver	Stop – caution alarm and service	None	Speeds in a visual display unit (authorized and target speeds)	Allowed and target speed –lamps	Warning dial	10 signal aspects Target speed value	
Interface between wayside and signalling eqt	Relay	Relay	Microprocessor encoders	Relays	Serial/parallel via beacon	Relay or SSI module output	Track circuit
SAFETY RELIABILITY							
Protection of information	Hamming distance: 2	None	Checked code Hamming distance: 4	Fail-safe filtering of the signals	Hamming distance: 4	None	Checked code redundancy No coded current transmission
Protection against wayside eqt failure	Automatic train braking	None	Coded microprocessors & complete redundancy	Fail-safe design and redundancy	Beacon duplication	Permanent magnet	Dual software
Protection against carborne eqt failure	Automatic train braking	None	Coded microprocessors & complete redundancy	Absence of information	Dual software	Robust design	Dual software Dual microprocessor
Protection against wayside removal	None	None	–	–	Info link between beacons	None	No coded current transmission
EXISTING APPLICATIONS							
Extent	90% of network	All traction units Main network	New TGV lines	South–East and Atlantic TGV lines	Main lines	98% appropriate routes	–
Quantity of carborne equipment	2667	–	300	400	4000	6000	153
Quantity of wayside equipment	20 000	–	600km double track	650km double track	6000 signals	20 000km	300km double track

Railway	15 FS	16 CFL	17 NSB	18 NS	19 BV
TYPE OF EQUIPMENT					
Manufacturer	Ansaldo SASIB	–	ABB Signal AB	GRS	ABB Signal – ATSS
Trademark	–	Crocodile	Ebicab	ATB	Ebicab – L 10 000
Year of introduction	1970	1973	1979	1966	1978
BASIC FUNCTIONS					
Transmission wayside to train	Inductive Continuous	Brush contact Intermittent	Inductive Intermittent	Inductive Continuous	Inductive Intermittent
Info displayed to driver	Lamp	No	Lamp – digital display	Allowed speed	Lamp digital display
Acoustic warning to driver	Buzzer	Yes	Buzzer	Gong, bell, buzzer	Buzzer
Speed check	Yes (external system)	No speed and	Comparison between actual speed and received information	Comparison between actual speed and received information	Comparison between actual received information
Acknowledgement by driver	Yes	Yes	No	40km/h only	No
Automatic braking	Yes	Yes	Yes	Emergency only	Yes
Data recording	Yes	Yes	Yes	No	Yes
Other information	–	–	Ebicab and L10 000 are compatible	–	Ebicab and L10 000 are compatible
PHYSICAL CHARACTERISTICS					
Physical layer	Track circuit	Trackside iron bar fed by 12–18V DC	Beacon	Track circuit	Beacon
Track to train transmission	1 frequency carrier 50Hz	(+) and (–) polarities	Inductive	75Hz carrier A.M.	Inductive
Train to track transmission	No	No	No	No	(2)
Power supply for wayside eqt	150V AC	–	Not required	Not required	Not required
Power supply for carborne eqt	24V or 110V DC	24–72V DC	24–72V DC	110V	24–72V DC
Wayside equipment data with dimensions (mm)	ZI bonds 370×615×355	5.2m long	540×400×38	Diverse –	540×400×38
• position	–	Axis of track	Axis of track	–	Axis of track
• installation	–	90mm above rail level	40mm above sleeper top	–	40mm above sleeper top

(2) Train class information only for level crossings operation

Railway	15 FS	16 CFL	17 NSB	18 NS	19 BV
Carborne equipment data with dimensions (mm)	– 525×440×286	Metallic brush –	Cabinet 400×380×270	Cubicle –	1 rack 340×500×350
Typical logic equipment	Relay –electronic	Relay	Microprocessor	Relay– microprocessor	Microprocessor
Speed measurement	No	No	Tacho generator	Tacho generator	Tacho generator
OPERATIONAL FEATURES					
Info transmission capacity	4 active codes	1 bit	4 × 8 bits telegram	5 speed codes	4 × 8 bits per transponder
Max. allowed train speed km/h	180	–	300	140	300
Info from track to train	4 speed values	Clear warning	Speed aspect signals	Max. allowed speed	Signalled speeds, restriction, target distance, gradient
Info from train to track	No	None	None	–	None
Info displayed to driver	5 signal aspects	None	Allowed speed and target speed below 70km/h – lamps	Max. allowed speed	Allowed and target speed, lamps
Interface between wayside and signalling eqt	Track circuit	Relay	Encoder	Track circuit	Encoder
SAFETY RELIABILITY					
Protection of information	Checked code redundancy	None	Hamming distance: 4	Checked code	Hamming distance: 4
Protection against wayside eqt failure	No coded current transmission	None	Beacon duplication	No coded current transmission	Beacon duplication
Protection against carborne eqt failure	–	None	Dual software	Fail-safe design	Dual software
Protection against wayside removal	No coded current transmission	None	Info links between beacons	No coded current transmission	Info link between beacons
EXISTING APPLICATIONS					
Extent	–	All network	50% of network	70% of network	60% of network
Quantity of carborne equipment	3000	120	350 3050km	1000	1020
Quantity of wayside equipment	3812km double track	232	7000 beacons	1746km double track	6270km line

10 High Speed Line Signalling System

J. PORÉ

Contents

10.1 Basic principles 283
    10.1.1 Traffic features 283
    10.1.2 High speed line signalling: on sections with speeds between 160 and 250km/h 285
    10.1.3 New very high speed lines (speeds ≥ 250km/h) 286
    10.1.4 Functional organization at different levels on very high speed lines 287

10.2 General operating modes on very high speed lines – speeds ≥ 250km/h 290
    10.2.1 Operating mode on the high speed lines 290
    10.2.2 Safety and availability aspects 294
    10.2.3 Maintenance 294

10.3 General signalling system description on very high speed lines: speeds ≥ 250km/h 295
    10.3.1 SNCF track to train transmission systems TVM 295
    10.3.2 DB AG speed control system LZB 297
    10.3.3 RENFE speed control system LZB 298
    10.3.4 FS speed control system 298
    10.3.5 Main technology used on high speed lines 298

Illustrations

Fig.10.1   *Wayside general architecture for SNCF high speed lines*   286
Fig.10.2   *LZB indications for cab signalling*   288
Fig.10.3   *Lille PAR*   289
Fig.10.4   *Marker on TGV line*   290
Fig.10.5   *TVM 300 and 430 curves*   291
Fig.10.6   *LZB indications for cab signalling*   293
Fig.10.7   *TVM cab signalling*   296
Fig.10.8   *LZB equipped AVE cab equipment*   299
Fig.10.9   *Main characteristics of very high speed and high speed line signalling systems*   302

10.1 Basic principles

10.1.1 Traffic features

In the 1960s several European countries considered that, on suitably upgraded railway lines, higher and higher speeds would be possible. In Germany, some suitable lines were upgraded to ABS (*Ausbaustrecken*) and, after carrying out substantial improvements, speeds of up to 200km/h were achieved.

Another example is the Great Western Main Line (GWML), from London to Bristol and South Wales in Great Britain, where, thanks to the wide gauge of the Brunel railway, high speed diesel trains have been able to run at speeds of up to 200km/h on the Intercity 125 service. This service has now been in operation for some 20 years.

In France, SNCF developed their red liveried flagship 'Le Capitole', which, from 1968, has linked Paris to Toulouse (750km) in just under six hours, reaching 200km/h along a greater part of the route. This experiment was followed shortly afterwards by 'l'Étendard' and 'l'Aquitaine', linking Paris and Bordeaux (650km) in just under four hours, thanks to an improved profile of the track, together with a maximum speed of 200km/h.

These developments, however noteworthy, were certainly not considerable enough in the long term, for either commercial or traffic purposes.

In several countries, especially France, Germany and Italy, the saturation of railway lines and other factors such as competition from road traffic, led railway administrations to look at newly built high speed lines, superimposed on the existing network. But from the very start, the philosophy and political aspects which were developed in the three countries by their railway administrations SNCF, DB AG and FS, were fundamentally different.

The SNCF high speed new lines network emerged as a result of the saturation of the Paris–Lyon line where, as traffic flow increased, quality of service declined. This prompted SNCF to promote the design of a new, fully separated line. The main objective was to divert the passenger flow onto a dedicated line in order to reduce the journey time by a significant amount.

The main features were:

- homogeneous traffic, using only dedicated electrical multiple units, powered at 25kV–50Hz, running at 270km/h and fully compatible with existing lines;
- a straight layout to reduce the journey and the civil work costs, with gradient up to 35‰;
- centralized control and monitoring of the traffic.

The first such line, commissioned in 1981, was genuinely successful and permitted a return on the SNCF investment of more than 10% per year, which was better than expected. Currently, about 120 train sets are in use at peak periods. Another major advantage of the TGV (high speed train) is its ability to perform a great number of journeys utilizing existing lines – for instance, Paris–Geneva and Paris–Chambéry.

A second high speed line came into revenue service in September 1989, bringing with it substantial improvements to passenger services to west and south-west France – the Atlantic line. The main features are identical to those of the earlier model except for the speed, which has been increased to 300km/h, with the headway cut down to 4 minutes. The train fleet consists of about 100 train sets.

More recently, SNCF has been engaged in construction works for the northern high speed line which will link Paris to Brussels and London via the Channel Tunnel in 1994. As with the western line, the initial speed will be 300km/h but with a provisional increase to 320km/h and a headway of 3 minutes. Moreover a loop line around the eastern outskirts of Paris will link the northern and southern lines, and the south-east line will be extended southwards, initially as far as Valence, on the River Rhône.

In parallel with the design of very high speed lines, SNCF has not neglected the improvement of traffic conditions on the existing network, with line sections authorized up to 200 or even 220km/h.

Where France or indeed Great Britain are both very much centred around the capital city, Germany on the other hand, has centres of population distributed more or less equally throughout. These centres are linked by railways, the majority of which were built during a time when Germany consisted of dozens of independent states, each jealously guarding its borders. The main arteries of traffic followed an easy countryside around borders to avoid substantial structures and neighbours' consent. After the Second World War, traffic and traffic patterns changed in type, volume and direction, all aggravated by road competition creaming off highly rated goods, speeding on newly built, generously designed highways, at little direct cost. The German railway administration Deutsche Bundesbahn (DB AG) decided to look into high speed railway lines in order to win back what had been lost.

In addition to the upgraded lines, where maximum speeds of 200km/h are possible, where existing lines were too circuitous and/or lacked potential for improvement, completely new high speed lines or NBS ('*Neubaustrecken*') were to be built to provide a running speed of up to 280km/h.

Any of these lines, however, had to be designed to carry conventional trains as well as special high speed trains, both for passengers and freight. The philosophy was, and still is, to create mutual benefit by the integration of all such lines into the existing network.

The Italian railways administration, FS, from the very start had an approach similar to that of DB AG, as a result of having several large cities or city centres in the two northern thirds of the country. Since then, the design and building of a comprehensive high speed network has involved many interconnections with the existing lines. One of the major reasons for this was the existing natural crossings. Indeed, the existing network, due to the hilly terrain, was not so direct, involving many curves, bridges and tunnels. The new lines, designed of course to be much straighter, were to cross the existing lines naturally.

From the start it was intended to include mixed traffic on the high speed lines – high speed passenger trains as well as standard passenger traffic and freight trains.

Following the lead of SNCF, DB AG and FS, several other European administrations considered introducing high speed lines, that is purpose built lines operating at very high speeds of 250km/h or more. However, to the present day, only Spain (with RENFE's Madrid to Seville line, opened in early 1992) has joined the group of countries with *very high speed lines*, with speeds equal to or higher than 250km/h and operating in revenue service.

Several other countries, such as Great Britain and Sweden, operate high speeds of 200, 220 and 230km/h, and Austria and Switzerland hope to achieve these speeds in the next few years. Some European railway administrations have plans to build very high speed lines or to upgrade existing lines for high speed and Belgium has a long-term project, which should start in 1996. In Great Britain, projects are also under discussion for very high speed lines.

In order not to confuse the reader, this chapter will only cover very high speed lines that are purpose built and can cope with speeds of 250km/h or more. In fact, according to UIC standards, lines with speeds of up to 220km/h are termed 'high speed' – above that, they are known as 'very high speed'.

Henceforward, references to lines in France, Germany, Italy and Spain will only be very high speed lines.

It is, however, worthwhile to indicate some of the features involved with high speed lines when these are upgraded sections of existing networks. These lines involve speeds higher than 160km/h but, generally speaking, lower than 250km/h.

The table in **Fig.10.8** shows the main characteristics of the very high speed lines existing networks, and also gives the main data for upgraded high speed lines in France, Germany, Italy and Spain, and in other countries as well.

The very high speed lines chief data, in France, Germany, Italy and Spain, are presented in the main characteristics table (see **Fig.10.8**), and amongst the most important features to point out are that:

- the maximum operating speeds range from 250 to 300km/h or more, with the theoretical minimum headways ranging from 5 down to 3 minutes;
- depending on the type of traffic, lineside signalling does not exist at all in France, except for very few shunting signals, or is only present because freight or nominal passenger trains run along the same lines;
- the line geography also depends much on the dedicated types of trains in France and Spain today, perhaps on the future DB AG line between Frankfurt and Köln. It may not apply in either Germany and Italy. Thus, the curve radius, the maximum gradients, and the maximum ton/axle weight, are much lighter in France and Spain (17 tons), than in Germany or Italy.

There are also historical reasons for differing choices about existing lines compatibility. Indeed, Spain's choice for the standard gauge for its high speed network means that today there is no possibility of interconnecting its high speed trains with the rest of RENFE's network.

On the other hand, France, Germany and Italy often run their high speed trains elsewhere. In these cases, generally speaking, high speed trains are able to run at up to 200km/h or even up to 220km/h on some sections of lines, on the upgraded sections of the main network.

10.1.2 High speed line signalling: on sections with speeds between 160 and 250km/h

Building a 100% new line is not only very costly but may also be difficult or indeed impossible for many reasons; for example:

- availability of space in built up areas;
- geographical conditions, for example in mountainous areas, etc.

A good compromise, perhaps, is only to adapt the existing lines to higher speeds by improving the tracks and points areas, rectifying sharp curves, and other such improvements. This develops the concept of upgraded lines where high speeds of between 160 and 250km/h can be achieved.

France, Germany, Italy and Spain have not only built very high speed lines but have also continuously improved their existing networks and have proposed higher speeds.

Other countries have also improved their lines to operate at speeds higher than normal. These countries include Austria, Sweden, Switzerland and Great Britain. Belgium and the Netherlands will jointly carry out similar projects during the next few years.

There now follows an analysis of some aspects of how high speeds have been achieved effectively on the main networks.

Firstly and dependent on the country, existing lineside signalling was generally kept as a basic system, but improved by adding another aspect – say from three to four or from four to five aspect signalling, as in France, Switzerland and Great Britain. The maximum allowed speed, with the improved lineside signalling, was increased up to 200–220km/h.

On the other hand, in Sweden, it is mandatory to have an automatic train control system in service for speeds higher than 80km/h.

Usually, independent of lineside signals and without any kind of ATP in operation, the maximum line speed is 160km/h in Belgium, France and Switzerland, and 100km/h in Germany. In Great Britain the maximum speed is 200km/h. However, ÖBB (Austria) and DB AG (Germany) use INDUSI as an intermittent train protection system, for speeds of up to 160km/h.

Another effect of high speeds on the normal networks is the choice of ATP, and whether it has to be of the intermittent or continuous type. For further details regarding the various standard automatic train protection systems in service in Europe today, see Chapter 9.

Intermittent ATP systems are in service or being installed in:

- Belgium (NMBS/SNCB): TBL 2 system for future applications
- France (SNCF): KVB system
- Sweden (BV): ATC2 system

An LZB continuous system has been chosen by DB AG in Germany, and in Switzerland (SBB) the ZUB system is planned.

In Great Britain (BR), except for pilot ATP systems being tested on the Chiltern Lines (SELCAB) and the Great Western Main Line (TBL) no ATP is in revenue service on the normal network.

As far as data capacity is concerned, the various systems, either intermittent or continuous, tend to have a rather large capacity, since they have to include a target speed, a target distance, the track gradient ahead, and other miscellaneous information. The capacity per information point seems to range from about 60 to 110 useful data.

Another interesting characteristic of high speed lines includes the total or progressive withdrawal of level crossings. Level crossings are not allowed on high speed lines in Austria, Belgium, France, Germany and Switzerland. In Sweden, they are accepted for speeds of up to 200km/h – all crossings are automatic, regardless of train speed. In Great Britain, level crossings are automatic up to 100km/h. They are manned, including close circuit TV for control up to 200km/h.

Depending upon the railway administration and the speed, there may be a requirement for a second driver to be present in the cab. This is more a regulation concern, directly connected with railway rules and procedures, than a requirement for technical reasons.

In Belgium, France, Germany, Spain, Sweden and Switzerland only one driver is present in the cab.

In Great Britain, the rule is one driver up to 175km/h and two drivers above that speed.

As far as tilting trains are concerned, their technological principle seems to be used far more to improve travel times on low traffic/low speed lines, than to increase speeds on high speed lines. In

Austria, Belgium, France, Spain, Switzerland and Great Britain, there are no tilting trains in service or being planned. In Germany, tilting trains are already in service, but not on high speed lines. In Sweden, tilting trains run on upgraded lines but for speeds higher than 160km/h, tilting trains are also allowed to proceed with extended signalling, the on board signalling data given to the driver by the ATP system being even more permissive than the lineside signalling information.

10.1.3 New very high speed lines (speeds ≥ 250km/h)
For France, Germany, Italy and Spain, the main data is summarized in **Fig.10.8** at the end of the chapter. What follows is additional to that.

Since mid-1993, about 2500km of new very high speed lines are in revenue service in France, Germany, Italy and Spain.

For practical and comprehensive reasons, all networks have considered bidirectional running on all of their lines. Passing lines, or more simple short passing loops exist in each country, but because of the design and individual railway philosophy, their frequency varies greatly: from a junction between both parallel tracks each 7 to 8km in Germany up to 20 to 25km elsewhere.

It has been suggested by railway personnel that loops and junctions should only be used when absolutely necessary. In Sweden, however, a country using upgraded existing lines for high speeds, the junctions are in frequent use. For example, they are used when routing a slow freight train over to the other track to make a free way for a high speed train. This feature also applies to countries with upgraded lines in high density traffic areas.

Signalling layouts also vary, and by way of example two signalling applications for France and Germany are analysed.

France – SNCF
In order to be economically viable, signalling and control systems should provide a smooth flow of traffic with a minimum of constraints, even in disturbance conditions. In the same way, high availability and consequently easy maintenance must also be achieved for these systems. To reach these goals, a centralized modular architecture has been adopted.

In normal operation, control and supervision of switches, routes and train sets are managed from a single control centre, one for each high speed line. This centre is called a PAR (*Poste d'Aiguillage et de Régulation*) – centralized traffic centre. The centre provides the following functions for traffic management purposes:

- automatic route setting with respect to train identification;
- general monitoring of data related traffic as well as the traction power supply;
- train radio communications.

All interlocking functions are achieved locally, the PAR involving no interlocking itself. In backup mode, local operation is possible from each signalling centre, located every 20 to 25km.

Ground equipment is grouped and controlled either from a signalling centre or from an intermediate centre. These centres are spaced at 10 to 14km intervals and gather all interlocking and wayside cab signal equipment data. In fact, very few active components remain at the trackside. This architecture, combined with hot redundancy for important devices, improves the availability and makes for easier maintenance.

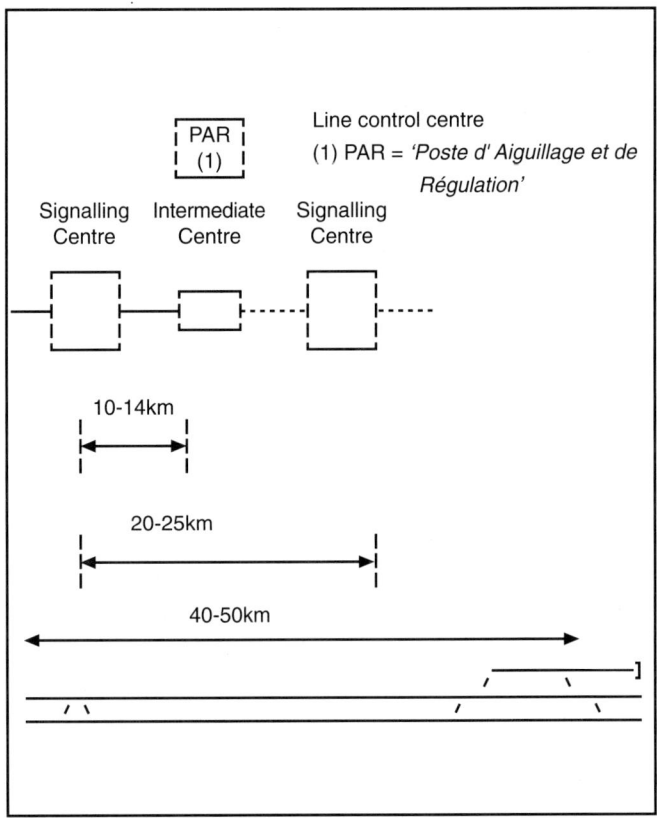

Fig. 10.1 Wayside general architecture for SNCF high speed lines

The wayside general architecture illustrated in **Fig.10.1** shows three levels: the central level, the local level and the track level. Standard communication is used to transmit data between the line control centre PAR and the wayside, signalling and intermediate centres, and trains. Three communication systems are used: a copper and/or fibre optic cable system based on a ring arrangement to prevent cut effect. One cable runs parallel with the high speed line, a second way is made through a cable via the classical line, a radio data transmission from wayside to train and a cellular radio (Radiocom 2000) from the on board phone, and connected to the national network.

Germany – DB AG

In Germany, control centres are situated at major stations or other strategic locations and control their own local equipment, together with the connected remote field interlockings.

A typical control centre handles 75km of line with seven field interlockings, controlling a total of 87 points, 77 main signals, 62 distant signals and 242 track sections.

The centralized electronic interlockings control the field stations via an extended bus for interchange of vital data. The line diagram, only showing the most important information, is depicted on an indication or mimic panel. From each of the major control centres, all passing loops and junctions in the area are fully controlled.

This, then, is a very different approach than the one adopted by SNCF, or indeed by FS in Italy and RENFE in Spain. In Germany, a control centre manages up to about 75km at a maximum, whereas in the other three countries, the whole line is managed from a single location with various local or decentralized signalling centres spread along the line.

10.1.4 Functional organization at different levels on very high speed lines

The high speed line systems of each country include different subsystems whose data is managed by the supervision system. However, at a higher level, each high speed line network in Europe has been designed with a different philosophy.

In France, to achieve an economic operation, a centralized architecture has been adopted. For example, the 278km Atlantic line is controlled under normal conditions by two people, one for the traffic and the other for the power supply control.

Nevertheless, at peak hours or in case of disruptions to traffic, dual manning is possible.

From the driver's point of view, two important factors have guided SNCF in its choice of high speed line driving aids, and, consequently, high speed line signalling principles:

- the higher the speed, the more difficult it is for drivers to perceive lineside signals properly;
- despite the railway's excellent safety record, human beings are more prone to failure than automatic devices.

As a result of these two factors, the following principles were adopted during research for new driving and signalling systems:

- a cab signal type signalling system had to be designed so that safety of very high speed trains would not be dependent on correct observation of lineside signals.
- trains are manually operated; drivers controlling acceleration, deceleration, coasting and braking; but a speed monitoring system has to be mounted to check train running in relation to the signals displayed and, if necessary, has to trigger emergency brake application.

This philosophy was decided by the SNCF management in order to keep drivers alert and in the best operational condition in case of emergency. These principles introduce the major differences between the functional needs of conventional lines and those of high speed lines.

On the other hand, in Germany, with the speed control system LZB, all LZB equipped leading vehicles are fitted with displays in the cab which indicate the actually permitted target speed and also the target distance. Within the displayed limits the driver can control the train either manually or let it run under automatic operation. It should be pointed out that lineside signals are only relevant for non-LZB trains.

This presents a fundamental difference of principle. Whereas with SNCF's speed control system TVM, the driver is always fully controlling his train's movement, with LZB, both in Germany and Spain, the control of the train may well be completely automatic as selected by the driver. In Italy, as in France, the train is more or less only under the driver's control as far as traction and service braking is concerned.

As far as the line control is concerned, it was

shown in Chapter 1 (**1.2**) that the German philosophy differs enormously from the others because of the fact that for each 50 to 75km, there is a line control centre.

Fig. 10.2. LZB indications for cab signalling

For DB AG, three levels for line management have been chosen:

- the vital operative level at field stations;
- control centres;
- dispatching level.

An unmanned individual field station interlocking is provided at each station, junction or crossover. It consists either of a remote controlled all relay hardware, or of equipment as an extended part of an electronic control centre. At stations, all through tracks are safeguarded by hard flank protection. All trackside elements are wire controlled.

In the control centres, in main stations or major junctions, all adjacent station and local equipment data is centralized. An indication mimic panel and its control computers manage the control centre process.

In front of this panel, two operators work at desks, each equipped with a keyboard, an input monitor and two colour VDUs. Each colour VDU can display any one of eight detailed stations layouts, which together cover all controlled interlockings. The display is safe, since all controls – particularly those bypassing safety levels – are executed by the operator on the basis of the displayed status. The control centre also houses the train describer system (TD) for the line, from which records and train routing information is derived. In addition, the TD system provides data to the dispatching level for train supervision. Printers are provided to log and record all events relevant for documentation, fault finding and maintenance. The operators normally supervise automatic operation of the line; however, they can override automatic operation at any time.

DB AG's third level consists of dispatching centres. They are situated at DB AG's regional headquarters. They receive data directly from the associated control centres at computer level. The task is to regulate traffic, co-ordinate connecting trains, ensure quality of service and provide information for staff and passengers, particularly in case of irregularities, and to provide both historic and realtime management information.

DB AG's interlocking is by relay or microcomputer based systems.

Each crossover, junction or station has its own remote or locally controlled interlocking for the home and starting signals with associated distant signals 1.5km in the rear. Simultaneously, each interlocking is the basis from which the LZB line centre derives all actual status information. Use is made of automatic route setting where only one route is possible, or of the train describer system based automatic train routing where more than one route is possible.

Interlockings on DB AG high speed lines also have to perform functions in addition to those of standard interlockings. They have to provide information for the related LZB line centres and they have to handle the particular functions associated with block sections described below and the starting signals covering these sections.

Track sections between stations are fully bidirectional, but have no intermediate block signals for economic reasons. They are replaced by marker boards (fictive block signals) which subdivide the line into block sections of between 1.5 and 2.5km length, each consisting of one or more track circuits.

The SNCF philosophy is rather different.

Each high speed line is controlled from a line control centre which fulfils the following functions for the whole line:

- automatic route setting;
- traffic control;
- remote control of the substations supplying power to the overhead lines.

These functions are housed together in the same room for the Paris–South-East line or for the Atlantic high speed line. On the North TGV line, signalling and power supply functions are controlled from two adjacent rooms.

Route setting is automatically controlled using a computer control module of similar design to that used in the SNCF's computerized signalboxes described in Chapter 5.

Fig. 10.3 Lille PAR

The traffic controller includes:

- a wall mounted mimic panel displaying the train describer, general monitoring data warning systems, hot box alarm, radio and various signalling indications signal status, track occupancy to be sure that everything is working well.
- four colour VDUs with their associated keyboard, interfaced with the computerized control and monitoring systems. The VDUs are able to display a magnified image of each station including route programming and emergency detailed data.

This way the traffic controller knows the exact situation of the route set, the train position, the train schedule discrepancies as well as hot box detection data or bridge monitor data which will involve automatic train stopping.

The traffic controller also has a telephone unit with a keyboard and VDU, displaying ground to train radio information and passenger commercial data and statistics processing indications. These functions are based on the use of the SNCF standard modules developed for the general network and described in Chapter 11.

As far as signalling equipment is designed, the different SNCF high speed lines involve slightly different solutions.

The signalling definition on the South-East and Atlantic lines are very much alike, the only difference being that on the Atlantic line the addition of one block section allows for a maximum speed of 300km/h instead of 270km/h. Thus, the speed steps selected for the deceleration phase are 270, 220, 160 and 0km/h and the minimal train spacing without any disturbance is seven block sections, taking into account the block section going free by the former train set. As lines are designated for bidirectional running, as soon as a train set has crossed a track crossover at 160km/h and can resume its normal running speed to 300km/h, train spacing remains the same, irrespective of the direction of running.

There is in France a specific mixture of high speed line and standard lineside signallings. It has indeed proved useful for high speed and conventional trains to proceed together on a section of line (for example Tours bypass, about 20km). Consequently, lineside signals identical to those used with normal automatic block signalling have been placed along the track for conventional trains running at up to 200km/h. For their part, high speed train sets receive signalling data, displayed to the driver's cab, in the same way as the one used on the high speed line. Block section layouts are the same with cab and lineside signalling.

The signalling definition on the North high speed line improves on this to achieve a 3 minute headway. Three different measures have been taken to effect this objective.

Firstly, the deceleration speed graduation has been adjusted with the 300, 270, 230, 170 and 0km/h sequence, providing a better sharing of the energy to be dissipated in each block section.

Secondly, the stepped speed control curve has been replaced by a smoother, more elaborate curve, more closely in line with the pattern normally followed by the drivers themselves.

Lastly, a specific indication in the driver's cab has been added in case the next block ahead presents a more restrictive speed order, so that drivers can start to brake in advance. This additional indication consists of the flashing of the speed display. The result is that block lengths can be reduced to 1500m on the flat, instead of 2100m on the South East line and 2000m on the Atlantic line.

It should also be noted that the SNCF driver's cab signalling display, designed in a fail-safe way, is complemented by lineside signs (markers) which indicate where speed orders have to be respected.

The continuous speed monitoring on SNCF high speed lines acts as a backup for the driver. If the train speed is higher than the maximum speed limit, an emergency brake is triggered. The emergency brake can be cancelled by the driver when the train speed becomes lower than the maximum authorized speed. Approaching the stopping point, the control curve is automatically released to a threshold value of 35km/h in front of the marker and maintained on the overlap block section.

Fig. 10.4 Marker on TGV line

Moreover, the new train sets for the North TGV line (called TGV–R or *réseau* – network) will also be able to run on the South-East and Atlantic lines. The train borne equipment is fully compatible with all other lines and can process the other types of TVM systems.

Italy's high speed lines are also centralized. On FS high speed lines, four main levels of line management exist:

- the network central supervision post controls the whole network;
- there are line central supervision posts for each line;
- satellite supervision posts, for each line section, up to 250km, also include vital functional management.
- there is a remote controlled peripheral post every 12km, which also handles vital signalling functions.

Solid state vital signalling provides local control, point operation, jointless audio frequency track circuit control, and continuous and intermittent automatic block function.

Finally, Spain's Madrid to Seville high speed line also has a single traffic control centre. The entire line can be controlled from this single centre, through the transmission network. The line has nine electronic interlockings (ENCE), eight of which (not the main one in Madrid–Atocha) manage 28 technical buildings, and each ENCE manages approximately 60km of double track.

The high speed line signalling is processed through the automatic train driving system CAT which, in Spain, is the equivalent of the German LZB, or continuous automatic train control. Each ENCE includes the CAT or LZB central computer corresponding to its zone. Therefore, there are eight LZB centres along the whole route.

10.2 General operating modes on very high speed lines – speeds ≥ 250km/h

10.2.1 Operating mode on the high speed lines

This section deals with the different operating modes and the detailed functions implemented along with the cab signalling system for all very high speed lines.

For clarity's sake the operating modes of SNCF and DB AG are discussed sequentially and illustrated with several speed control curves, which are a good indication of how the trains are controlled.

France – SNCF
SNCF speed control is based on the TVM system, which stands for *transmission voie machine* and means track to train transmission. The technical aspects regarding train detection, transmissions and both the wayside and trainborne organization and characteristics are further described in Chapter 2.

Several sequences of operating modes are illustrated in **Fig.10.5**.

Fig.10.5 shows a stopping sequence as used on the Atlantic line, together with the indications displayed in the driver's cab and the speed control curve. Crossing points and junction points are proceeded at 160km/h.

The second diagram in **Fig.10.5** shows a stopping sequence as used on the North line. Flashing indications down to 170km/h are displayed to show that a more restrictive speed has to be enforced on the next block section. Crossing points and junction points are proceeded at 170km/h.

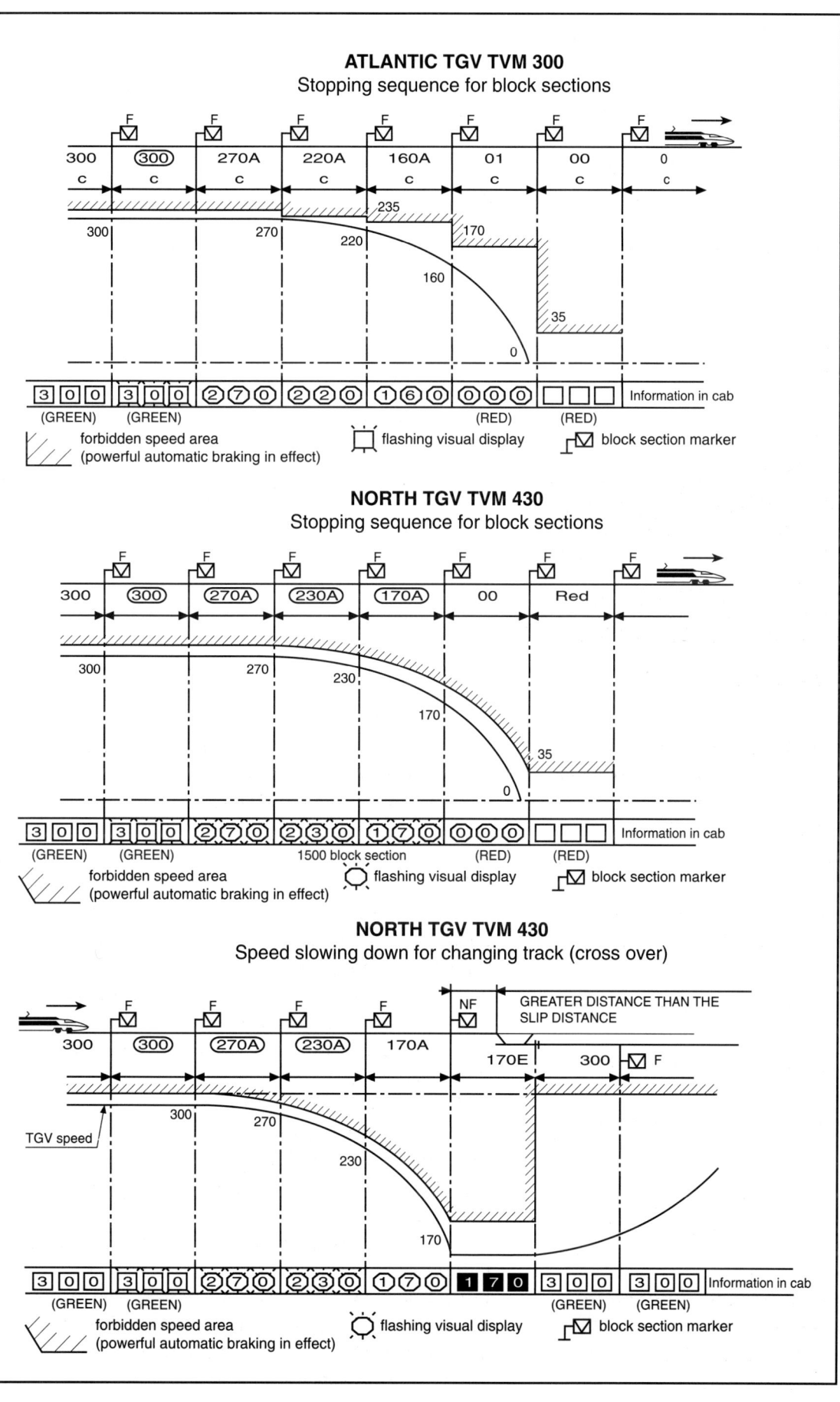

Fig. 10.5 TVM 300 and 430 curves

The last diagram shows the same items for a restricted speed applied for changing from one track to the other. Steady indications are displayed on the running of the point block section in the same way as the ceiling speed of the line. As it appears the distance between the tip of the switch and the reference marker backwards is greater than the overlap length.

Other configurations can be met such as transitions from a conventional line to a high speed line or vice versa, or from former higher speed lines to a more recent high speed line and vice versa.

Other protections are available. Strictly speaking, these are not part of the signalling requirements, but they contribute to the safety concept and are connected with it. The first of these is the hot box detector, as described in Chapter 13. There is a hot box detector every 25km of the SNCF high speed line. This equipment will bring a train that is suspected of overheating to a halt.

The second such protection system is the falling vehicle detector. This consists of electrical wires positioned in such a way that they will be broken if a road vehicle were to fall on the track from a bridge, triggering an alarm and consequently bringing trains in the vicinity to a halt. Falling vehicle detectors are also described in Chapter 13.

Additional driving aids are provided by the signalling system. The first of these is the overhead power supply protection which isolates the power supply by controlling the circuit breakers at the limit between electrical sections, and lowers the pantographs when changing from one power supply voltage to another.

A second such driving aid ensures that no absolute stopping point is passed, by the triggering on of an emergency brake, in the event of a train overrunning a stopping point. The third driving aid involves several facilities such as the identification of radio channels, tunnel entrance and exit indications, data logging, etc.

The nominal operating mode of the SNCF high speed lines achieves a very high level of availability. Statistics show that, after more than 12 years of TGV operation, late arrivals due to signalling equipment failure are insignificant.

In respect of the failure of either a track circuit or the track to train transmission systems, or, in the worst scenario, a cab equipment failure, the sight driving mode with caution has to be brought into service. Accordingly, the train may then have to run on sight, or other special procedures may have to be followed.

Germany – DB AG

Operating modes on high speed lines in Germany are very much paralleled by those in Spain. The only major difference between the German and Spanish systems is the fact that the RENFE line, using standard gauge tracks, cannot of course accept any train other than the TGV sets or specially adapted rolling stock, such as Talgo trains, hauled by standard gauge locomotives.

In Germany, the majority of trains running on high speed lines are LZB equipped, controlled and operated.

On the basis of clear track sections being available ahead of a train, interlocking information, line gradients, permanent or temporary speed restrictions, and the actual position and speed transmitted from the trains, the target distance is transmitted individually to each train. Braking capability, train length and maximum train speed are stored on board the locomotive, for each particular train. From all available data, the maximum permissible speed is computed permanently. If the speed of the train exceeds the permitted speed, the service brakes will be applied.

This means that successive LZB equipped trains may follow each other at intervals governed by the block section length plus the service brake distance plus a safety margin. At a speed of 250km/h, this global distance is about 10 000m.

From the above it can be seen that lineside signals are not really required for LZB controlled trains and, in practice, they are only provided along DB AG high speed lines to protect the point areas. Conventional trains, with locomotives not equipped with LZB, may also use the high speed lines. The drivers of such trains rely on the remaining lineside signals, observation of which is ensured by the conventional INDUSI intermittent inductive automatic train protection system. INDUSI is also fitted to all LZB trains as a fall back level. Due to the limited number of signals on high speed lines, conventional trains can only follow each other at signal to signal distance, plus an overlap. Marker boards (fictive block signals) are irrelevant for conventional trains.

INDUSI, which is the DB AG standard automatic train protection system is described in Chapter 9.

As a result of these regulations, a conventional train can only follow an LZB train on green signals, as it would follow a previous train on any other line. Some block sections thus extend over approximately 7km. Conversely, an LZB controlled train can follow a

HIGH SPEED LINE SIGNALLING SYSTEM

Fig. 10.6 LZB indications for cab signalling

Reference Mark	Name
a	Windscreen Wiper
b	Circuit-breakers, Signalling and Headlamps
c	Intermittent Lights
d	Sanding facilities
e	Dead-Man
f	Emergency Brake
g	Assistant Brake
h	Clock
i	Brake Control
j	Brake Cylinder
k	TFA (Spanish abbreviation for Assistant Train Brake Pipe) Manometer
l	Pressure Gauge General (Manometer)
m	Pilots Signalling Failures
n	ASFA (Spanish abbreviation for Signalling Warning and Automatic Brake)
o	Power Lever
p	Manually-controlled Switchogroup
q	LZB
r	Programmed Speed, Motor Ammeters, Line Voltmeters
s	BL (Desk Key)
t	Reverser
u	Speed/Brake Selector
v	Driving Assistant System (in Spanish: SIAC)
w	Passengers Public-address System/Train Intercom
x	Control Pantographs
y	Train/Ground Radiotelephony
z	On/off Lighting Switch Control
aa	Dead-Man

conventional train as it would follow another LZB train. In order not to restrict high speed line capacity unduly, on LZB controlled trains (and this is the most significant deviation from conventional operation practice) *cab indications have priority over remaining lineside signals.*

With an LZB train approaching, starting signals (block entry signals) are suppressed. All unlit signals, however, have their INDUSI track transponders active in order to prevent conventional trains inadvertently overrunning such a signal. On board trains where LZB is active, the INDUSI system is suppressed. Consequently, trains where LZB is inoperative run only as conventional trains.

Line capacity is about 12 LZB controlled high speed trains per hour which means a headway of about 5 minutes. The theoretical minimum at 250km/h is from 2.5 to 3 minutes. If a mixture of LZB controlled and conventional trains is encountered, the line capacity is of course significantly reduced.

For these reasons, DB AG tries to avoid the mixture of all categories of train at all times. It has been found preferable only to run high speed trains and the more conventional LZB passenger trains at 200km/h during the daytime. At night, freight trains are permitted to run on DB AG high speed lines, which makes for shorter trips and, consequently, very efficient journey times.

To ensure safety on DB AG high speed lines, various other safety devices have been installed, such as hot axle box detectors and wind velocity meters. These are situated at strategic places along the lines. In case of very strong lateral winds, the speed of certain trains may be restricted. The direction of the draught inside tunnels is also accurately monitored in order to manage the evacuation of passengers in case of fire and/or smoke efficiently.

Finally it should be noted that in Italy too, where a mixture of high speed trains and conventional passenger and freight trains can be contended with, train traffic, as on the DB AG lines, is very much influenced by the presence of trains running at lower speed.

10.2.2 Safety and availability aspects

As already noted, safety and availability objectives are a major requirement when considering high speed operations and a high level of service quality.

For SNCF high speed lines, safety concerns mainly interlocking and cab signalling systems and must be achieved with fail-safe techniques based on either oriented process for discrete component conception, ORE A 119 recommendation or probability process for computerized conception of hardware and data coding redundancy.

On DB AG high speed lines, in order to prevent trains from coming to a stop inside a tunnel after an emergency brake application, the driver can override the braking action until the train emerges. High speed trains on DB AG high speed lines are also sealed to prevent uncomfortable and potentially dangerous sudden changes in pressure. Non-sealed trains must reduce their speed if they meet other trains in a tunnel. This should also be controlled later by LZB.

DB AG freight trains must reduce their speed over viaducts in case of an excessive lateral wind. At present, the trains concerned are individually informed by radio of such a situation. Later it is anticipated that LZB will take over this function.

The availability aspect for the SNCF system, mainly concerns the cab signal system, because there are no more lineside signals. A high level of availability is achieved by hardware redundancy and two fully independent subsets for each device so that no single failure will lead to a cab display indication showing sight driving with caution over more than two block sections and, exceptionally, three sections.

10.2.3 Maintenance

There is not much to say about maintenance on high speed lines. Indeed, because of the innovative nature of these lines, maintenance has always been a consideration from the very start of each project.

Regular preventive maintenance, inspections and immediate repair of minor defects are most important on high speed lines. Predictable time interval based services are carried out at times of lower train frequency, particularly at night and during the weekends. Line closure for the benefit of individual services is discouraged in favour of co-ordinated simultaneous work by all involved services.

It is worthwhile to note two specific aspects concerning maintenance, as designed by SNCF:

- A computer aided maintenance system called SIAM, from a PC AT microcomputer in each signalling centre, enables the recording of information concerning operation of the track to train transmission system and of track circuits. This system can be remotely interrogated and can thus efficiently help preventive and corrective maintenance, by restoring, on request, every event

or a part of an event that occurred during a determined period.
- In respect of the protection of staff working along the tracks, elementary protective zones, called ZEPs, corresponding to sections of track on the site, have been implemented. When these are made active, they inhibit the setting of any route giving access to the zone(s) concerned. These controls are carried out from the central control room, the PAR.

10.3 General signalling system description on very high speed lines: speeds ≥ 250km/h

The signalling systems of all four railway administrations covered in this chapter – SNCF, DB AG, RENFE and FS – are fundamentally different. These are described one after the other.

Fig.10.9 at the end of the chapter shows the different high speed line signalling systems available and summarizes the various comparisons that can be made.

10.3.1 SNCF track to train transmission systems TVM

10.3.1.1 General
In the mid-1970s, SNCF decided to equip the high speed network with a cab signalling system, since it was often awkward for a driver to observe lineside signalling under difficult conditions such as fog or rain with speeds higher than 200–220km/h. Moreover, should only a lineside signalling system have been chosen, it would have been necessary to increase the number of light indications displayed during a braking sequence for which at least four block sections would have been required.

When this decision was adopted, the track to train transmission bearer support was also chosen. With a choice of rails or a cable laid between the running rails, SNCF opted for a transmission mode through the rails, which also enabled broken rail detection, and avoided the disadvantages of a cable, especially during track maintenance operations.

10.3.1.2 TVM 300
For the South-East high speed line, SNCF has developed a continuous track to train data transmission system from the UM71 audio frequency track circuit.

The basic frequencies of this track circuit are 1700, 2000, 2300 or 2600Hz. The basic frequency F assumes alternately the values F+10Hz and F−10Hz at a rate varying from 10.3 to 29Hz, according to the information to be transmitted. This modulating frequency (f) can assume 18 different values. It is then possible to transmit one out of 18 data, which correspond to the speed rates to be displayed in the driver's cab.

The UM71 track circuit – also used on conventional lines equipped with automatic block system installations – has the advantage of not using any insulating joints. It is therefore possible to use long welded rails, with the attendant well known advantages of comfort and track maintenance.

The continuous data transmission system is completed by intermittent data transmission facilities which consist of loops of cable, 10m long, laid in the track, and able to transmit one out of 14 frequencies in the range from 1300 to 3700Hz.

The intermittent transmitted data is especially concerned with:

- activation/deactivation of track to train transmission at the entry to and exit of a high speed line;
- ensuring that no absolute stop signal is passed;
- radio channel changes;
- isolating the traction power;
- lowering the pantograph(s).

TVM 300 is used on the South-East Paris to Lyon high speed line (maximum speed: 270km/h) as well as on the Atlantic high speed line (maximum speed: 300km/h). Respectively, this system enables a 5 minute and 4 minute headway between two trains.

10.3.1.3 TVM 430
Despite the highly satisfactory performance and reliability of TVM 300, this system was not considered efficient enough to cope with the anticipated volume of traffic expected on the North high speed line.

The technical departments of SNCF were therefore asked to develop a signalling system characterized by a 3 minute headway at speeds of up to 300km/h. The TVM 430 project was subsequently launched at the end of 1987.

In addition to the constraints of a 3 minute headway, other objectives have been developed:

- the possibility of increasing the line speed. Indeed, the speed of 320km/h has been reserved for the North line and the system capacity provides for future speed rates of 350km/h and more.
- the high flexibility in available speed rates. Increasingly the new high speed lines are coming into city centres, such as in Lille, for instance, or into the airport areas of Roissy near Paris and Satolas near Lyon, where different speed rates may be required.
- downward compatibility.
  the trains equipped with TVM 430 system will have to run without constraint on existing high speed lines, equipped with the present TVM 300 data transmission.
- flexibility outside SNCF new lines.
  TVM 430 system is an open system whose numerous parameters can be determined by the network user, especially the speed rates – for example the Channel Tunnel.
- a best balance of deceleration rate could be necessary in each block section, by modifying the speed rates of the stop sequence [in km/h]:
  TVM 300: 300 270 220 160 0
  TVM 430: 300 270 230 170 0
- systematic information given to the driver: a flashing illumination of the cab display indication warns the driver that, in the following block section, the information will be more restrictive, enabling him to anticipate his action.
- a better definition of speed control curves: replacement of a control curve with varying speed steps by a parabolic control curve, thanks to the wayside data transmission concerning the length and profile of the block section.

All these improvements have enabled SNCF to achieve 1500m long block sections on level ground instead of the usual 2000m.

UM71 audio frequency track circuits are also used as a support for the TVM 430 track to train data transmission system. The transmission capacity of the TVM 430 system which uses digital technics, is of 2^{27} data, whereas TVM 300 system capacity is only one out of 18 available data.

TVM 430 system is made up of two subsystems:

- the wayside system;
- the trainborne system.

The information is exchanged through a one way track to train transmission. In both systems, the coding consists of modulating the frequency of the track circuit.

With TVM 430, the signal modulates its carrier frequency with the sum of n sinusoidal signals, out of 27 possible signals. The presence or the absence of these signals characterize the information to be transmitted.

The transmission therefore has a 27 bit capacity, 6 bits being reserved for the transmission protection by coding; 3 bits are used to define the user network name or, rather, the type of mission of the train. 18 functional bits are available for each network for speed, distance to go and profile information.

On the SNCF North high speed line and for the new lines to be built in the future, the functional bits are used as follows:

- 8 bits for speed data;
- 6 bits for distance to go;
- 4 bits for profile.

From this code the trainborne system deduces:

- the speed data to be displayed to the driver and the

Fig. 10.7 A & B TVM cab signalling

flashing illumination of the indicators, when required;
- all data necessary for working out the speed control curve.

An intermittent transmission is associated to TVM 430 as it is with TVM 300. The integration of transmission loops TVM 430 system gives a global transmission capacity of 2^{28} data bits per direction of traffic.

The wayside TVM 430 equipment is especially efficient at coping with the spacing function. The function of the wayside equipment is to work out the code to be transmitted depending on the different signalling conditions, for example stop indication and speed restriction. The function of the TVM 430 trainborne system is to decode the signal transmitted through the track, to display the speed indication to the driver, to supervise the train speed and to process the trainborne automatisms.

10.3.2 DB AG speed control system LZB

A major reason for the development of a new continuous speed control system for DB AG was the need to increase the maximum allowed speeds of 140–160km/h, without the need to alter the existing infrastructure and with an increase of line capacity in mind. The new system had to be applied using the existing interlockings, block sections and signal aspects. On the other hand it had to work as a stand alone system. It also had to be able to handle different types of train, as well as mixed traffic.

The continuous automatic train control system in operation with DB AG is called *Linienzugbeeinflussung* (LZB). It is a vital speed control system based on modern technology, including cab signalling, and the central monitoring of the speeds of all trains within a specific area.

The principal characteristics of the LZB system are:

- continuous data transmission between the track and the trains, in both directions;
- continuous control and monitoring of train speeds;
- the use of vital multicomputer systems, both at the LZB line centre and on board the trains.

LZB was first used in the mid-1960s. The LZB system, based on free programmable components and as described in the following paragraphs, has been in scheduled operation since 1977.

The LZB system involves a line centre in which are available all fixed line data, geographical, gradient, speed profile, as well as all variable data, route, temporary speed restriction, from the interlocking(s) and the operator terminal.

In addition, the line centre continuously receives information from all equipped trains in the area. This can be divided into specific train characteristics, such as the length of train, braking capability, and variable data on train position and actual speed.

On the basis of this information, a control command is computed for each individual train. The control that is computed differs in Germany and Spain. The German LZB transmits from the line centre, a control command every second.

The information of the signal aspects – respectively the conditions of the block section ahead of the train – permits an extended line visibility and thus an appropriate increase of speeds along the line.

When all trains within the system controlled area are equipped with LZB, the conditions for driving by moving block (LZB block) can be fully met and, consequently, a higher frequency of trains is possible. This also results in a considerable reduction of the number of fixed signals required along the line.

The ground equipment information is transmitted to the train in a telegram form. The telegrams contain the following data:

- the length of the available braking distance, indicating the point where the braking action has to be initiated, based on each individual train's available braking capability;
- the target to be indicated in the driver's cab, ie the distance to nearest restriction and the permissible speed at the restriction;
- any additional operating information.

On the basis of this information, the LZB on board equipment continuously calculates the maximum permissible speed, controls the indicators in the driver's cab and monitors the speed to be followed.

This data that is evaluated by the LZB system also permits full automatic train operation (ATO). In the event that the train exceeds the maximum permissible speed, braking will be initiated automatically.

Similarly, two types of return telegram are sent from the trains to the line centre:

- when a train announces itself for the first time after entering an LZB controlled area, it transmits its specific data to the line centre;
- as soon as the train has been registered by the LZB

line centre system, the cyclically sent telegrams will include the actual positioning of the train, the braking characteristics, actual speed and internal operational data.

For an uninterruptible exchange of trains between adjacent LZB line centres, additional data lines are provided.

A special feature of the LZB system is the use of operator terminals at the line centres. It is possible to exchange data via this operator terminal which is operated by the signalman. Key data inputs for this operator terminal include:

- temporary speed restrictions; and
- a requirement for status of operating conditions of the system, including peripheral equipment.

Key data outputs include:

- information from the trains controlled by the LZB; and
- information regarding possible failures in the system, the equipment, the transmission lines elements connected to the system.

It is also worthwhile to note that on lines without LZB, the INDUSI functions integrated in the LZB trainborne equipment are automatically activated and the train is protected in accordance with the operating programme.

10.3.3 RENFE speed control system LZB

The RENFE speed control system is very similar to that operated by DB AG. The only major difference is that the RENFE line can only accept standard gauge vehicles. Thus in Spain, there is no real mixture of traffic as there is in Germany.

10.3.4 FS speed control system

The train detection on the Firenze–Roma line is carried out by coded track circuits. Two carriers (50 and 178Hz) and four modulating frequencies (75–120–180–270 codes per minute) are used. Each block section has a length of about 5400m and includes four track circuits, each having its own coded current according to the allowed speed value; the message transmitted from track to train is therefore a speed value. Each block section has a block signal preceded by a warning signal, 1350m in advance.

On future FS high speed lines, train detection will be carried out by audiofrequency digital track circuits, providing the transmission of track to train messages of an appropriate number of bits. This system will, in fact, be analogous to the SNCF TVM 430 system. The information supplied to the train will include track circuit identity, target distance and other auxiliary vital data. Furthermore, every 12km, an intermittent track to train transmission system, at present under development, will be installed to increase the amount of information supplied to the train and to act as a back up system. Lineside signals are not going to be installed. Block section length will be 1500m.

10.3.5 Main technology used on high speed lines

Various types of computer technologies and fail-safe techniques are used on the different European high speed lines.

On SNCF lines, UM71 track circuits assume three functions – train detection, track to train transmission and broken rail detection. Processing is carried out by microprocessors. Safety is ensured by duplicating the microprocessors and comparing their results with a third microprocessor using the encoded microprocessor technique, and by the code rereading transmitted in the track.

Availability is ensured by duplicating the processing units and efficiently regrouping the interface equipment, block section by block section, geographically, in order to limit the consequences of any failure.

It is also worthwhile to note that equipment on the SNCF high speed lines is concentrated in the signalboxes; trackside equipment is kept to a minimum and no power has to be distributed along the track.

The TVM 430 trainborne system architecture is built around three subsystems processing each of the following functions:

- the signal is decoded by two microprocessors performing in parallel a demodulation, then a discrete transformation of Fourier, on a sliding time division sample of the signal;
- a coded monoprocessor compares the result of these two decodings and displays the speed indication on the driver's cab display;
- speed control and trainborne automatisms are processed by a specific microprocessor.

With the DB AG speed control system LZB, train detection is ensured by audiofrequency jointless track circuits, type FTG–S, while the track to train transmission is made through a specific cable. As for processing, the LZB line centres are equipped with a two out of three computer system.

The central logic unit then consists of three independent microcomputers. Additional elements of the equipment are duplicated for reasons of availability. For the LZB on board equipment, a two out of three computer technique has also been chosen.

Fig. 10.8 LZB equipped AVE cab equipment

INDICATOR	TYPE	MEANING	MAXIMUM SPEED (STANDING ORDERS V_{cons})	INDICATION PRECAUTION	SPEED LIMIT
T (white on blue)	VIEWER	~ it is lit if transmission LZB is correct ~ it turns off if transmission LZB is not correct	From 0km/h to 270km/h	Transmission fault of LZB. Viewer T with flashes	From 0km/h to 270km/h
S (white on blue)	VIEWER	~ it is lit if LZB is on service ~ it is off if it is out of service	Regardless of speed	Does not intervene	Regardless of speed
FIN (black on yellow)	VIEWER	~ end transmission LZB in operations and end of route with LZB. Two indications: ~ With flashes from 1700m before LZB end signal and until confirming with RELEASE button which passes to : ~ End permanent illumination until 50m in front of LZB end signal	V_{cons} < 270km/h V_{cons} < 270km/h	END with flashes V in permanent red END permanent V in permanent red V turns off when V real is ≤ 40km/h	40km/h 40km/h
A (black on white)	VIEWER	~ with no application in the preset exploitation. It was used during the test to warn entrance into tunnels			
OE (white on red)	VIEWER	~ it lights up when it goes over the stopping point ~ it lights up with flashes indicating stop because the train that ran by the adjacent track saw a risk	V_{cons} ≤ 270km/h V_{cons} ≤ 270km/h	OE with flashes	0km/h 0km/h

INDICATOR	TYPE	MEANING	MAXIMUM SPEED STANDING ORDERS (V_{cons})	INDICATION PRECAUTION	SPEED LIMIT
EL black on white and blue	VIEWER	~ it lights up permanently in the neutral area of the 25KV 50Hz /catenary exchange phase ~ it lights up with flares in the catenary exchange area of 25KV 50Hz to 3KV c.c.	$V_{cons} \leq 270$km/h $V_{cons} \leq 270$km/h	Besides, a sound mechanism operates intermittently Besides, a sound mechanism operates intermittently	$V \leq 270$km/h $V \leq 270$km/h
V white on red	VIEWER	~ it lights up in the braking sequence of slackening off speed: ~ at 1000m after beginning the braking curve ~ during the braking curve ~ and until $V_{real} \leq V_{limit}$ then, V turns off ~ it lights up permanently when LZB ends, according to sequences: ~ at beginning of the 1000m when starting the braking curve ~ and until $V_{real} \leq 40$km/h then, V turns off ~ it lights up permanently in transmission failure ~ it lights up with flashes when the standing orders speed overruns 5km/h ~ if overrunning the real speed to V_{cons} is more than 5km/h it lights up permanently and the emergency brakes actuate	$V_{cons} \leq 270$km/h $V_{cons} \leq 270$km/h $V_{cons} \leq 270$km/h $V_{cons} \leq 270$km/h $V_{real} = V_{cons} + 5$km/h $V_{real} > V_{cons} + 5$km/h	 ~ END with flashes and then passes to permanent ~ viewer T lights up with flashes and C/I lights up permanently ~ viewer E lights up and P lights up with flashes	For example: $V = 100$km/h or $V = 0$km/h $V = 40$km/h $V = 0$km/h in Partial Block. $V = 40$km/h in Total Block $V \leq 270$km/h $V \leq 270$km/h
P black on yellow	VIEWER	Brake pressure. It lights up permanently when brakes operate in case of failure of transmission of the LZB, both in Total Block and Partial Block It lights up when: ~ overrunning of the standing orders speed is more than 5km/h ~ when starting the putting into service phase of LZB	 $V_{cons} \leq 270$km/h	~ viewer 80 lights up ~ viewers E and V light up ~ viewers C/I, 60 and R light up with flashes	 $V \leq 270$km/h
R black on white	VIEWER	~ overrunning. It lights up with flashes when the system needs an overrunning authority to continue running in the cases of: ~ putting into service of LZB ~ failure of transmission of LZB in Partial Block	 0km/h 0km/h	~ viewers C/I, 60 and p light up with flashes ~ viewer C/I lights up permanently	 0km/h 0km/h

HIGH SPEED LINE SIGNALLING SYSTEM

INDICATOR	TYPE	MEANING	MAXIMUM SPEED STANDING ORDERS (V_{cons})	INDICATION PRECAUTION	SPEED LIMIT
80 white on black	VIEWER	~ It lights up, permanently to go into LZB after switching Alert and Release		Viewer off	~
		~ it lights up permanently when there is a loss of LZB with Partial Blocking after switching Overrun	$V_{cons} \leq 270$ km/h	Viewer T off and Viewer P lighted	0km/h. Then until 80km/h
		~ it lights up permanently for end LZB	-	Viewer T off	40km/h. Then until 80km/h
		~ it lights up with flashes when is a loss of LZB with Total Block to pass to ASFA. It is confirmed by switching Release and viewer 80 lights up permanently	$V_{cons} \leq 270$ km/h	~ viewer C/I lighted and viewer T off ~ viewer T off and viewer P lighted	40km/h. Then until 80km/h
60 white on blue	VIEWER	~ it lights up with flashes to go into LZB. After switching Alert and Release, 60 turns off and 80 lights up permanently	~	~ C/I, P and R, light up with flashes	~
C/I black on white	VIEWER BUTTON	~ it lights up with flashes to go into LZB. It turns off after pressing C/I to insert the train's information	~	~	~
		~ it lights up permanently when LZB transmission fails, either in Total or Partial Block. C/I turns off when switching Release in Total Block, or Partial Block	~	~ T lights up with flashes	~
		~ it is used as a button to check out the system	~	~	~
E white on red	VIEWER	~ emergency brake. It lights up permanently when the emergency brake is operated. P turns off when $V_{real} = V_{cons}$	$V \leq 270$ km/h	~ V lights up permanently and P lights up with flashes ($V >> V + 5$km/h)	$V \leq 270$ km/h
OP 40 black on white	VIEWER	~ precaution order in 40km/h. With no application in exploitation	~	~	~

Fig. 10.9 Main characteristics of very high speed and high speed line signalling systems

Very High Speed Lines (V ≥ 250km/h)	SNCF [1,2]	SNCF [3]	SNCF [4]	DB AG	FS	RENFE
1. BASIC PRINCIPLES						
Very high speed lines today (speed ≥ 250km/h)	Yes	Yes	Yes	Yes	Yes	Yes
• ATC name	TVM 300	TVM 300	TVM 430	LZB 80	BACC	CAT (or. LZB 80)
• ATC associated with other system (backup)	KVB, Crocodile			INDUSI		ASFA 200
• Maximum operating very high speed (km/h)	270	300	300[6]	250[16]	255	270[17]
• Total very high speed lines length (km)	417	282	558	426	262	471
• Additional length planned (km)	> 1500	–	–	800	> 1000	600
• Minimum curve radius	3200m	4000m	4000m	7000m	5400m	4000m
• Maximum gradient	35‰	18‰	25‰	12.5‰	18‰	12.5‰
• Ton/axle (maximum)	17	–	–	21	20	17
• Minimum headway (12)	5′	4′	3′	3′	5′	3′
• Homogeneous traffic on very high speed lines or not	Yes	–	–	No	No	Yes
• On board compatibility with existing lines signalling	Yes	–	–	Yes	Yes	Yes
• Bidirectional running	Yes	–	–	Yes	Yes	Yes
• Passing loops: side track (km)	40 to 50	–	–	20 to 25	48	50
• Connecting switches: junction between both track (km)	20 to 25	–	–	7 to 8	24	25
• Rail traffic management	Yes	–	–	Yes	Yes	Yes
– control centre	Whole line	–	–	Up to 75km	Whole line	Whole line
– local centres (one for x km)	20 to 25	–	–	7 to 8		60

Very High Speed Lines (V ≥ 250km/h)

	SNCF [1,2]	SNCF [3]	SNCF [4]	DB AG	FS	RENFE
• Interlocking	Relays	Relays	Relay PRCI computer	Relay + micro computer	Relays computer	Relay + micro
• Lineside signals (except for shunting)	No	–	–	Yes[22]	Yes[22]	No
• Level crossing on very high speed lines	No	–	–	No	No	No
• If any: associated warning system	–	–	–	–	–	–

2. GENERAL OPERATION CONDITIONS

• Information displayed to driver	Yes	–	–	Yes	Yes	Yes
allowed speed	Yes	–	–	Yes	Yes	Yes
distance to go	No	–	–	Yes	No	Yes
• ATO	No	–	–	Yes	No	Yes

Very High Speed Lines (V ≥ 250km/h)

	SNCF			DB AG	FS	RENFE
	1, 2	3	4			

3. GENERAL SYSTEM DESCRIPTION

	SNCF 1,2	SNCF 3	SNCF 4	DB AG	FS	RENFE
• Train detection (TC = track circuit)	Jointless TC	–	–	Jointless TC	TC with joints	Jointless TC
• Track to train transmission	Coded TC	–	–			
	+ intermittent transmission			Cable loops B3 pose	Coded TC	
– number of codes transmitted	1 of 18	1 of 18	27 bits (21 for data, 6 for coding)	Up to 83 bits and coding	1 of 9	Up to 70 bits
– principle	Frequency modulated in frequency	–	–	Frequency modulated (frequency shift keying)	50Hz+178Hz amplitude modulated	Frequency modulated (frequency shift keying)
• Train to track transmission	No	–	–	Yes	No	Yes
• Transmission along the line: fibre optic	No	Mixed	Yes	No	Yes	Yes
• Line parameters	Speed	Speed	Speed distance gradient	Speed distance gradient	Speed distance gradient	Speed distance gradient
• Train parameters	–	–	brake ability max. speed[13]	brake ability train length max. speed	–	brake ability train length max. speed
• Vital logic/computer: – ground	–	–	(2 out of 2) x 2	2 out of 3	Coded micro-processors	2 out of 3
– trainborne	Hybrid circuits intrinsic safety	Hybrid circuits intrinsic safety	Coded micro-processors[14]	2 out of 3	Coded micro-processors	2 out of 3

HIGH SPEED LINE SIGNALLING SYSTEM

High speed lines (speeds between 160 and 250km/h)

	SNCF	DB AG	FS	RENFE
1. BASIC PRINCIPLES				
• High speed on existing lines	Yes	Yes	Yes	No
• ATC name	KVB	LZB	BACC	
• ATC associated with other system (back up)	Crocodile	INDUSI		
• Maximum operating high speed (km/h)	220	200	230	
• Maximum speed without ATC (km/h)	160	160[24]	150	
• Total high speed lines length (km)	785	800	1000	
• Additional length planned (km)	?	?	600	
• Minimum curve radius[18]	1563/1724	2484	1231/2154	
• Maximum gradient	8%	12.5%	8%	
• Ton/axle (maximum)	22.5	22.5	20	
• Minimum headway[12]	3′	2′	5′/8′	
• Homogeneous traffic on high speed lines or not	No	No	No	
• Lineside signals on high speed lines	Yes	Yes	Yes	
• On board compatibility with existing lines signalling	Yes	Yes	Yes	
• Bidirectional running	Yes[19]	Yes	Yes	
• Passing loops: side track (km)	≃ 30	3 to 30	Various	
• Connecting switches: junction between both tracks (km)	≃ 12	2 to 20		
• Rail traffic management	Yes[19]	Yes	Yes[19]	
– control centre		up to 30km		
– local centres (one for x km)		2–15km		

High speed lines (speeds between 160 and 250km/h)

	SNCF		DB AG	FS	RENFE
• Interlocking	Relay or relay + computer	Relay or computer	Relay		
• Level crossing on lines with high speed trains	No		No	Yes	
• If any: associated warning system	–		–	half barrier / full barrier	

2. GENERAL OPERATION CONDITIONS

	SNCF	DB AG	FS	RENFE
• Information displayed to driver				
– allowed speed	000[20]	Yes	Yes	
– distance to go	No	Yes	No	
• ATO	No	Yes	No	

3. GENERAL SYSTEM DESCRIPTION

	SNCF	DB AG	FS	RENFE
• Train detection (TC = track circuit)	TC: HVITC or UM 71	TC or axle counters	TC	
• Track to train transmission	Intermittent transmission	as for ≥ 250km/h	Coded TC	
– number of codes transmitted	24 bits		1 of 4 / 1 of 9	
– principle	4.35mHz carrier / 50kHz AM		50Hz + 178Hz amplitude modulated	
• Train to track transmission	No		No	
• Line parameters	speed distance gradient		speed	
• Train parameters	brake ability train length max. speed			
• Vital logic/computer: – ground / – trainborne	Yes / No (KVB)		Coded microprocessor	

High speed lines (speeds between 160 and 250km/h)

	ÖBB	BV (ex. SJ)[7]	SBB[11]	BR[11]
1. BASIC PRINCIPLES				
• High speed on existing lines	Yes[15] Yes	Yes[8]	Yes[15]	Yes
• ATC name	LZB 80	ATC 2	ZUB 100	No[21]
• ATC associated with other system (back up)	INDUSI	–		No
• Maximum operating high speed (km/h)	200	200	200	200
• Maximum speed without ATC (km/h)	160	160 or 80[9]	160	200
• Total high speed lines length (km)	25	605	–	1000
• Additional length planned (km)	140	> 1000	300	500
• Minimum curve radius	–	3000	3000	–
• Maximum gradient	–	10%	25%	–
• Ton/axle (maximum)	22.5	18.3[10]	22.5	–
• Minimum headway[12]	[23]	3′	2′	3′
• Homogeneous traffic on high speed lines or not	No	No	No	No
• Lineside signals on high speed lines	Yes	Yes	Yes	Yes
• On board compatibility with existing lines signalling	Yes	Yes	Yes	Yes
• Bidirectional running	Yes	Yes	Yes	Yes
• Passing loops: side track (km)	≃ 8	25	30	–
• Connecting switches: junction between both tracks (km)	≃ 4	12	7	20
• Rail traffic management	Yes	Yes	Yes	Yes
– control centre	8	–	–	Up to 100km
– local centres (one for x km)		150		

High speed lines (speeds between 160 and 250km/h)

	ÖBB	BV (ex. SJ)[7]	SBB[11]	BR[11]
• Interlocking	Relay or Computer	Relay + Computer	Relay + Computer	Relay + SSI
• Level crossing on lines with high speed trains	No	Yes	No	Yes
• If any: associated warning system		Full barriers + obstacle detectors + ATC protection	–	CCTV
2. GENERAL OPERATION CONDITIONS			–	
• Information displayed to driver	Yes	Yes		No
– allowed speed	Yes	Yes		No
– distance to go	Yes	No		No
• ATO	Yes	No		No
3. GENERAL SYSTEM DESCRIPTION			–	
• Train detection (TC = track circuit)	TC with joints	TC with joints		TC
• Track to train transmission	Cable loops	Intermittent transmission		No
– number of codes transmitted	**Chap. 9** LZB	24 bits		–
– principle	Telegram	4.35mHz carrier 50kHz AM		–
• Train to track transmission	Yes	Yes for level crossing		No
• Line parameters	**Chap. 9** LZB	speed distance gradient		–
• Train parameters	**Chap. 9** LZB	Brake ability train length max. speed		–
• Vital logic/computer: – ground	2 out of 3	–		–
– trainborne	2 out of 3	2 out of 3		–

Notes:

[1] Paris–South-East TGV: Paris to Lyon.

[2] For information concerning general aspects, please read this column.

[3] Atlantique TGV: Paris to Le Mans (Nantes) and Tours (Bordeaux).

[4] North TGV (Paris to Lille and the Channel Tunnel or to Belgium) and future other lines (around Paris, around Lyon, etc).

[5] TGV trains are able to run up to 220km/h, instead of 200km/h, for normal trains, on existing lines, equipped with an intermittent speed control system (KVB).

[6] Line and signalling designed for speeds up to 320km/h.

[7] No very high speed lines; figures are given for tilting trains in standard (possibly upgraded) lines.

[8] Tilting trains.

[9] 160 for conventional trains with ATC, equipped with soft bogies; 80 for any train without ATC in function. (140 is max. speed for older trains with ATC, equipped with stiff bogies).

[10] 22.5 ton/axle for freight trains on the same line.

[11] No very high speed lines; figures are given for 'high speed trains' (diesel or electric) running on conventional lines.

[12] Effective minimum headway used on revenue service; *not* theoretical headway!

[13] These parameters are used in the TVM 430 on board equipment to allow heterogeneous traffic to run through the Channel Tunnel; these parameters are not accessible to the driver.

[14] • Ground: DDMR (Dual Duplex Modular Redundancy) using the coded microprocessor technics for comparisons.
• Trainborne: DSMR (Dual Simplex Modular Redundancy), also using coded microprocessor technology.

[15] Today's maximum speed is 160km/h.

[16] Maximum allowed speed 280km/h; max. operating speed: 250km/h.

[17] Maximum allowed speed 300km/h; max. operating speed: 270km/h.

[18] 1563m (220km/h); 1724m (220km/h).

[19] Not systematic.

[20] With KVB, a '000' indication is displayed when speed limit goes from 'maximum speed' to '0'.

[21] Conventional BR AWS (Automatic Warning System) in use.

[22] Lineside signals only in stations; no block signals. LZB cab signalling is valid for LZB trains.

[23] In Austria, ÖBB considers that the minimum headway has no fixed value; it could even be less than 3 minutes in case of short block distances.

[24] Vmax without LZB or LZB imperative (INDUSI only).

11 Rail Traffic Management

A. EXER

Contents

11.1	Introduction	312
11.2	**Traffic supervision**	313
	11.2.1 Operational concept	313
	11.2.2 Operational planning	316
	11.2.3 Disposition	317
	11.2.4 Control	318
	11.2.5 Safety	318
11.3	**Timetable**	319
11.4	**Train description**	319
	11.4.1 Introduction	319
	11.4.2 Summary of technical solutions	320
	11.4.3 Summary of functionality of train description	320
	11.4.4 Functional description of signalling information and stepping	321
	11.4.5 The operator's control	322
	11.4.6 Fringe boxes	323
	11.4.7 Link of train describer systems	324
	11.4.8 Fault logging, alarms and warning messages	324
	11.4.9 Automatic editing of train numbers	324
	11.4.10 System configuration	324
	11.4.11 Safety and other aspects	325
11.5	**Train recording**	326
11.6	**Operation**	326
	11.6.1 Introduction	326
	11.6.2 Display – visualization of the process	326
	11.6.3 Control functions – manual route setting	327
	11.6.4 Automatic route setting	330
	11.6.5 Shunting operation	332
11.7	**Passenger information**	332
	11.7.1 Visual information	333
	11.7.2 Voice communication	333
	11.7.3 Other systems	334
11.8	**Centralized traffic control**	334
	11.8.1 Traffic supervision system (super regional centres)	334
	11.8.2 Remote control centres (regional centres)	336
11.9	**Other management and information systems**	342
	11.9.1 Freight traffic information system	342
	11.9.2 Vehicle identification system	342

Illustrations

Fig. 11.1	Typical structure of rail traffic management	312
Fig. 11.2	Levels of operational organizations of European railways	313
Fig. 11.3	Block diagram of the hierarchical structure	313
Fig. 11.4	Information flow through the hierarchy	319
Fig. 11.5	Concept of berth allocation	321
Fig. 11.6	Status of berths	321
Fig. 11.7	Stepping of train description	323
Fig. 11.8	Operator's control desk – Winterthur station – SBB	323
Fig. 11.9	Typical train describer system configuration on SBB	325
Fig. 11.10	Examples of records on SNCF	326
Fig. 11.11	IECC network	329
Fig. 11.12	Block diagram for automatic route setting using timetable data and train description on SBB	330
Fig. 11.13	Configuration of shunt route request system on SBB	332
Fig. 11.14	Typical arrangement of a passenger information system	333
Fig. 11.15	Traffic supervision system (RZÜ), Frankfurt, with four high density monitors, a dialogue monitor and a digitizer	335
Fig. 11.16	Typical configuration of a remote control centre	337
Fig. 11.17	Configuration to suit availability requirements on SBB	339

11.1 Introduction

Rail traffic management embraces the subjects of:

- planning and co-ordination of traffic operation;
- informing the customers, such as passengers and those who wish to transport freight;
- setting up of timetables and recording of completion and adherence to the timetable;
- planning and dispatching of trains;
- operating the control centres at all levels, such as: super regional centres, regional control centres and local stations;
- ensuring traffic operation even under irregular conditions, for example technical failure and severe weather conditions.

This chapter is divided into:

- traffic supervision
- timetable
- train description
- train recording
- operation
- passenger information
- centralized traffic control

It gives a general view of each of the topics and, where appropriate, addresses the particulars of different countries with regard to their operational needs.

Rail traffic management is usually organized and structured in such a way that the more general or global the activities are, the more centralized they are dealt with. The planning functions are on the top, the supervisory and control functions are near to the process. This leads to a hierarchical structure which is visualized in **Fig. 11.1**.

This approach has been adopted by several railway administrations and introduced into their operational concepts: **Fig. 11.1** indicates the levels of operational organizations that are currently being applied by European railways. The responsibilities that are assigned to each level may differ from country to country, as may the range of operational functions. The individual operational functions of the railways in relation to the hierarchical level are addressed below.

General management

- Planning and co-ordination of traffic operation.
- Providing guidelines for the setting up of timetables.

Super regional centres

- Setting up of timetables and recording of completion and adherence to the timetable.
- Verbal co-ordination of traffic according to the timetable and actual traffic flow in situations where global aspects have to be taken into account, such as delays, special trains, peak load due to holiday seasons etc.

Regional centres

- Informing the customers, such as passengers and those who wish to transport freight.
- Planning and dispatching of trains in co-ordination with staffed stations.
- Operating the control centre.
- Ensuring traffic operation under irregular conditions, for example under technical failure and in severe weather conditions.
- Feed back to super regional centre as required by the conditions.

Stations

- Informing the customers, such as passengers and those who wish to transport freight.
- Planning and dispatching of trains.
- Ensuring traffic operation under irregular conditions, for example under technical failure, and in severe weather conditions.

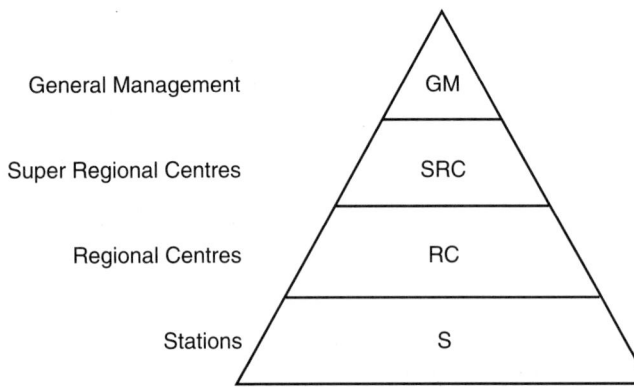

Fig. 11.1 Typical structure of rail traffic management

11.2 Traffic supervision

11.2.1 Operational concept

Passenger and freight transport are usually two business areas. The strength of each varies between countries. They are mostly of equal importance in terms of traffic operation. This also applies for most private railways. The railway network has to be shared by both transportation activities with some exceptions, for example BR multitrack lines and TGV lines in France.

The traffic management is determined by the operational concept. The operational concept will therefore be outlined for each individual country, in order to provide an understanding of the solutions adopted by the railways concerned.

Fig. 11.2 (left) Levels of operational organizations of European railways

Railway	Super Regional	Regional	Station
ÖBB	Y	Y	Y
NMBS/SNCB	Y	Y	Y
SBB	Y	Y	Y
DB AG	Y	Y	Y
RENFE	N	Y	Y
SNCF	N	Y	Y
BR	Y	Y	Y
FS	P	Y	Y
CFL	N	Y	Y
NSB	N	Y	Y
NS	Y	Y	Y
BV	N	Y	Y

Y Yes, operative or in the process of installation
P Planned
N No, not required or not yet planned

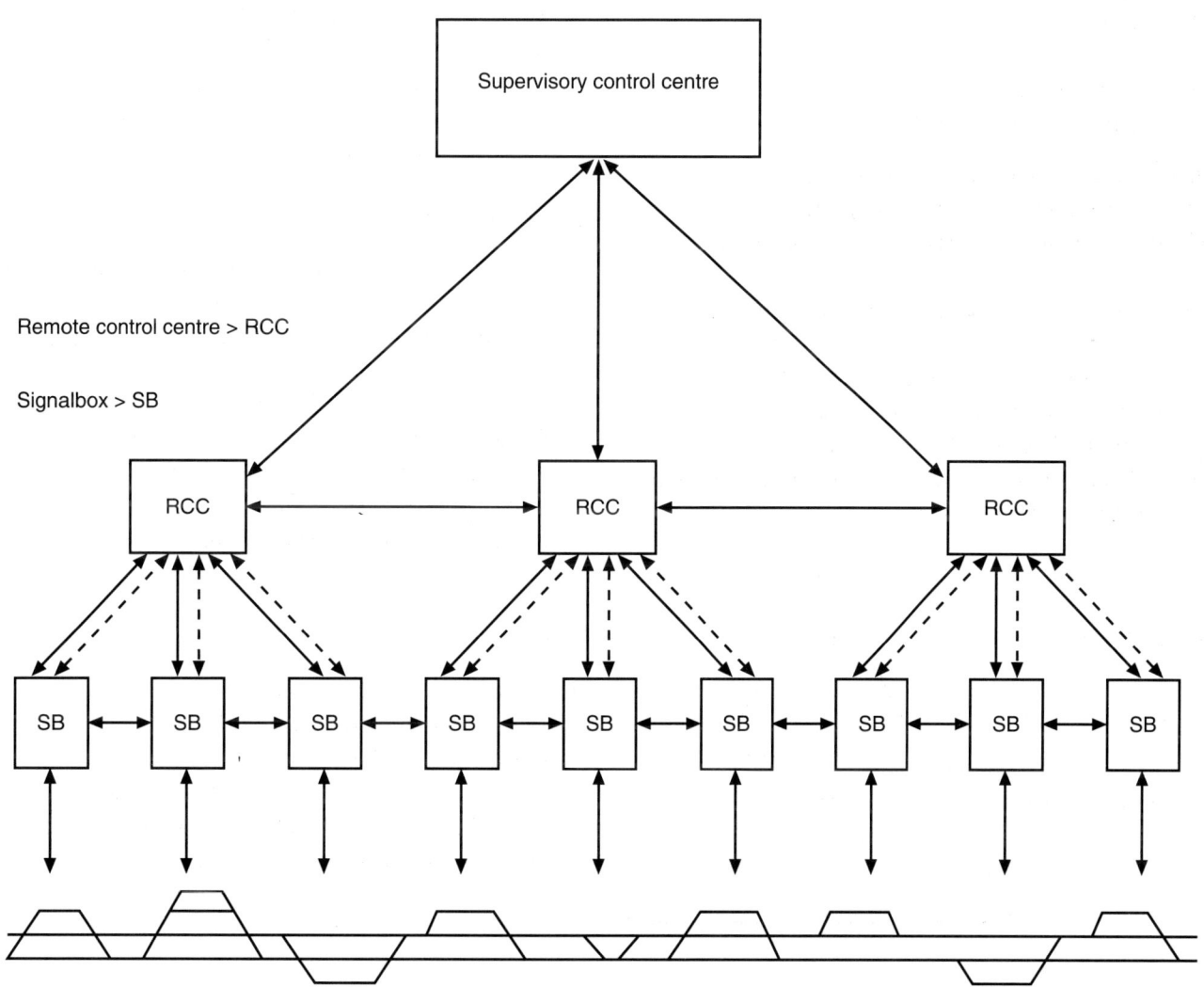

Fig. 11.3 Block diagram of the hierarchical structure

ÖBB

The Austrian railway network is divided into four regions. Each region has its supervisory control centre using a computer aided train control system for the supervision of the main lines. The train controllers make decisions in conflicting situations by telephone instruction to the signalmen. That is, there is no facility for direct control, thus the signalman remains responsible.

These control centres also have train radio equipment for direct communication with train drivers.

NMBS/SNCB

The NMBS/SNCB network is divided into five regions. Each region has its own supervisory control centre with regional dispatching, which is linked to the remote control centres by telephone links. Only a few centres are equipped with train graph.

There is one supervisory control centre for the entire network, with central dispatching of main lines. This computerized system directly controls the high speed lines, > 300km/h, and communicates with the control centres on the main line through the communication network.

The responsibility of the regional dispatchers is thus reduced to the secondary lines.

SBB

The network of SBB is divided into three regions. The traffic is mixed, using the network as required – that is there are no dedicated lines and tracks to business areas.

The passenger transport is based on a timetable concept which provides in general for at least one train every hour on the same minute in the same direction. This timing is backed by a node concept. The Swiss rail network has been split into nodes, normally at important junctions in order to provide for running time between nodes within an hour. The correspondence at the nodes is normally at the hour including an allowance for changing trains. This concept is referred to as the spider principle, with arrival before the hour and departure after the hour. Intercity trains connect the nodes, fast trains the cities and towns. Regional trains provide the majority of passenger services.

International trains are also scheduled on an hourly basis as far as the co-ordination with railways across the border permit and the respective national timetables take this into account.

The overall timetable allows for additional trains in peak seasons.

The freight concept is based on Cargo Rail (*Wagenladungsverkehr* – carriage load and full train load), Cargo Domicile (*Stückgutverkehr* – part load) and Cargo Combi (*Huckepack-und Grosscontainer-Verkehr* – Piggy back and container). There are also train services available for domestic transport, import, export and transit.

DB AG

The expansion of the offered services and the continuously rising quality requirements in passenger and freight traffic require a centralized supervision and dispatching of a considerable number of trains. On DB AG this is ensured in two levels:

- high quality, long distance passenger and freight trains are supervised and dispatched from super regional operational control centres;
- regional control centres monitor the entire traffic on the partial network.

The quantitative and qualitative requirements with regard to customer and market oriented train services can only be met by means of appropriate operation and communication technology. In order to ensure reliable supervision of the main lines DB AG introduced computer aided train supervisory systems, where the data of the actual train movements is collected and displayed on graphical displays. By means of time distance graphs the dispatchers of the supervisory control centres are able to locate and manage conflicting trains at an early stage.

The following developments are planned for the next few years:

- automatic detection and management of conflicting train movements.
- automatic transmission of instructions from the supervisory centres to the operative level.

In addition, links of certain dispatching tasks between the various operational levels are planned in order to concentrate into multifunctional units. The aim is to operate trains over more than 100km from one operational centre.

RENFE

In the RENFE organization there are two general offices which deal with customer services, passenger and freight transport and railway infrastructure.

In the general office for transport services there are offices and business sectors responsible for dedicated tasks. These are classified by:

- passenger services:
  - high speed;
  - intercity;
  - suburban.
- freight services:
  - load;
  - combined transport;
  - parcels.

In the general office for infrastructure there are offices and business sectors responsible for dedicated tasks, such as:

- infrastructure management:
  - planning and co-ordination;
  - maintenance of infrastructure;
  - traffic.
- infrastructure maintenance:
  - signalling systems;
  - projects for new installations;
  - direction of new works;
  - normalization;
  - technical specifications;
  - new products;
  - maintenance facilities.

SNCF

SNCF is divided into 23 regions. Each region has a traffic management centre from where the traffic of its region is supervised. In some cases there are also traffic control centres. The high speed lines are monitored by unique centres – PAR (*Poste d'aiguillage et de Regulation*).

BR

In Great Britain, BR is divided into business sectors with each responsible for particular types of service. The main business operations include:

- InterCity, mostly trunk passenger services;
- Network SouthEast, commuter services in the south-east of the country;
- European passenger services;
- parcel and postal services, rail express parcels;
- Regional Railways, all other passenger services;
- trainload freight, block freight trains.

Services are organized on a route basis with profit centres responsible for operations of one or more routes. Management and supervision is organized on a profit centre basis for both infrastructure and services.

Control of the railway is exercised by a line of route controllers who are in contact with the signalboxes on the route concerned and who have access to real time national running information.

The number of signalboxes on a route varies depending upon the length of line and age of the signalling. The future trend, however, is to concentrate control into a minimum number of strategically placed signal control centres.

FS

Many lines of the FS network are operated by centralized traffic control, with others being operated by systems that cater for co-ordination and supervision of local operation. FS is going to link all these systems with regional centres and the regional centres with a supervisory control centre. This integrated centre will be provided with facilities for planning and supervision of traffic, and for evaluation of results from it in order to propose appropriate changes.

CFL

The CFL network consists of only one region. All main lines but one carry mixed traffic. One main line and four secondary lines carry only freight traffic. National passenger trains are scheduled on an hourly basis as far as possible. The radial lines connecting Luxembourg City to the bordering network also carry international traffic. The timetable of those result from a compromise with the bordering railways.

There are both national and international freight trains which are concentrated or issued in most cases to or from the national marshalling yard located in Bettembourg near the French border.

Traffic supervision is normally done by remote control centres. Alternatively, stations are responsible for the traffic.

There is one supervisory control centre in Luxembourg City; its main tasks are:

- the co-ordination of traffic in case of irregular conditions, such as important delays, accidents, severe weather conditions;
- the setting up of timetables of special or urgent trains.

NSB

The NSB network is divided into four regions, each with one to three control centres, eight in total. The control areas are not restricted to the regional boundaries. NSB is divided into several business sectors for different services, such as:

- IC trains;
- regional trains and commuter services;
- trainload freight and block freight trains.

Four centres are responsible for all operational planning and for the operation at the eight control centres. Each control centre has functions for all main lines within the assigned area. These control centres are also provided with both train radio equipment and conventional mobile telephone system (NMT) for direct communication with the train drivers.

NS

NS is preparing for a substantial increase in the scope and complexity of the timetable. On the principal main lines the timetable concept is already in operation. This provides for 15 and 30 minute intervals for local trains, 30 minute intervals for Inter City trains and additional fast trains at peak hours. By 1996 this will become the NS three train system – that is Inter City, Inter Regional and Suburban.

The traffic control service unit will be contracted by the passenger and freight business unit for managing the timetable. The capacity management unit will be in charge of capacity allocation.

The traffic control service unit will carry out its work at two levels, disposition and process supervision. Disposition on NS is handled by 17 traffic control centres. This number will be reduced by co-ordination with the main traffic control centre in Utrecht.

BV

- BV stands for Swedish National Railway Administration in charge of fixed installations. The Swedish railway network is divided into five BV regions, each with one to three remote control centres. The control areas for these remote control centres are not restricted to the regional boundaries.
- SJ stands for the Swedish State Railways in charge of traffic operation. SJ are responsible for schedules, all operational planning except network and for operating the 11 remote control centres.

Each control centre has centralized traffic control functions for all main lines within the area. These control centres are also provided with both train radio equipment and conventional mobile telephone system (NMT) for direct communication with the train drivers.

SJ is divided into several business divisions with dedicated tasks:

- traffic management division.
- passenger services division for:
  - SJ traffic;
  - contracted traffic;
  - Stockholm regional traffic;
  - parcels and post;
  - marketing.
- freight services division for:
  - marketing;
  - car load traffic;
  - system train transports;
  - continental traffic;
  - iron ore traffic.
- rolling stock division for:
  - supplies;
  - development;
  - workshops.

11.2.2 Operational planning

Operational planning is mostly the responsibility of the upper level of the operational hierarchy and covers the long-term range in terms of years. The scope of the planning normally encompasses many aspects of the business, including an assessment of future requirements and the provision of future resources. These include:

11.2.2.1 Customer

- Marketing passenger service/transport.
- Marketing freight service/transport.

11.2.2.2 Timetable

- Yearly planning broken down to minutes, deadline for request of modification, review by the public on the level of municipality.
- Basic timetable containing all trains including train orders for special trains and provisions for alternatives in exceptional situations.
- Generation of graphic timetables.
- Generation of official timetable (booklet) for the public transport.

11.2.2.3 Network

- New lines, extension of existing lines, for example additional track on the open line.
- Reconstruction, track renewal.
- Limitation of work sites, co-ordination with timetable.
- Scheduling of open lines.

11.2.2.4 Rolling stock

- Circulation of rolling stock, including empty carriages.
- Purchase and maintenance.
- International and national exchange of rolling stock with all relevant administrations.

11.2.2.5 Tractive units

- Circulation including return trips.
- Purchase and maintenance.
- International and national exchanges with all relevant administrations.

11.2.2.6 Personnel

- The personnel of the railways are of different educational and professional background.
- Some may start at the beginning with an apprenticeship, others may come from industry or from universities.
- Recruiting, educating, training.

11.2.2.7 Energy

- Electricity.
- Gasoline, fuel.

These general considerations will not be discussed further, other than to confirm that they are of great importance as part of rail traffic management.

11.2.3 Disposition

Disposition is understood to mean action in the short term – from hours to one year. Instructions are prepared as required based on real time information, regional to super regional, without direct control of process.

Irregularities are picked up, short-term measures are planned and transferred to the operational control levels. Disposition also takes the market into consideration, and several other aspects, such as:

11.2.3.1 Customer

- Rapid changes of demand, for example sporting events requiring spontaneous dispositions.

11.2.3.2 Timetable

- Deviations to target timetable, plus or minus.
- Arranging of trains according to train orders provided by base timetable.

11.2.3.3 Network

- Arranging modifications of track utilization.
- Undertaking unplanned track repairs, fixing of consequences to timetable and arranging the required track repair.

Today, efficient and economic rail traffic management requires powerful, computer based systems. Most railways use such systems or plan to use them on the levels of super regional and regional centres with distinctive scopes of disposition assigned to each level (see **Fig. 11.2**).

In general, the main objective of the disposition is to minimize the effects of irregularities and to make decisions on measures to be taken in order to meet this objective. Events that have to be looked at are:

- considerable delays;
- line prohibition;
- important faults;
- special customer requirements.

The function of disposition is divided into overall and final disposition, and carried out by the respective centres.

For *overall disposition* at the super regional centre with no direct control but the tools for the overall disposition and for the direction of the dispatchers at the regional centres, the required information is:

- actual timetable data;
- actual position of trains;
- actual data on availability of the network;
- actual data on freight traffic, demand, load, destinations;
- actual data on passenger traffic.

Based on this information the supervisory control system will draw the actual timetable in relation to the planned timetable, which will be available on a graphic display and serve as a means for decision making.

The output of the system contains:

- updated track occupancy plans of the stations within the area;
- forecast of train moves;
- early detection of conflict situation where decisions will be required to set priority;
- early detection of difficulties (conflict) to maintain correspondence;
- listing of trains for which decisions will be required;
- graphical display of train orders.

As soon as the difference between given and actual timetable reaches the threshold the supervisor will be requested to take care of this situation. His measures can be new train orders, shift of crossing points, and change of train orders for which he can recall checklists from the system. His decisions have to be transmitted to the next level, the regional centre, since he is not in a position to take direct control.

Final disposition at the regional centres is where the staff are primarily involved with train dispatching according to the timetable and with the instructions received from the super regional centres. Their daily business is supported by computerized supervisory control systems providing, for example, detailed timetables and exact information on train positions. But of even more importance are the automatic train routing facilities.

11.2.4 Control

The process is controlled according to the target timetable and dispositions within an assigned area, automatically or manually, with the area being usually regional or local. This function is performed according to real time information and accounts for irregularities. However, in addition to train routing the control function covers all other aspects such as:

11.2.4.1 Customer

- Monitoring of customer flow;
- Passenger information, for example in stations and on trains.

11.2.4.2 Network

- Track assignment.
- Planning and carrying out of shunting.
- Planning and carrying out of track maintenance.

11.2.4.3 Personnel

- Planning taking into account staff leave for holiday and other purposes.
- Activation of job assignment.
- Emergency alterations to train crew due to late running.

11.2.4.4 Energy

- Switching of transmission lines.
- Switching of catenary.
- Selecting of traction voltage and frequency at border stations.

The control of routes is usually initiated according to the actual progress of trains, either manually or, more often, automatically. For this, information is continuously collected at various stages from various sources.

Examples of such information are:

- actual timetable data, containing programmed route data, where required;
- orders from super regional or regional centres;
- actual status of wayside objects;
- actual train description;
- actual track engagement;
- identification of loco in shed tracks.

The data received is continuously processed to an exact display of real time information of all train movements and stored for other purposes, such as listings of train progress, forecasting, reporting etc. This data is also processed in combination with the continuously updated timetable data for the utmost possible automatic control of operation.

11.2.5 Safety

Safety is provided by the signalling system. Route control is only performed as long as all safety conditions are met. This also applies for all normal operational commands which will be executed only if allowed. In cases where a dispatcher or a signalman has to intervene, bypassing part or all of the interlocking, he has to take over the responsibilities that require predetermined procedures. These procedures are enforced in such a way that no such intervention is made without all concerned being aware of the actions to be taken. The range of such possible interventions depends on the operational conditions of each national railway. The details of the individual railways in terms of operation are addressed in **11.4**.

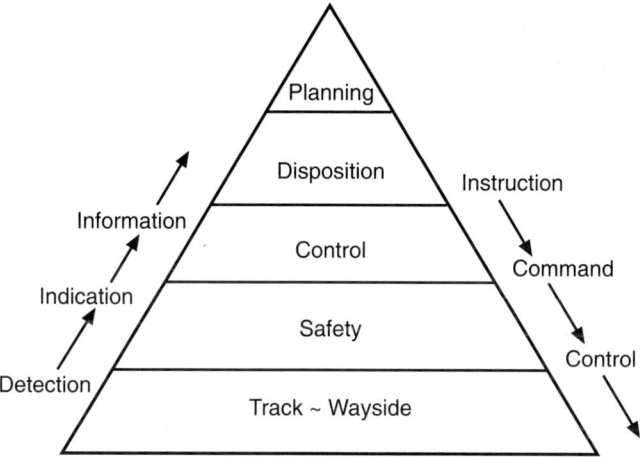

Fig. 11.4 Information flow through the hierarchy

11.3 Timetable

The timetable is a means of planning the railway operation correctly in order to meet the target for safe, reliable and economic transport. The concepts, however, may differ from country to country because of their specific operational requirements.

The timetable is usually documented in three ways:

- graphically, for use at stations and for track maintenance.
- tabular, as a service timetable for drivers.
- paperback for customers/passengers.

Timetable generation software is available, which in some cases assists and in other cases does the work of generating timetables. In its simplest form, the timetable generator converts a manually produced timetable into a form suitable for computer storage and compatible with the train describer database. However, much more complex and sophisticated timetable generators are available which work from basic information such as the number of trains, the headway requirements and the track topology; and these automatically generate a full timetable including crew assignments.

The timetables usually contain:

- detail planning per centre;
- planning per station;
- planning of resources;
- planning of track utilization in stations and on the open line;
- planning of rolling stock;
- planning of personnel.

As a matter of fact it is unlikely that an ideal timetable will ever be published – ie a timetable that suits everybody. A timetable is a compromise taking into account the needs and wishes of customers balanced with the economic utilization of available line capacity and rolling stock. The timetable is reviewed and updated periodically.

11.4 Train description

11.4.1 Introduction

The purpose of train description is to identify the train at a particular place and to keep track of its progress. This identification allows the operators to follow the progress of trains and enables them to detect the origin, destination and characteristics of trains. As far as unambiguous information can be derived from the train description and, if necessary, with the help of timetable information, the train description can be used, to a great extent, as a means for automation. All depends upon the setting up of train identification by inserting the appropriate description at the location of entry into a controlled area, from where the information has to be stepped according to the progress of the train.

Train describer systems are important tools for efficient traffic management and provide the essential backbone for traffic supervision, automatic train routing and passenger information systems. Modern technology allows for sophisticated application of all information.

The train description (train numbers) can take the form of a numerical or an alphanumerical designation system, depending on the country and its use.

Train numbers are of several digits (system of allocation according to railway rules) and serve the purpose of:

- identification of passenger, freight, Intercity, direct or regional trains;
- discrimination of origin and destination;
- identification of running direction according to international agreements (UIC), ending number pair for east to west and south to north.

Typical examples of designation systems are:

BR

position	type	signification
• first character	numeric	train classification
• second character	alpha	train destination
• third character	numeric	train number, character one
• fourth character	numeric	train number, character two

- for example display 1A27 may be the 27th express train, destination London

SBB

All train descriptions are numeric from one to five digits. Concept and application are predetermined in the appendix to the operational rule book (no. 310.3), with examples as follows:

number	signification
• 1 to 199	Euro–City trains (EC)
• 200 to 499	International fast trains
• 500 to 999	Intercity trains
• 3000 to 8999	Regional trains (suburban, agglomeration)
• 14 000 to 19 999	S-Bahn trains (agglomeration)
• 23 000 to 25 999	Service trains
• 40 000 to 49 999	International freight trains of various classes

- all these descriptions take the places of second to sixth digits, the first digit being used or as spare for automatic routing purposes.

11.4.2 Summary of technical solutions

- Electronic equipment using discrete components.
- Computer systems.
- Display by means of light emitting diodes (LED), liquid crystal displays (LCD) or video displays which may be either monochrome or colour.
- Set up of train numbers either manually at the origin of trains or automatically by computers according to timetable information and route setting.
- Automatic stepping from signal to signal with the progress of the train by monitoring the state of the signalling.
- Recording at predetermined locations of place, time and train number with print out on request.

11.4.3 Summary of functionality of train description

- Display of train numbers of a certain area at information/control panels, video displays or other systems.
- Keep track of train moves by appropriate stepping of train numbers, also under special operational conditions, such as: group signals; train entry into engaged tracks; simultaneous departures of trains from the same platform track to the opposite direction.
- Exchange of train numbers between control areas.
- Facility for simplified automatic train routing by means of the first digit.
- Transmission of criteria for automatic route setting to the remotely controlled stations.
- Interface with other systems, for example passenger information, supervisory control centre, train graph.
- Log of train moves at determined spots onto printers.

A track layout is divided into signal and block sections, where normally only one train can be accommodated in one section at a time. In every section a berth is provided for the display of respective train numbers. A train number is assigned to each train, which has to be set up at the starting point of the train. By the progress of trains the numbers will be moved from berth to berth in accordance with signal and route information.

Train describer systems offer a range of alternatives for the display of train numbers:

- LED or LCD on panels or mimic diagrams.
- video display with simplified geographical arrangement according to the track layout.
- tabular display of train numbers combined with their station and berth identification.
- display on other systems, such as a colour video monitor.

The train numbers are usually moved from one berth to another when the train passes a signal – ie signal, track circuit and point information are required. Route information is required for the stepping which can also be used for other purposes. For example, on SBB the address information is complemented by the information as to whether the train has passed the signal or not as soon as the route is set. A particular berth can thus have three states:

- empty;
- train not yet passed the signal 'ROUTE';

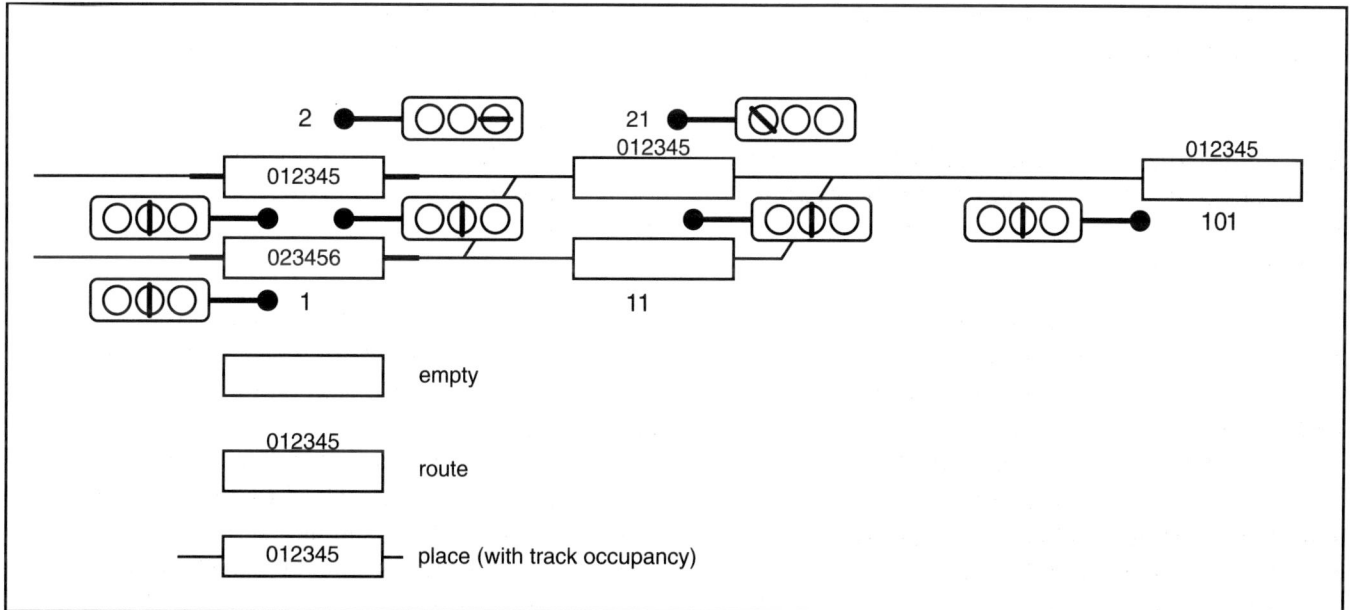

Fig. 11.5 Concept of berth allocation

Fig. 11.6 Status of berths

- train passed the signal 'PLACE'.

This different information can be of further use for automation purposes.

11.4.4 Functional description of signalling information and stepping

The train describer receives information regarding the position of trains from the signalling system interlocking. Track circuits or axle counters detect the presence of a train in a given section of a track. The interlockings hold information concerning the status of the track circuits or axle counters, either occupied or not clear, the position of the points, routes that have been set, and the status of the signals. In relay interlockings, the information relating to each of the track circuits, points, routes and signals is made available to the train describer system by relay contacts. The train describer system interfaces with these contacts by a parallel interface. The information is then transferred to a serial code and processed by the train describer system. Electronic interlockings usually provide a serial stream of the signalling information.

The train describer system (TDS) holds a database of the status of all the signalling information. The database is updated in real time, as train movements occur. The train describer only monitors the status of the interlockings – it has no direct control function, although it can be used to provide input for automatic routing. The train describer system also holds a database containing train number berths.

A berth is a storage area allocated to, or associated with, a given signal section, and holding the train description itself. When the TDS recognizes a change of state of any of the signalling inputs, it will process the change. The TDS will follow a certain algorithm relating to the signalling information and certain conditions that need to be satisfied in order to generate a step. The change of the state recognized will be processed with the other related conditions by the TDS, according to this algorithm. The TDS will then step the content of the originating berth associated with the given signalling information, to the destination berth. The panel display or video display unit (VDU) will be updated automatically to show the signalman the new position of the train. The movement of train numbers from one berth to the other is known as *stepping*.

In the BR four digit TDS, the description step is usually initiated by *route set and first track occupied*. If a route has not been set and the signal is passed at danger, the description will remain in the rear berth. This can be regarded as admissible evidence of a SPAD (signal passed at danger). Some older TDS required *signal cleared and first track occupied*, which is even better from the evidence point of view. However, if the signal has to be passed at danger by a succession of trains, ie during a failure, the signalman would have to reinstate the description for each one. The above applies to controlled signals. For automatic signals in route relay areas, there is no signal indication brought back to the signalbox. The description would step on first track occupied, unless it is already occupied by the previous train.

Some countries use modern technology so that the stepping is carried out in two different steps. In the first step the receiving berth is marked with an address *route*, indicating that the route to the destination berth has been set, the train is still in the originating berth. With the actual progress of the train the berth will be updated with the information *place* that the train has passed the signal in rear and thus entered this section. The step *route* is carried out as soon as the route has been set and the associated signal cleared, this as far as routes are set and signals cleared. When the train progresses, the step *place* is triggered as the train occupied the first track circuit after the signal.

An empty berth will be marked *blank* at the time of clearing its associated signal to either direction and the destination berth will receive the same information. As soon as a train number is inserted into the originating berth, its information will immediately be transmitted to the destination berth as route indication. If the train progresses without a train number in the starting berth – ie the train occupies the track circuit after the signal – then a dummy number will be generated in order to inform the signalman of missing train description. The stepping is illustrated in **Fig. 11.7**

11.4.5 The operator's control

The signalman communicates with the train describer via an operator's control unit (OCU), normally in the form of a keyboard and a VDU, interfaced to the TDS by an appropriate data link. A TDS may have several OCUs, of which there is normally one per operating position. The TDS will store a number of OCU maps, depending on the number of OCUs and the individual requirements. A given map can be called up from any of the operating positions.

Usually the TDS does not require manual operation, the train numbers normally being transmitted between the operational control centres via local area networks (LANs) or other appropriate data transmission systems. When setting up new trains or for ending trains or under failure conditions, however, manual intervention by an operator may be necessary. Therefore, the following functions can be performed at an OCU:

- *interpose a description*: This is the process of entering a train description into a berth address. Interposing is one of the main methods of entering train numbers into the system, especially when a train is set up, a train enters the area without a train description, the following train is given a new description.
- *cancel, ie clear a berth contents*: Either cancel a berth address or a certain train number, when, a train has been removed from the system by a shunt move, two trains are combined into one.
- *recall the contents of a berth*: The train number contained in a berth is displayed in a predetermined area of the display.
- *recall the location of a train*: The train number is displayed in a predetermined area of the display.
- *edit a train description*: Modify or replace a train number already in the TDS if, the train number in the system is incorrect, a dummy number shall be replaced by a correct one, a train changes its description, for example a returning train.
- *manual stepping of train description*: When automatic stepping could not take place because of, shunting moves, moves using auxiliary signals, moves to verbal order, any failure of stepping.

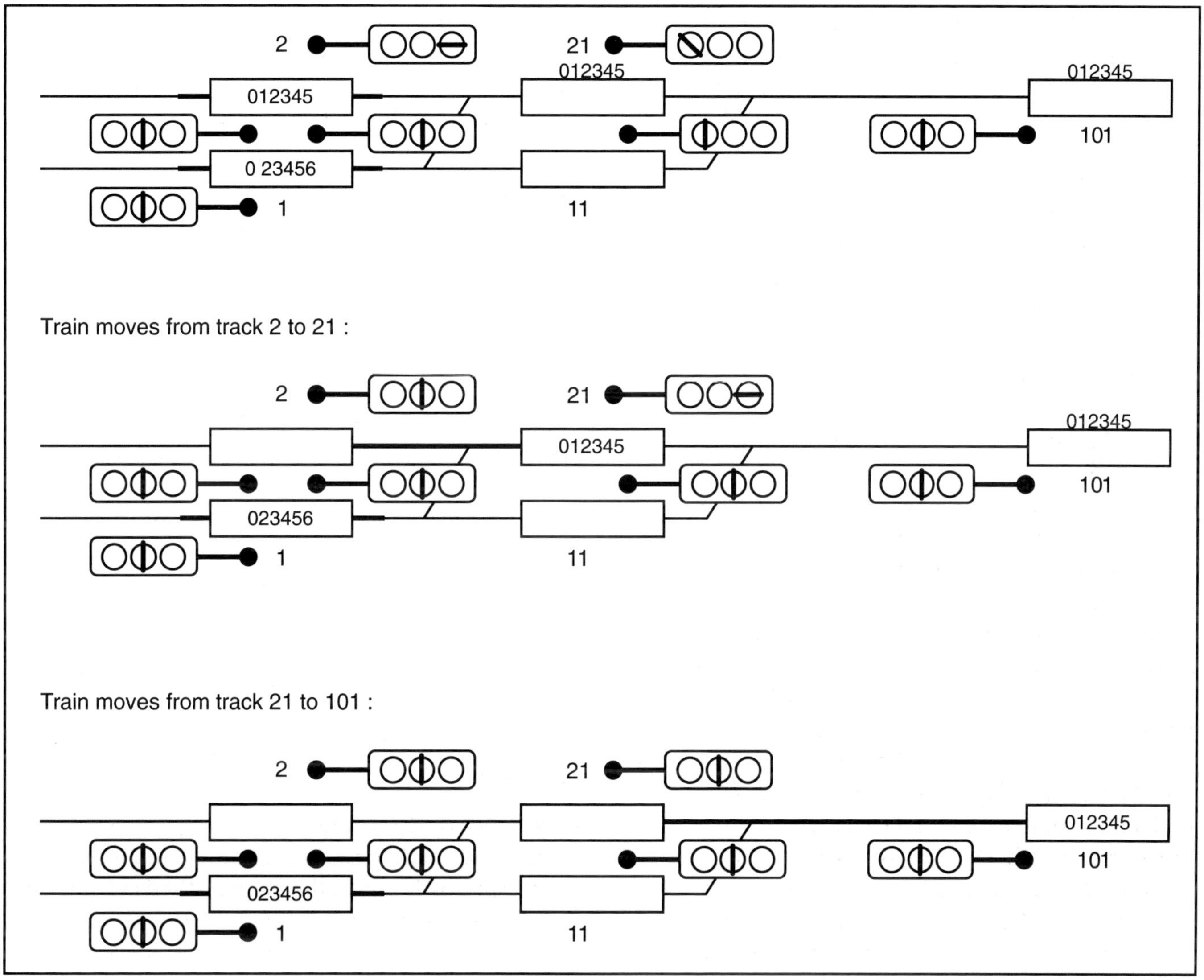

Fig. 11.7 Stepping of train description

Fig. 11.8 Operator's control desk – Winterthur station – SBB Photo: Siemens Integra

11.4.6 Fringe boxes

Fringe or outpost signalboxes are located at the boundaries of a controlled area, and from a train describer viewpoint, act as remotely connected main box OCU. The fringe box signalman enters or interposes train descriptions as trains enter the controlled area.

The fringe box OCU is usually a VDU with keyboard, connected to and driven from the TDS at the main box by a serial communication channel. Train numbers set up by the fringe box signalman are transmitted to, and automatically displayed on, the main box control panel or VDU in approach berth displays. The TDS will step the train numbers as the routes at the main box are set and the train moves into and across the area.

11.4.7 Link of train describer systems

Each train describer system covers an area assigned to it. At the boundary of each area a communication link is required in order to transmit the train description from one area to the other. Depending upon the system of the adjacent area, modifications on the train description may be required; for example a superimposed routing information for automatic train routing. Overlapping of berths between two adjacent areas is usually provided for – its extent depends on the operational requirements.

Communication between the adjacent TDS is via serial communication channels. When a train is going to leave the area of one TDS, its train description is automatically sent to the adjacent TDS via the communication channel. The train description is entered into the approach berth at the receiving TDS and stepped as the train moves through the new area.

Serial communication interfaces are required in many of the links to a train describer system. Standardization of the interfaces simplifies maintenance and reduces the required spares holding. The software protocol used by the communication link is also standardized. A fixed protocol, to be applied across all links, enables one manufacturer's TDS to talk to another manufacturer's TDS, or platform indicator system, etc.

11.4.8 Fault logging, alarms and warning messages

The train describer is a monitoring and information system, and the controlling computer is capable of identifying system operating problems and equipment fault conditions. The TDS provides both the signalman and the system maintainers with visual and audible indications of fault and warning conditions.

The operator is informed of the presence of a problem and the nature of it, but his alarms are limited to operational activities.

The maintainer's displays are more comprehensive. The maintainer receives detailed printouts from fault printers connected to the TDS. These printouts include an indication of the priority level of each fault, ranging from those that are purely advisory, to those that demand immediate attention. They identify the nature of the fault or warning and the time of occurrence. Maintainer's displays cover equipment faults as well as operational faults.

The signalman's or operator's alarms are normally presented at his control unit (OCU). The alarm is a flashing display on the VDU accompanied by an audible warning. The operator must acknowledge the alarm by pressing an acknowledge button in order to silence the audio warning and to steady the VDU display.

The maintainer's alarms are presented on fault printers and on a separate equipment fault alarm panel. The maintainer is normally only concerned with equipment faults rather than operator's warnings and alarms, although he is informed of all identified problems.

The TDS monitors both itself and all of the equipment in the system, so immediate action can be taken to correct a problem.

11.4.9 Automatic editing of train numbers

On some networks, such as NMBS/SNCB, the train numbers can be automatically replaced by the corresponding number according to the timetable. This is carried out, for example, by the integrated electronic control centre (IEEC) after the arrival of a train at its destination. Similar arrangements are available for splitting or joining trains.

11.4.10 System configuration

With a train describer system using single hardware configuration a failure of any unit will lead to the complete loss of all facilities. In situations where, for operational reasons, the total loss of the train describer is unacceptable, a more complex configuration of the hardware modules is required. A suitable arrangement is one where automatic reconfiguration of the hardware modules takes place whenever a failure occurs. Self-healing configurations of this nature, which normally provide for uninterrupted operation in the event of a major hardware failure, inevitably involve duplication of equipment. Each computer has its own dedicated interface system for receiving signalling information and train descriptions including those of the operator's control unit and fringe boxes. The output system – displays and logging printers, etc – however, is driven by an auto switch unit, which takes output from the online system only. Both systems will normally be running and performing all of the functions of a train describer, but only one output system will be connected by the auto switch. In the event of a fault affecting this online system, or its input interface, the outputs will be switched automatically to the standby system. An equipment fault alarm will be raised automatically. Automatic switching is performed under the control of a

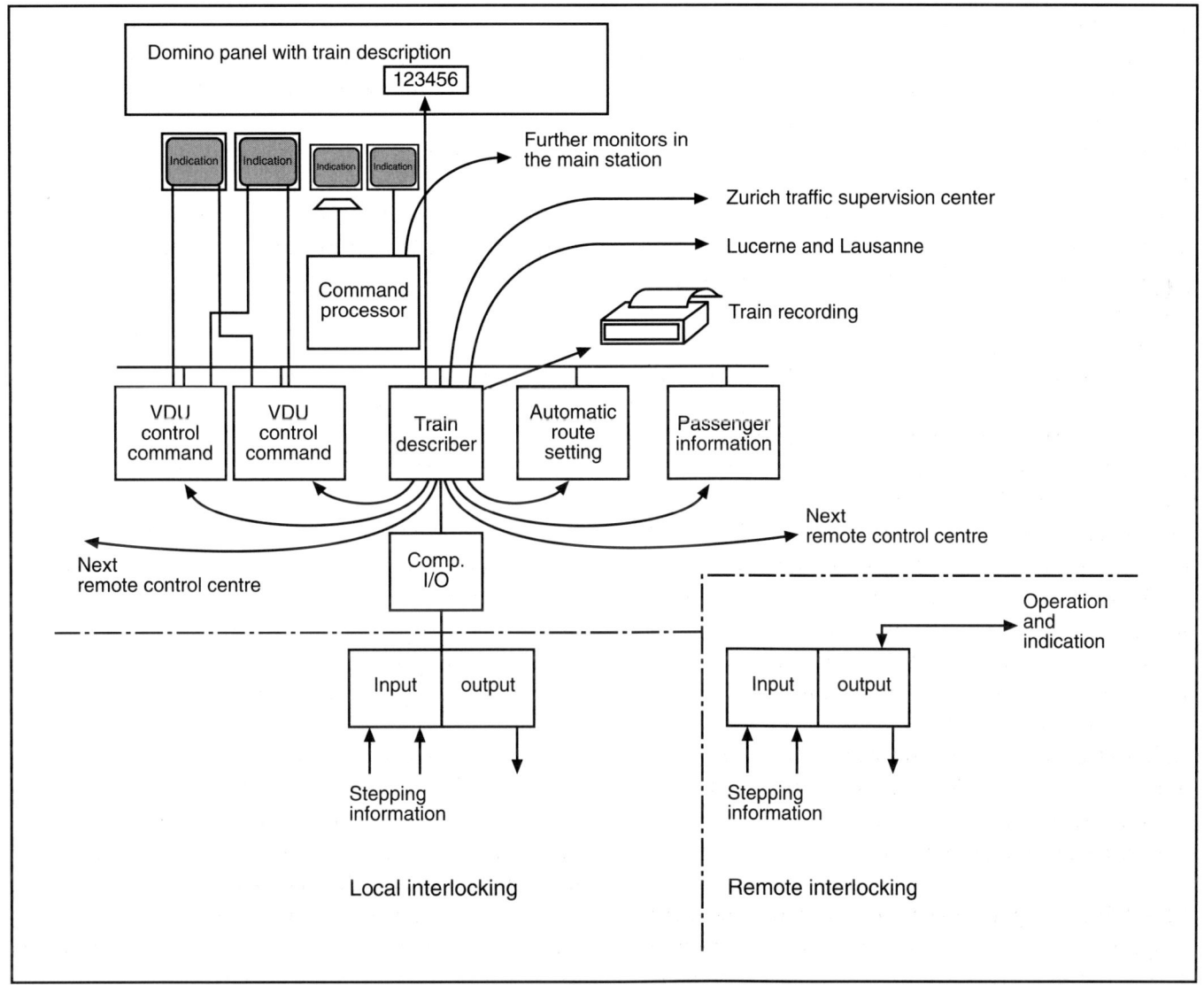

Fig. 11.9 Typical train describer system configuration on SBB

watchdog timer. Each of the computers is required to report to the watchdog timer at intervals of a few seconds. They only do so provided the systems they are controlling are fault free.

Should the online system fail to report, then provided the standby system is healthy, control of the train describer is switched to the standby computer. If the cause of the problem necessitating switch over was the computer itself, or if one of the computers had to be disconnected to repair its subsystem, that computer will need to be updated with the current state of the train describer when power is restored. This is carried out by a serial data link between the two systems, known as update link. The maintainer can force the memory data of the online system to be downloaded to the restored standby, thus allowing normal operation to be resumed. Update is accomplished in a few seconds. Other duplicated configurations are possible depending on the degree of protection against equipment failure required by the railway authority. All such configurations are designed to increase the system availability and improve reliability.

11.4.11 Safety and other aspects

The train number is utilized as a means for information and automatic control which can be carried out only if safety conditions are met – ie there are no safety requirements but requirements on reliability and availability.

Railway operating staff, in remote sites away from the signalbox, can be informed of train movements and other relevant information, on printers and VDUs at appropriate locations. These VDUs and printers are driven directly from the TDS. Terminals may be equipped with interrogation facilities which enable the information provided to be increased by asking the computer for specific data. The complexity of the train reporting network depends upon the railway layout, the extent of the control area and the particular operational requirement.

11.5 Train recording

Statistical information can also be produced as a permanent record of train movements for operating staff. Such a permanent record can be provided on a train graph recorder or computer log printout, using information derived from the basic train describer function.

Graphical recording of:
- actual train moves;
- planned train moves;
- comparison of actual with planned moves.

Method:
- by stepping of train numbers;
- track occupancy.

Means for recording:
- manually on paper;
- *automatically on*:
  - monitors;
  - plotters;
  - printers.

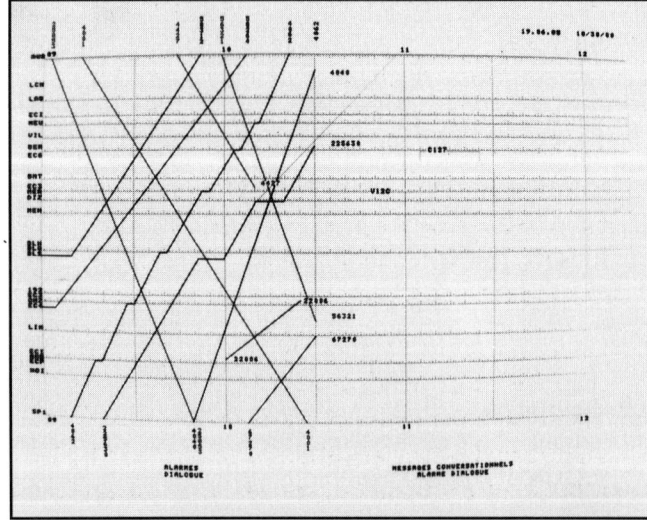

Fig. 11.10 Examples of records on SNCF
Photo: GEC Alsthom

11.6 Operation

11.6.1 Introduction

Operation in the context of rail traffic management means all aspects that are related to the interface between operators, supervisor, dispatcher, signalmen, etc, and the process control. A distinction has to be made as to whether the signalbox should be operated as stand alone or as part of a remote control centre or centralized traffic control.

The stand alone system of a relay interlocking is usually equipped with a panel where a mimic diagram and the means for manual control with push buttons or switches are arranged. Push buttons and indications are arranged geographically, where the control and indications are combined; in other cases switches are arranged separately. Commands are carried out by two button operation, entry–exit principle.

Commands bypassing part or all of the interlocking require predetermined procedures, such as breaking of seals of emergency buttons with recording of time and command carried out. Checklists *must* be consulted. This kind of operation is known as *critical operation*.

Large stations are equipped with a vertical mimic diagram including all means for manual operation ('Panorama') and a separate desk type panel is provided for regular operation with push buttons or keyboard. Regular operation means route setting and other repetitive commands which have to be carried out frequently. The configuration depends on the specific requirements and can range from regular operations at the panel to a full duplication of all operational means at the panel and vertical mimic diagram.

The stand alone system of an electronic interlocking is mostly equipped with a set of video display units and a keyboard or/and a keypad.

11.6.2 Display – visualization of the process

Two distinctive systems are currently in use:

- conventional display on mimic diagram using mosaic type or plain panels in desk or wall form.
- colour video display.

As long as the operator or signalman has no facility to insert commands that may bypass the safety functions, the display does not need to be vital. But as soon as the operator or signalman can do this, arrangements have to be made towards a safe display.

Various systems are in use:

- conventional panel with mimic diagram and push buttons.
- panoramic display panel with or without possibility of manual operation at the display panel.
- video display for indication of commands.
- colour video display for local area and/or remote controlled area.

The colour video display needs to be fail-safe where the operator or signalman can send critical commands. This can be achieved by a dual processor system generating the information to be displayed. The display unit is switched to use alternately one of the two identical information channels at an interval of 1 second. If on one channel the information is different from the other one, it can be detected by the operator since the display will flash where the difference is. Thus, the system complies with safety requirements. Normally signalbox information, train numbers and catenary status information are displayed.

As an example, SBB is using the following concept:

Each interlocking function displayed is represented as a graphic symbol. All functions are described by Karnaugh diagrams, which include exact definition of every possible state and its appropriate representation in terms of symbol and background colour. A Karnaugh diagram is a method of representation of all 2^n possible combinations and logical variables of a process element. In colour video technique this method is used to describe all possible element states for the process visualization of each element. The diagram in table form is evaluated online by the computer system.

According to the intended operation various pictures can be selected:

- overview of several stations for supervision of trains on a big scale.
- standard close up of station for the operation of train and shunt routes.
- full close up of station for operation including critical and emergency commands.
- main signal close up of station with display of main signal designations.
- shunt signal close up of station with display of shunt signal designations.

By the issue of commands the appropriate picture will be selected automatically and displayed – no blind operation is possible with this system.

The colour video display system is based on a two out of three computer system with manual switching over facility.

FS uses a fail-safe concept based on a three computer system where the video memories are compared.

Full graphic displays associated with appropriate means of operation, using, for example, Windows-X technique are in use on various European railways or else development is currently in progress.

11.6.3 Control functions – manual route setting

Means are provided for the operator to send commands to the system he controls. These commands consist of route related or individual commands. There are various systems of manual operation, such as push button, push and pull button, two buttons to be pressed simultaneously, two buttons to be pressed one after the other, switches, keyboard using standard arrangement or purpose designed with dedicated function keys; and other means such as mouse, light pen, digitizer and trackerball are also in use.

The commands may be summarized as:

- interlocking commands: route setting, route release, points switching, block and auxiliary functions;
- train describer commands;
- automatic train routing commands: enabling, disabling, data input, modifications of parameters;
- passenger information system commands: data input, modifications of parameters;
- shunting operation commands;
- level crossing commands.

Conventional operation is carried out at the control panel by means of push buttons. According to the command, simultaneous operation of entry–exit buttons is required, whereas for other commands the individual button has to be activated together with a common control button, for example point control, route release.

At larger stations or at remote control centres,

keyboard operation including other means like mouse, light pen, trackerball as part of an integrated operator's desk, are provided. The integrated workstations provide for all the functions required, such as:

- data terminal and colour video display for the visualization of the process and the dialogue;
- keyboard operation – easy access to all controlled stations, various kind of manual interventions;
- train describer – for actual information of the operator or signalman on train location;
- automatic train routing system – triggers route setting automatically according to the programmed data and the actual progress of the trains;
- means for updating and modifying of programmed data;
- passenger information – platform indication for the announcement of train departure time and destination;
- shunting operation – voice communication or radio controlled.

The keyboard operation system enables operation of the interlocking and other functions of a control system to be carried out by standard keyboard and/or other standard means. It is implemented on the command processing computer and, together with the colour video display system, represents the actual interface between the operators and the process.

The commands on SBB, for example, are entered as strings of code words with parameters. Frequently used code words are kept short, seldom used commands and critical commands have long code words. Various facilities assist the operator. For example each code word is associated with a dedicated mask where the operator has to fill in the gaps. Incorrectly entered commands are rejected, an explanatory text is displayed and the cursor is positioned just below the incorrect parameter value.

In addition to the summary of commands given above, the following commands are normally included for standard signalbox operation:

- control of block functions;
- control of line prohibition;
- control of track prohibition;
- critical commands which may bypass safety conditions;
- furthermore there are some voice proceed commands which also bypass safety conditions, which, together with the critical commands

require a strongly defined procedure to be followed by the operator using a set of check lists which are designed for this purpose only.

For large schemes, BR, and to some extent NMBS/SNCB and other railways, are using an *integrated electronic control centre* (IECC). The BR concept, for example, consists of two local area networks on which are configured a number processor modules. One network deals primarily with signalling functions, the other with information and interfaces. The segregation is designed to ensure control over the level of data flowing in the signalling network, in order to maintain quick response times.

The various processor modules comprise solid state interlocking (SSI), automatic route setting (ARS), gateway system (GWS), timetable processor (TTP), external communication system (ECS), IEEC system monitor (ISM) and signalman's display system (SDS), and remote relay interlocking interface (RRI).

It is SDS that forms the man–machine interface. One is provided for each signalman's position up to a maximum of three. The signalman sits at a workstation in which the SDS is housed together with up to five high resolution graphic monitors (128 characters by 48 lines), a keyboard and a trackerball device. The graphic monitors depict the layout under control using a special character set. Both area overview and detail screens are available, plus a general purpose display for alarms and simplified adjacent area real time maps. Command inputs and output messages are also handled by the general purpose display, and communications facilities can be incorporated into the workstations.

Route setting is normally carried out automatically by the ARS package in response to the approach of trains, based on data passed from the timetable processor. ARS hold operating strategies which cater for out of sequence and late running. They are designed to minimize the overall delay in the event of a conflict occurring. The signalman is able manually to set routes ahead of ARS and has control to remove certain trains or areas from ARS operation. Manual route setting is required for unscheduled movements and for some position light signals.

The trackerball and associated control buttons are used to set routes on an entrance–exit basis. This is carried out by positioning a cursor over the signal symbol from where the route is to commence, followed by pressing a *set* button. The cursor is then moved by operation of the trackerball to a position over the symbol of the signal at the end of the route. The *set* button is again pressed and when the route correctly sets the relevant portion of line changes from a grey colour to white. Track circuit occupation is shown by the line colour turning to red. The train description number is displayed in line with the red occupied indication.

A cancel button is provided for use in conjunction with the trackerball for manual cancellation of routes. Similarly three control buttons are used for individual point operation in association with positioning the cursor over the symbol of the points to be moved. The trackerball is used for many of the other routine commands in conjunction with a set of icons depicted in the lower portion of each graphic display.

All signals, points, level crossings and others are dynamically displayed on the graphic monitors. Equipment numbers are displayed optionally upon request by the signalman. The setting up of train descriptions and interrogation of the system are carried out through the keyboard which is also available as a fall back mode of operation in the event of trackerball failure.

The gateway system – GWS – interfaces between the signalling and information networks, regulating the exchange of traffic between them. The various timetables applicable to particular days of the week are held by the timetable processor. The relevant version is passed to the ARS processor in 12-hour portions and is available to assist in the driving of platform information systems. Information passing in and out of the IECC to other computer systems such as those at adjacent signalboxes, stations and central offices is handled by the external communications system processors. Their main task is to provide translation of message formats to and from the connected systems. Fault reporting and system monitoring are handled by the ISM as is time synchronization of the complete system with reference to the Rugby radio clock or the Dresden clock in the event of loss of signal from Rugby. A technician's interrogation facility of keyboard and monitor acts as a window to the system via the ISM.

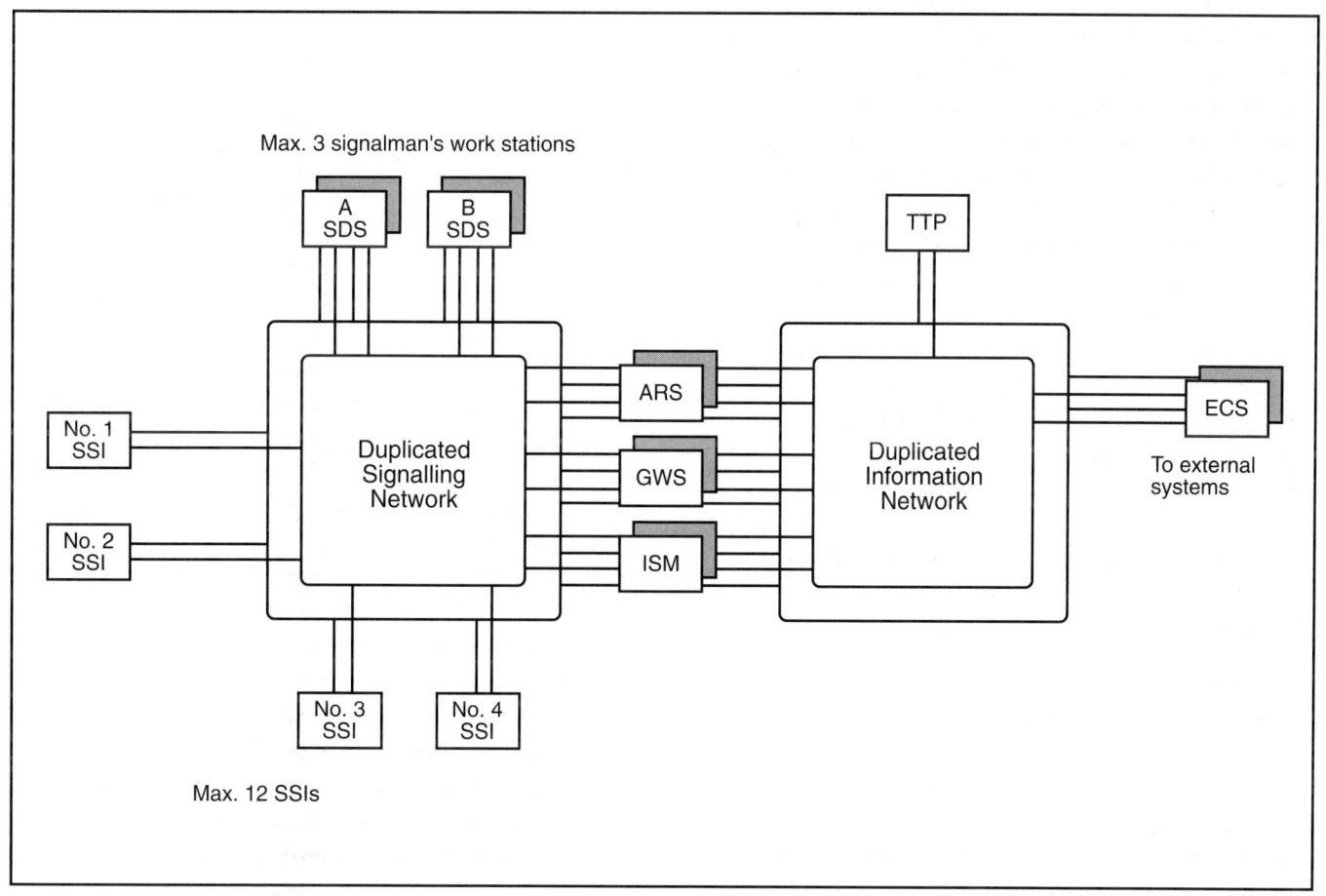

Fig. 11.11 IECC network

Electronic man–machine interface 'Videopult' for relay interlocking

In Austria, ÖBB are using Videopult for large schemes. Videopult consists of similar functional components as the IECC where the station layout is displayed on up to five colour VDUs. This is accompanied by a panoramic mimic diagram as a second display medium. The VDUs need not be of a fail-safe concept. Commands are given by the operator via a light pen, the output of the Videopult control the relays of the interlocking system. Like the IECC concept, Videopult is the basis for further facilities such as automatic route setting, train logging, automation of shunting operation, optimization of traction load, monitoring and control of catenary sections.

An electronic interlocking is equipped with the same man–machine interface using a fail-safe VDU concept. The station layout is displayed on up to five monitors, according to the size of the station.

11.6.4 Automatic route setting

A distinction should be made between the terms *automatic route setting* and *automatic train routing*, where the former refers to automatic setting of routes according to stored information and to the actual train moves, and the latter refers to the overall function of train routing consisting of several routes based on train description and database information according to several timetables depending on days of the week, season, etc. Automatic train routing requires the automatic route setting facilities.

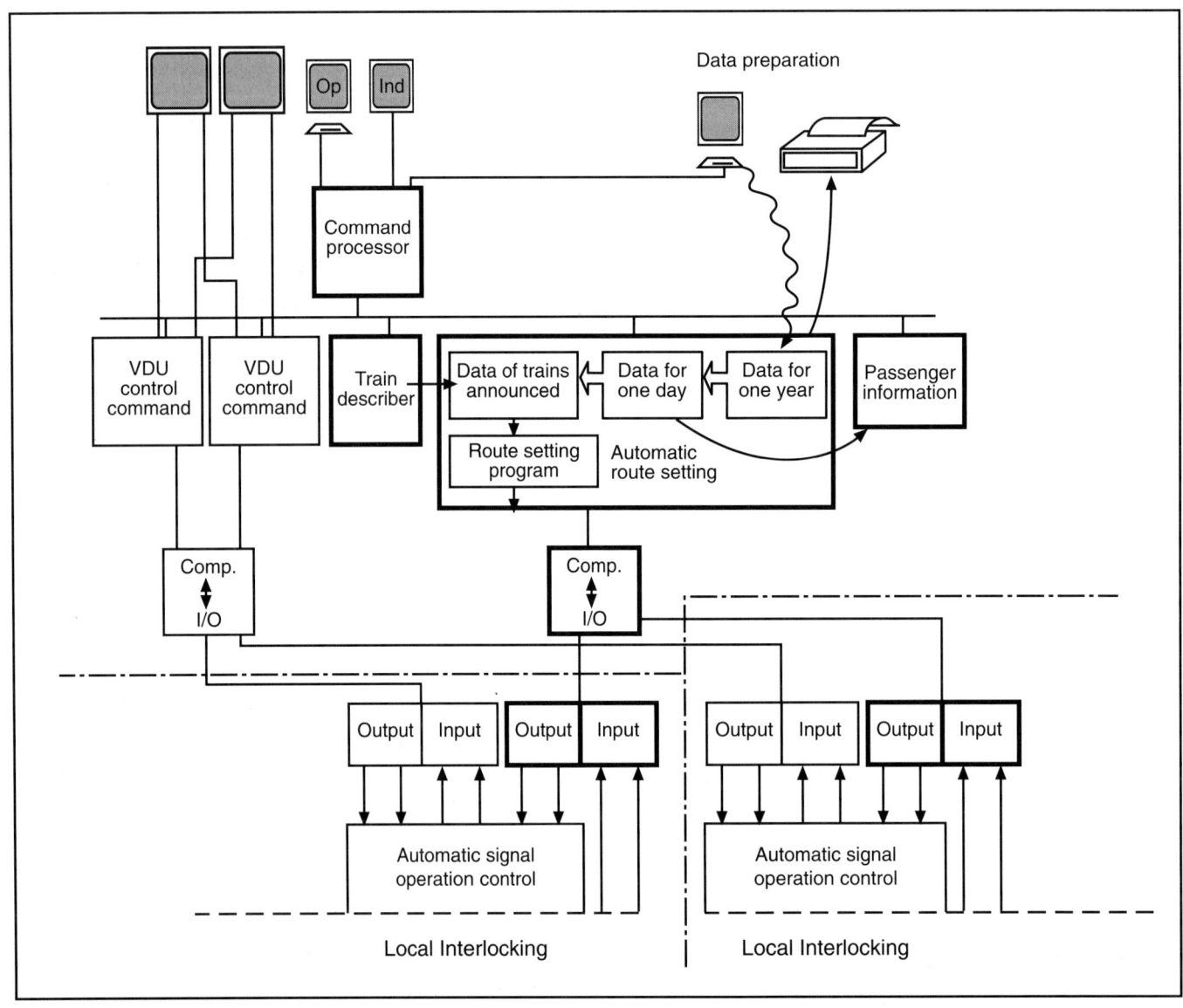

Fig. 11.12 Block diagram for automatic route setting using timetable data and train description on SBB

Provision can be made so that all regular operational commands for train routes of a determined area can be carried out by the function of automatic route setting. Thus, the control of normal operation of train movements can be automized to a large extent, relieving the dispatchers of routine operations.

The train describer holds information on the location of trains within a controlled area. If a timetable is fed into the train describer system, the location of trains can be compared to intended destinations for those trains, thus providing the capability of automatic route setting and of comprehensive passenger information systems.

Automatic route setting on SBB, for example, is arranged in various grades of implementation with several other European railways also using automatic facilities which differ somewhat from that discussed below:

- preselection of routes into blocked line sections, engaged by a preceding or opposing train.
- automatic operation of a station by through switching – ie points locked in normal position where trains can either stop at or run through the station, signals being cleared and replaced to danger automatically by train moves. These modes of operation require a procedure which proves at switching on the automatic mode that tracks are clear, required points in normal position locked. No manual intervention is possible before automatic mode is switched off by emergency signal replacement. The modes of automatic operation depend on the interlocking technique in use:
  - SD: *Stellwerk Durchschaltung* → signalbox through switching, electromechanical
  - DG: *Durchgangsbetrieb* → through running operation, relay interlocking without automatic block release
  - ADG: through running operation, relay interlocking with automatic block release
  - AB: *Automatischer Betrieb* → automatic operation to a limited extent at stations with small relay interlocking. These modes have to be switched on or off by the signalman in charge.
- in stations with signalled and interlocked shunt routes predefined signals can individually be switched to automatic operation, ASB (*Automatischer Signalbetrieb*) without the need for points to be locked. These signals are cleared and replaced automatically according to the progress of trains and based on a defined program. The signalbox is available for any additional operation required during the ASB mode, for example shunting operation, and other, unprogrammed, train moves. The signals in ASB mode can also be controlled manually in order to reroute a train or to shift a train crossing from one station to the other or just to set a route for an unprogrammed train.
- automatic train routing is based on ASB or AB as described above. In addition to that, the program is flexible and follows the data provided by the first digit of train describer or the programmed data according to the train numbers. The data for all trains is then stored into the train data files for a timetable period.

The route setting becomes effective as soon as the train approaches the location where the signal has to be cleared if in ASB mode. Overriding of the automatic system is possible at any time since the operator always has priority, as long as the route has not been triggered, by temporary alteration of the automatic routing parameters in the computer or by manual preselection of the route. Timetable conflict situations can be detected even before they become visible to the operator and will be handled accordingly.

A typical system for automatic train routing provides for following functions:

- *Automatic route setting*
  - preferred route selection
  - alternative route selection
  - automatic return to the preassigned route after alteration, as soon as permitted by the track layout

- *Hold*
  - at conflicting points in order to wait for passing train with higher priority
  - programmed hold at the end or beginning of a train route

- *Conflicting points*
  - wait for crossing train
  - wait for connecting train
  - enforced succession of trains

- *Type of trains*
  - starting trains
  - terminating trains

- joining trains
- separating trains
- shuttle service trains
- returning trains

- *Miscellaneous*
  - triggering of level crossings
  - altering of train numbers
  - assigning first digit to existing train numbers at predetermined locations

11.6.5 Shunting operation

Shunting route request:
- shunt radio – voice communication;
- shunt route request system – data communication;
- fixed installation;
- mobile equipment.

Setting of shunting routes:
- by signalman or dispatcher;
- consent by dispatcher;
- by shunting route request system.

Locomotive depot:
- administration;
- monitoring of locomotives.

11.7 Passenger information

The public, in particular the passenger, is informed by a train departure indication system and, where appropriate, by a train arrival indication system. This information is provided by large splitflap indicators and/or video monitors distributed over the station area. Audio announcements can be made from a central office and unmanned stations can also be

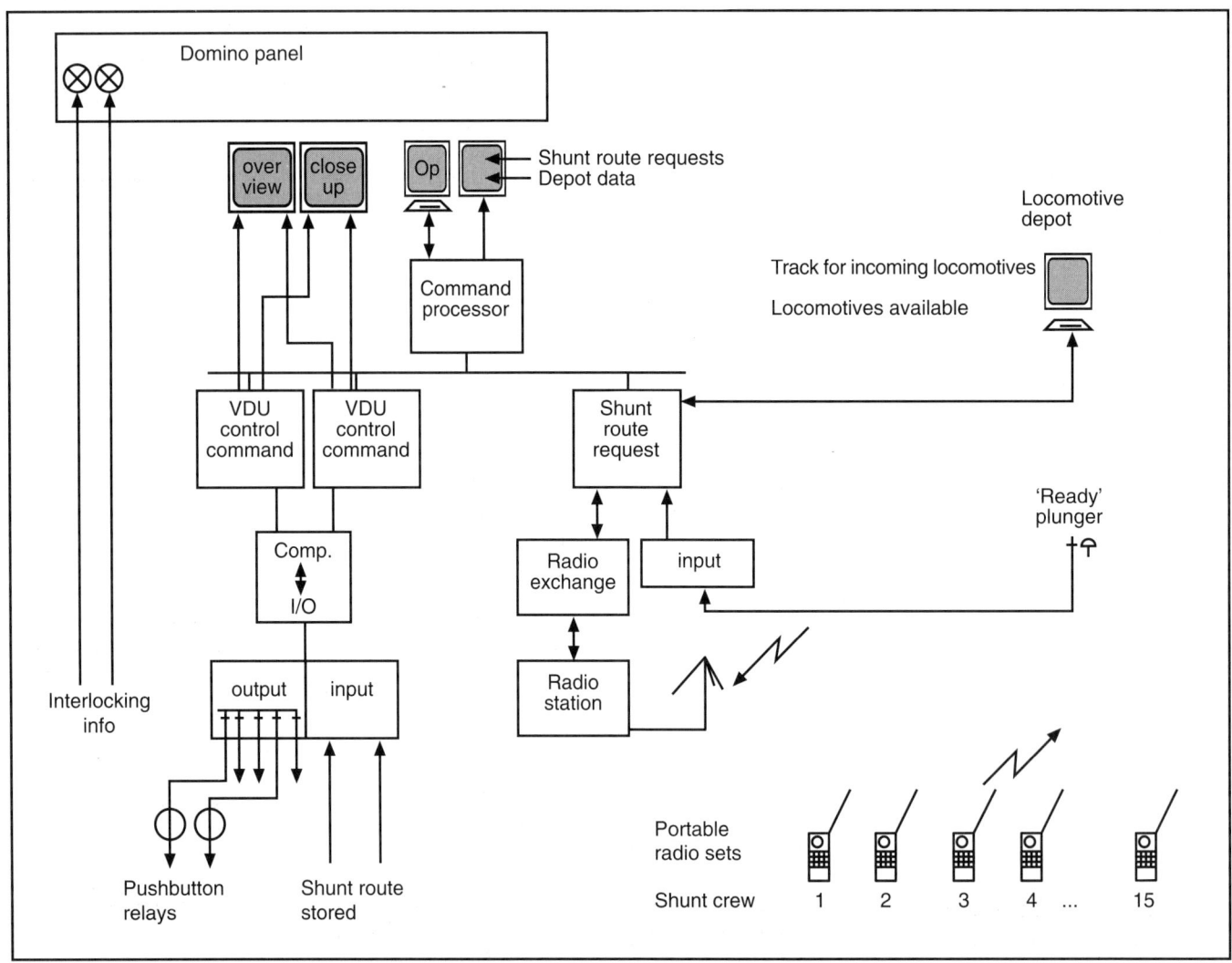

Fig. 11.13 Configuration of shunt route request system on SBB

addressed. In train information is also provided by an internal public address system generated by train attendants, voice recorders or radio link.

11.7.1 Visual information

Indication of train departure serves the purpose of informing passengers and directing them to the appropriate departure platform. Especially at bigger stations, an adequate system of information is an important facility, together with junctions where passengers have to change in order to reach the final destination and at stations with dense traffic. The system should provide customers with facilities to assist them to locate their train under confusing conditions; for example, delays, change of platform, etc.

On platforms there are indicators assigned to each track with the next departure displayed up to 30 minutes in advance. Overviews are placed at central locations or can be observed from monitors distributed along pathways.

In the Zurich area, for example, there are over 250 indicators distributed over 11 stations and forming a system which will be continuously extended according to need. The system consists of a fully redundant system which operates in a hot standby mode. By means of a local area network the data is exchanged between the different systems. The passenger information system receives its information from a common automatic train routing database twice a day. The train describer system sends information on actual train progress spontaneously for checking and clearing the platform indication.

Manual input is possible at the integrated operators desks of the remote control centre or at a station in order to define known track changes or any other remarks which can be useful for traffic control. In normal conditions the system operates automatically leaving the staff time for supervision and checking functions.

11.7.2 Voice communication

In addition to the visual information system there are loudspeakers placed all over the public areas of stations. These are not only for regular information to passengers when boarding, changing or leaving trains, but also, and more importantly, in case of emergencies

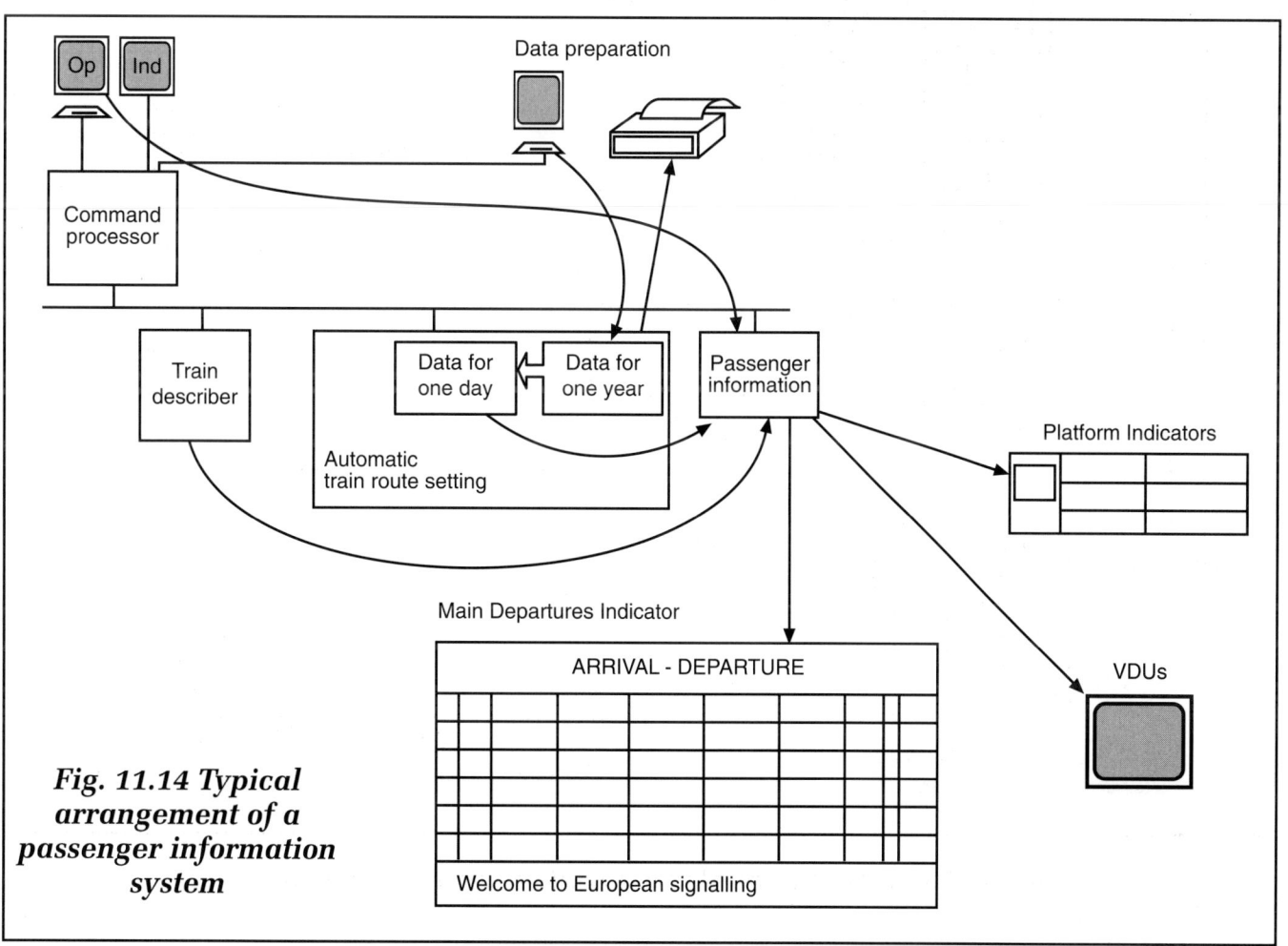

Fig. 11.14 Typical arrangement of a passenger information system

and other public announcements that relate to transport. The loudspeakers are usually driven by redundant amplifier systems and connected to the radio communication centre.

11.7.3 Other systems

11.7.3.1 Information and distress call column

These are locations where a customer can ask for information. Assistance in all kinds of emergencies is normally available at unstaffed stations.

The communication link is automatically established with the supervisor for the public at the supervision centre office. There, decisions have to be made accordingly.

11.7.3.2 CCTV closed circuit television

CCTV is placed at locations where supervision of passenger flow is not possible by train attendants or where the circumstances require special measures. This system is also linked to the supervision centre.

11.8 Centralized traffic control

Centralized traffic control is a distinctive subject, and subdivided into:

- super regional centres, and
- regional centres.

11.8.1 Traffic supervision system (super regional centres)

The purpose and key objectives of a traffic supervision system are:

- operational management facility to ensure the required high performance in terms of quality and stability of the timetable. It should further ensure the optimal utilization of fixed equipment, rolling stock and personnel.
- provision of customer service to ensure a high degree of punctuality of passenger and freight operation. In case of failures the effects should be kept to a minimum and the customers should be informed in time.
- the analysis of the reported data should be transferred into updated target planning data which should provide for conflict reduction.
- the optimized work flow based on a clear distinction of tasks on the three levels of operational control, under both normal and fault conditions.

The traffic management organization can be split into two functions:

- operational planning.
- operational control.

These functions can be subdivided into:

- *Operational planning*:
  - medium to long term;
  - annual timetable and planning activities down to two days hence short term;
  - planning for the day to day activities.
- *Operational control*:
  - supervision;
  - control.

The traffic control system embraces those aspects of the operational organization that are concerned with traffic supervision. Thus, the supervisory control system is the tool of the disposition and, as such, it is clearly separated from the control tasks. However, on the control level functions of disposition have to be handled as well, but are limited to the extent that these dispositions have no negative effect to the area beyond the controlled area.

For example, on SBB the organizational unit of traffic control is part of the division's production control at each of the three regions – Lausanne, Lucerne and Zurich. In order to meet the increasing demand of customers in terms of punctuality and shorter travelling time of passenger and freight traffic, in the 1950s SBB set up a system of traffic supervision and control in Lausanne, Lucerne and Zurich. The objective was to provide for the highest possible degree of adherence to the planned timing, for which the super regional control centres are responsible. Increasing traffic and improving train services made the mostly manual train reporting and scheduling system unsuitable, which led to the actual concept of computer based, supervisory control system. Especially, train supervision, preparation of actual timetables, location of trains en route and update of graphic timetables occupied supervisors to the extent that they did not have time available for the disposition tasks originally assigned to them. These demands can be met today using automated reporting systems and computer based comparison between target and actual timetable.

The staff of the three super regional centres in Lausanne, Lucerne and Zurich are responsible for the planning and co-ordination of operations and for information for passengers. Together they cover the whole SBB network. For example, in Zurich where a computer based train monitoring system has been provided, this system consists of:

- traffic controller workstations;
- auxiliary workstations;
- computers;
- LAN local area network.

The system provides for:

- monitoring the lines of the supervision area;
- train scheduling;
- scheduling of utilization of rolling stock;
- personnel assignment;
- information on track situation;
- time–travel diagrams of actual moves and projection to immediate future moves;
- information on track conditions;
- information on reasons for delays.

The data for target timetables is taken from 'System Fahrplan' (SYFA), whereas the actual data is transmitted from the remote control centres via a PIDN (process information data network). This data is processed by the supervisory control system and displayed on various monitors or printed on tables.

The DB AG and ÖBB approach to the traffic supervision is similar, since they both use a 'RZÜ' synonym for computer aided train supervisory system (*Rechnergestützte-Zugs-Überwachung*), which is considered to be on the planning level. The RZÜ systems monitor the real time traffic flow, and with the integrated timetable processing system the complete timetable data is produced. These systems are designed and used within large regions. The RZÜ relieves operators (supervisors) from tedious drawing activities, provides them with actual information on the rail traffic, which allows them to focus on events requiring their full attention.

Other benefits are reduced number of staff and more economic traffic management. A RZÜ centre is usually divided into several monitoring areas, each equipped with its own work place. The work place consists of:

- high resolution, full graphic colour monitor, for train graph curves and track network diagrams at eye level;
- operator's control unit in the form of a keyboard, a digitizer, a VDU, telephone and train radio equipment.

The train radio equipment is not part of the RZÜ; it is used for direct communications to signalmen and train drivers.

Fig. 11.15 Traffic supervision system (RZÜ), Frankfurt, with four high density monitors, a dialogue monitor and a digitizer. Photo: Alcatel SEL

Normally, two centre monitors are used for the display of train graph curves, and the outer monitors display the track layout of the whole area assigned to one operator. The train graph displays provide information on train moves of the last 30 to 40 minutes (review), forecast of the next 40 to 50 minutes (preview), track closing, bidirectional working and additional information as appropriate for efficient supervision. The time axis is arranged vertically and the distance horizontally. Review and preview are separated by the horizontal time line which moves every minute. The picture will be scrolled once in 10 minutes in order to provide an almost even split between review and preview on the graphical display. The curves are displayed in different colours, brightness and shapes; for example review brighter than preview, colour for class of trains. With the help of the digitizer the operator can move curves or can even design new ones as required.

The track diagram displays the actual status of the lines and the most important station tracks, in terms of:

- station name with phone number of the signalman;
- track designations and signal designations between the stations (home signals);
- signal locations with indication of normal running direction;

- location and class of platform (access, crossing facilities);
- level crossings;
- actual location of trains with their train description and delta time values as applicable;
- set routes (green track line when signal cleared);
- status of track availability, for example prohibited for electric traction, etc.

The colour coding provides additional information, for example a station name and the associated signals assigned to an operator are in one specific colour. Close up of specific track areas can be called as detailed track display in order to show additional information such as available length of tracks and catenary sections. The supervisor can communicate with the system through the operator's control unit by keyboard or digitizer.

Several other working places are integrated within a supervisory control centre, such as:

- timetable for editing *ad hoc* changes;
- locomotive scheduling of a certain area.

11.8.1.1 Traction voltage supply

Another important issue of traffic control is the energy supply. The traction supply system is monitored by a traction control centre providing for a 24-hour service, such as:

- co-ordination and check of all switching orders placed by various services;
- check of requests with regard to availability of supply;
- co-ordination with operation and power stations;
- systematic approach for fault detection;
- improving the availability by centralized reporting and analysing failures;
- switching within the required timing without delays;
- protection of personnel by accurate supervision of all switching parts;
- improving the availability by accurate information of traffic;
- co-ordination in irregular conditions;
- centre point for other services such as the fire brigade.

11.8.1.2 Locomotive management

The locomotive management plans and monitors the required traction units in co-ordination with the other regions and railway companies – ie the economical utilization of traction units according to requirements and taking into account maintenance and optimal engagement of personnel.

11.8.2 Remote control centres (regional centres)

Information and automation are key elements of any kind of centralization. The objective of a remote control centre is to ensure safe traffic operation according to a given timetable. The area of a remote control centre is normally limited by the workload of the dispatchers and the reasonable number of workstations in one big office.

By using state of the art technology the dispatchers can be relieved of most of their routine work in order to allow them focus on the more important aspects, such as advanced planning and detection of arising conflicts. Integrated workstations provide for all the functions required, such as:

- data terminal and colour video display for the observation of the process and the dialogue;
- keyboard operation – easy access to all controlled stations, all kind of manual interventions;
- train describer – for actual information of the dispatcher on train location;
- automatic train routing system – triggers route setting automatically according to the programmed data and the actual progress of the trains;
- means for updating the modifying of programmed data;
- passenger information – platform indication for the announcement of train departure time and destination;
- shunting operation – voice communication or radio controlled.

The area of a remote control centre usually extends from a station of a node function to an optimal number of adjacent stations. The optimal number is found by assessing the distances to adjacent nodes in relation to the traffic to be handled and also in consideration of the commercial importance of the respective node.

Centralization – remote control of interlocking:
- manual route setting;
- manual control of individual objects;
- critical commands;
- emergency commands;
- call on signal.

RAIL TRAFFIC MANAGEMENT

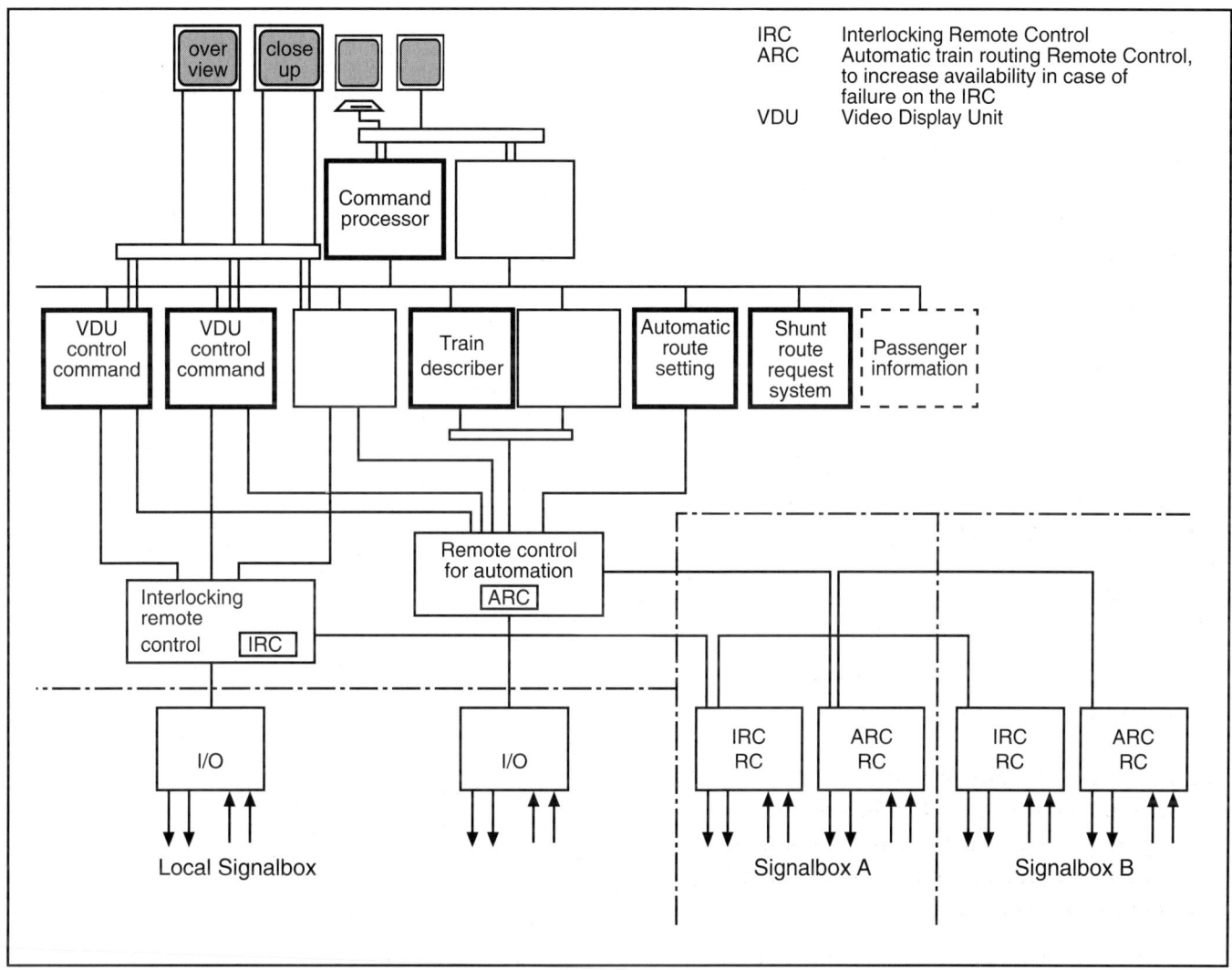

Fig. 11.16 Typical configuration of a remote control centre

Automation
- automatic route setting by train number or first digit of train number;
- safety aspects of prohibited track sections and signals not in automatic mode.

Transmission equipment
- protocol;
- reliability;
- redundancy or diversity;
- safety by arrangements in the interlocking, interfaces to the remote control and on the remote control.

These are the typical elements taken into account when designing a remote control centre. Specific approaches by individual countries are described below.

ÖBB
Remote control centres used to cover one or two adjacent stations only. With the present operational concept all high performance main lines with dense traffic, high speed and mixed traffic will be equipped with 'BOS–*Zentralen*' – operational control centres (BOS–*Betriebsoperationssystem*) installed on lines. The stations on a line over a length of 40 to 60km will be controlled from one centre which provides for an extended supervision area for detecting and handling conflicts more efficiently. The concept is based on 50 such control centres – one is currently in operation and five others under construction.

The realization of this complex concept will be carried out in stages and requires the following functional modules:

- train description;
- train supervision;
- automatic shunting;
- automatic train routing;
- remote control;
- interface to existing signalboxes;
- catenary control;
- optimization of power consumption – energy control;
- train control.

This modular set up provides the signal engineer with the appropriate configuration of operational control systems, taking into account and optimizing requirements and economic aspects. On ÖBB there are several systems in operation, such as:

- three different relay interlocking systems; and
- two different electronic interlocking systems.

Irrespective of the make of a local signalbox provided, all stations will be controlled through one common man–machine interface – the EBO (*Einheitliche Bedienoberfläche*). The EBO provides for uniform operation procedures, independent of the signalling equipment, and appropriate ergonomic working places for the operators. The primary means of operation is the mouse, which permits operation on all screens irrespective of the number of visual displays. The operation is supported by menus. A menu is a list of actions organized on a screen, which can be activated by pulling down the menu and choosing the appropriate command.

The track and signal layout is represented over one or more colour video displays in the form of a mimic diagram where the symbolic is designed to facilitate the association with reality. The maximum is approximately 200 points over five screens on one working place. The operation and indication are fail-safe by appropriate arrangements.

NMBS/SNCB

An entire signalbox or part of it can be remote controlled by another signalbox or by a remote control centre. Indications and commands are identical either for local or remote operation. Optical synoptic panels (OSP – mimic diagrams) are used. The remote control system consists of microprocessors and the OSP is driven by output relays on printed circuit boards. The output relays are monitored in order to ensure the safety of the vital information. Remote control is also possible by means of EBP, the Belgian integrated electronic control centre.

SBB

The total number of remote control centres in Switzerland is in the order of 50, including SBB and all other railway companies. A typical approach on SBB is to tailor the extent of tasks to be remotely controlled and monitored to the assumed requirements of operation. There are stations that are entirely under the mode of remote control and others that are only monitored, with few facilities for manual intervention by the dispatcher in the centre. The first decision is concerned with the size of station layout to assign for remote control, which may range from just the main station tracks (for example on a double line, two tracks) to the complete station layout including most of the sidings. There are three possible modes of operation:

Local operation
Only local operation by the signalman is possible, with full information of the signalling status of the dedicated station area to the remote control centre – ie the centre is limited to monitor the station only.

Mixed operation
Train routes are controlled from the remote control centre; shunt routes are controlled locally, with full information of signalling status at the centre.

Remote operation
The complete dedicated station area is remotely controlled. No manual intervention by the local signalman is possible, except for emergency replacement of signals.

The mode of operation has to be switched over from one mode to the other by a request and consent operation by both the operator at the remote control centre and the signalman at the station concerned. The remote control is often associated with automatic train routing which leads to considerable reduction of staff. In the event of failures of such systems it would not be easy to staff the station immediately, although arrangements are made for such occurrences. In order to avoid train delays in such events, a standby control

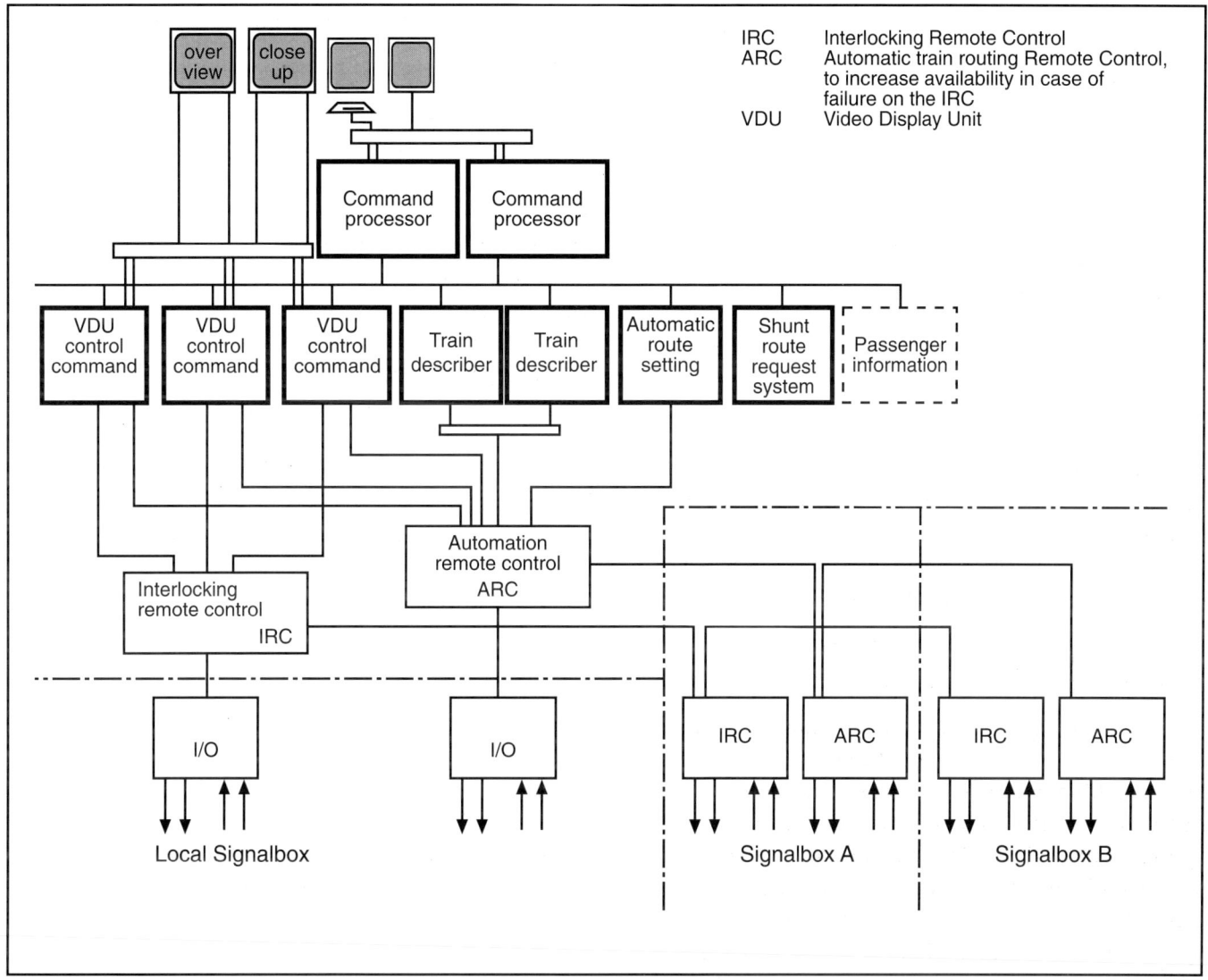

Fig. 11.17 Configuration to suit availability requirements on SBB

by an independent communication link is often provided. This is in the form of the train describer transmission link, which is an entirely separate connection and where the switching facilities for automatic route setting is superimposed.

DB AG

Almost all main lines of DB AG are supervised by computer based systems. These supervisory centres (RZÜ – *Rechnergestützte Zugüberwachung*) include train describer systems and man–machine-interfaces, which consists of dialogue monitors and high resolution video displays where overview diagram and time distance graphs are represented.

One of the first remote control centres was installed in Saarbrücken in the early 1970s. On DB AG more remote control centres will be installed in the near future.

Each remote control centre will cover more than 100km of lines with several stations, which are equipped with relay or electronic interlockings. Several operators in each centre will share their tasks according to traffic.

The remote control centres are usually equipped with:

- train description;
- train supervision;
- automatic train routing;
- remote control;
- man–machine interface;
- passenger information system;
- platform supervision system.

High resolution colour displays are used for time distance graphs and detailed diagrams (ISA – *Integrierte sichere Anzeige* → integrated fail-safe display). Terminals and digitizers are provided for the remote operation.

Remote control centre computers are connected by local area network (LAN) which caters for high flexibility in terms of expansion and variation. Data exchange over LAN uses international standards such as MMS (manufacturers message specification) between computers within a remote control centre and to other remote control centres.

RENFE

Seventeen remote control centres cover about 2670km of RENFE's network. Except on high speed lines the traffic on the remotely controlled sections is mixed. The stations are provided with local control facilities.

The most recent remote control centres are provided with modern facilities, such as train describers, train recording and automatic route setting. The stations are either equipped with relay or electronic interlocking systems.

The remotely controlled sections are also provided with other facilities like train to track communication, hot box detection and hot brake detection.

SNCF

SNCF is also following the tendency to focus on centralization and automation, although the network will still rely on the signalman for the final decision. With the design of a modular concept for all traffic management aspects, all control and supervisory functions, both for signalling and traction supply, will be concentrated at fewer places.

BR

BR uses the following systems for transmission of data between the stations and remote control centres:

- *Direct wire*: Where distances are short, the direct wire method has often been used. This consists of a micro core cable carrying individual functions over separate conductor. Typically, the cable consists of 360 cores each of 0.4mm diameter laid around a central 56/0.3mm diameter common return core.
- *Frequency division multiplex (FDM)*: Beyond a certain distance (10km) other methods of circuit provision become more economical than direct wire. One method uses discrete frequency tones for each function, in the range of 400 to 900Hz, superimposed on a single pair of wires. This FDM system can be used with both non-vital for up to 50 channels and, with less frequencies, 16 vital tones. By using this method, it is possible to extend the range of an interlocking and a number of such installations are in existence.
- *Time division multiplex (TDM)*: The most effective way to pass large amounts of data between control centre and remote interlocking is to use TDM techniques. Each function is allocated a time slot in a serial data stream, which is transmitted via a modem over a pair of wires at high speed. Both point to point and multidrop are used depending upon interlocking size and traffic density. Point to point systems connect the control centre (office) to just one interlocking (field) and can be duplicated for availability. Multidrop configurations connect a number of field stations to one communications bearer, which originates from the office. Each station on a multidrop system is addressed in turn. Serial to parallel multiplex interfaces are provided at both field and office as part of the equipment. Some systems employ panel processing as an addition to telemetry and may be connected to the control panel without the need for relay circuits.

FS

FS uses centralized traffic control systems for both low traffic and high traffic lines including the existing high speed line.

There are three possible modes of operation:

Local operation
Only local operation by the signalman is possible. The remote control centre is provided with full information of the signalling status if not hidden by a failure.

Local operation by permission of the dispatcher
This is a particular mode where operation is local but subject to the permission by the dispatcher of the remote control centre. The remote control centre is provided with full information of the signalling status if not hidden by a failure.

Remote operation
The entire station area is remotely controlled. No manual intervention is possible locally and the station is not staffed.

On low traffic lines the remote control centre is equipped with a mimic diagram and a keyboard. On main lines there are video control units instead of mimic diagrams and train describer systems with other facilities are made available for the assistance of the dispatchers.

The transmission systems between remote control centres and stations use dedicated telephone lines which are duplicated.

CFL

Fifty per cent of all signalboxes are controlled by six remote control centres. The controlled stations are equipped with relay interlockings and automatic block systems. The operators are supported by simplified automatic route setting.

The possible modes of operations are:

- local operation similar to that of SBB;
- remote operation without any possibility of manual intervention by a local signalman.

NSB

The typical configuration of regional centres is almost identical as that shown in **Fig. 11.16**, without the automatic route setting. The train describer system is integrated in the central command system. All commands from the control centre are sent through one transmission channel.

Eight remote control centres cover approximately 2400km of NSB network, mostly single lines, each covering an area of up to 300km. Only commands that are permitted by the interlocking are transmitted and executed – ie no bypassing of safety conditions. A route command that cannot be executed immediately will be stored by a route setting memory in the interlocking.

Provisions are made to protect working staff or maintenance work by locking commands for points, signals, track sections, in order to prevent the operator from unintentional route setting.

Possible modes of operation are:

- local and remote operation, similar to SBB and BV;
- automatic in local mode: an approaching train sets its route including facility for automatic crossing at all stations on single track lines;
- automatic route setting in remote control mode.

Several automatic functions are made available to relieve the operator from considerable routine work:

- all level crossings automatic on CTC lines, without CCTV supervision;
- in case of transmission failure the system falls back to automatic operation in local mode, thus providing smooth traffic flow in degraded mode, automatic crossing at stations on single track lines is also provided.

NS

The signalboxes are currently controlled by 30 control centres. This number will gradually be reduced to approximately 20. There are some remote control systems that have been installed previously using TDM (time division multiplex) or some FDM (frequency division multiplex) equipment, whereas recent installations of electronic control centres use modern communications techniques.

Signalboxes that are remote controlled do not have local control facilities except for shunting moves at some locations.

BV

The typical configuration of remote control centres in Sweden is almost identical to that outlined in **Fig. 11.16**, without automatic route setting. The train describer system is integrated in the central command system. All commands including those for automatic route setting are sent through one transmission channel. In case of transmission failure a local automatic mode will be switched on at the field stations.

BV have 11 remote control centres each covering up to 300km in either direction. Thus, approximately 4500km, mostly single, lines are covered for remote control. Only commands that are permitted by the interlocking are transmitted and executed – ie there are no commands bypassing safety conditions, nor will they be accepted by the local interlocking. A route command that cannot be executed immediately will be stored by a route setting memory in the interlocking.

Various TDM (time division multiplex) transmission systems are in use, such as relay transmission, electronic transmission using phase shift keying (1000 baud (bits per second), integrated electronic systems 1200 baud rate with central processor for monitoring functions).

Provision is made to protect working staff or maintenance work by means of locking commands for points, signals and track sections, in order to prevent the operator from unintentional route setting.

Possible modes of operation are:

- local and remote operation, similar to SBB;
- automatic in local mode: an approaching train sets its route including facility for automatic crossing at all stations on single track lines;
- automatic in remote mode: a comprehensive train describer function.

Some automatic functions are made available to relieve the operator of a considerable amount of routine work:

- all level crossings automatic on centralized traffic control lines, without CCTV supervision;
- fleeting and automatic route setting;
- in case of transmission failure the system falls back to automatic operation in local mode, thus providing smooth traffic flow in degraded mode, automating crossing at stations on single track lines is provided for;
- train describer commands are transmitted through the normal channel.

11.9 Other management and information systems

11.9.1 Freight traffic information system

ÖBB use a freight traffic management system where trains carrying freight cars are identified and the data recorded on a central computer system. The data consists of train and car number, weight, start and destination, receiver and classification of the freight. Terminals for data input and output are provided at all stations where required.

NMBS/SNCB and BV/SJ have a similar freight traffic management system, where car classification includes vital information of dangerous cargo which is referred to especially in the train consist list, extended by the international Rio number.

11.9.2 Vehicle identification system

All vehicles are provided with a label or tag, which can store specific data, such as vehicle number and vehicle related information. At determined locations there are reading facilities which enable electronic identification and localization of the vehicle, even on high speed lines. This data, together with identification of the location and time stamp, is transmitted to a data processing centre. It is intended to connect this system to a European freight information system.

12 Level Crossing Protection

P. MIDDELRAAD

Contents

12.1	Introduction	344
12.2	Characteristics of level crossing protection	344
12.3	Main forms of level crossing protection	344
	12.3.1 Manually controlled full barrier crossing	344
	12.3.2 Automatic open crossings with road signals	345
	12.3.3 Automatic half barrier crossing	345
	12.3.4 Automatic doubled half barrier and full barrier crossing	345
	12.3.5 Open level crossing without road signals	346
	12.3.6 Forms of level crossing protection control	346
12.4	Level crossing protection in Europe – road side	346
	12.4.1 Selection criteria	346
	12.4.2 Road signals and audible warnings	347
	12.4.3 Normal operation sequence	349
	12.4.4 Minimum open time	349
	12.4.5 Second train warning	349
12.5	Level crossing protection in Europe – railway side	349
	12.5.1 Technology used by European railways	350
	12.5.2 Measures taken to avoid overlong warning times	351

Illustrations

Fig.12.1 Automatic level crossing on NS 352
Fig.12.2 Automatic half barrier level crossing with separate barriers for the adjacent cycle path on NS 352
Fig.12.3 Automatic half barrier level crossing on ÖBB 353
Fig.12.4 Automatic half barrier crossing on FS 353
Fig.12.5 Level crossings in Europe 354
Fig.12.6 Normal operating sequence in seconds 355

12.1 Introduction

The preceding chapters have illustrated that there is hardly any standardization in European railway signalling systems and interlocking technology. The main reason for this is that there is no real need for standardization since it is an exception for trains to cross the border of their country. National regulations for the safety of running trains therefore work out quite satisfactorily.

Level crossing protection, however, not only serves the safe running of trains but, and perhaps more importantly, it also serves in the need to safeguard road traffic when crossing the railway line at a level crossing. It should be evident that this last reason for level crossing protection requires international regulation since car drivers frequently cross national borders.

Such regulation does indeed exist. The use of European road traffic signs was agreed in 1955 with the Geneva agreement signed by European governments. The use of road traffic signs at level crossings was laid down in the Vienna Treaty of 8 November 1968. Moreover UIC has issued leaflets 760, 761 and 762, which provide regulations for railways with regard to the forms of level crossing protection and the rules of application. Thus there is a basis for a harmonized European level crossing protection system. However, the regulations gave railways such freedom of action that this chapter cannot just deal with the European level crossing protection but must describe all existing European variations as well.

Nevertheless it is obvious that the only viable solution for the safe working of train and road traffic is not to mix them at all. Every effort has been made in Europe to abolish existing level crossings and to replace them with flyovers or subways. Naturally, all new high speed lines are free from level crossings from the design outset.

12.2 Characteristics of level crossing protection

Level crossing protection is the consequence of having level crossings on a railway line. Its aim is to avoid collisions between trains and road traffic. From the point of view of the railway, the requirement to be met by such protection is quite simple: it has to stop all road traffic before the passing of a train. From the point of view of road traffic, however, the protection must stop the train in order to safeguard road traffic when crossing a railway line. In other words, level crossing protection can only be the compromise in the conflict of interests between the two parties.

Depending on the situation, different forms of level crossing protection are therefore employed, in order to serve the interests of both parties as effectively as possible. It will become clear that those interests are not only created by reasons of safety but also by operational and economical reasons.

12.3 Main forms of level crossing protection

12.3.1 Manually controlled full barrier crossing

The manually controlled full barrier crossing can probably be considered the basic solution for level crossing protection, and the one from which other solutions are derived. The installation is provided with barriers which can be lowered in order to close off the whole road. The installation can be completed with road traffic signals and/or audible warnings.

A local control is provided in order to prevent road vehicles from being trapped between the barriers. A signalman or crossing keeper ensures that the barriers are lowered and that the crossing is clear before a train is allowed to approach. Controlled signals are therefore provided on each rail approach interlocked with the level crossing.

The advantage of employing full barrier level crossings is the high degree of safety that can be achieved. In principle a train can only pass the level crossing if the road traffic has cleared it. However, it has to be taken into consideration that this high degree of safety fully depends on the crossing keeper, who therefore has to complete his task with vigilance. On the other hand, the advantage of local control is that a train can be stopped in an emergency – in which situation an automatic level crossing is of no help in averting a possible disaster.

The main disadvantages of this type of level crossing protection are the high staffing costs, in addition to the very long closing times and possible train delays if the crossing is not closed in time. The high costs could be reduced to some extent by the use of remote control. The long closing times and possible train delays, will, however, limit the application to level crossings where other forms are not feasible; for example, too many tracks in the crossing. Most existing level crossings of this type will be replaced sooner or later by a subway or flyover, if not replaced by some other form of level crossing protection.

12.3.2 Automatic open crossing with road signals

The automatic open crossing can be considered to be an attempt to protect a level crossing without the disadvantages of the manually controlled full barrier crossing. This form of level crossing protection is controlled automatically by the approach of trains. In the basic form the installation is provided with flashing red road signals on each nearside road approach only and completed with the St Andrew's Cross fixed on the same pole as the signals. The application of white flashing lights in order to show that no train is approaching is allowed.

Further options are audible warning and additional road signals. The minimum warning time before the fastest train arrives depends on the length of the crossing but must be at least 20 seconds.

From the safety angle, it is clear that automatic open crossing cannot achieve the high degree of safety achieved by manually controlled full barrier crossings. Although the signal engineer, by using fail-safe techniques, will guarantee that the warning to road traffic will be given in time, much will depend on the road user who has to obey this warning and stop. The absence of a physical barrier, in order to force the road traffic to stop, will therefore affect the overall safety of this level crossing protection.

To discourage road users from taking risks it is therefore necessary to keep closing times as short as possible, since road users would be less inclined to take a risk if they were certain about the train being at the crossing in a very short time.

Apart from the risk caused by road users who deliberately ignore the warnings, a second train approach, instantly after the first train has passed, is a point of concern. A road user who is not aware of the still red flashing road signals could be tempted to start immediately after the passing of the first train. A barrier would prevent him from starting too early.

Another risk is that of road vehicles not clearing the crossing in time. This can be caused by traffic congestion or by crossing the railway line with a too long and/or too slow vehicle.

Therefore this type of level crossing protection can only be seen as a solution for level crossings with a low traffic density which is too small to provide half barriers but where a form of protection is required.

12.3.3. Automatic half barrier crossing

This form of level crossing protection is the improved version of the automatic open crossing. The additional half barriers improve safety by forcing vehicle drivers to stop and preventing them from starting too early in the case of a second train approach.

In the basic form barriers are provided on each nearside road approach only. The installation is at least provided with flashing red road signals. Further options are audible warnings, red barrier lights and additional road signals. The minimum warning time before the fastest train arrives depends on the length of the crossing but must be at least 20 seconds. The time interval between the beginning of showing the red flashing lights and the beginning of lowering the barriers must be at least 5 seconds, while the lowering of the barriers must take at least 7 seconds.

Since pedestrians and cyclists can still easily cross and vehicle drivers can do so with some effort by zigzagging around the barriers, deliberately ignoring warnings, it is still necessary to keep closing times as short as possible in order to discourage them from taking risks.

In places with high road traffic densities and nearby road crossings, additional traffic lights can be used in order to diminish the risk of road traffic not clearing the crossing in time. The traffic lights are coupled to the level crossing installation in order to regulate the traffic in such a way that road vehicles can easily clear the level crossing.

Generally speaking, the automatic half barrier crossing offers an optimal result, being a compromise between safety and performance.

12.3.4 Automatic doubled half barrier and full barrier crossing

This form of level crossing protection is a further development of the half barrier crossing by closing off the whole road. The basic form is similar to the half barrier crossing, but it is provided with additional half barriers at the exit side of the crossing or full barriers. In order to minimize the risk of trapping vehicles between the barriers, additional measures have to be taken. In the case of doubled half barriers the additional half barriers have to be lowered with some delay. When the full barriers are employed the time interval between the beginning of showing the red flashing lights and the beginning of lowering the barriers has to be increased.

The automatic doubled half barrier or full barrier crossing will be employed in situations where the highest possible degree of safety is desired. This protection is recommended for higher speeds above 140km/h up to 200km/h on existing lines with level crossings.

12.3.5 Open level crossing without road signals

Although this form of level crossing protection does not offer much protection, it is included to complete the overview of forms of level crossing protection. The regulations only mention the application of St Andrew's crosses.

12.3.6 Forms of level crossing protection control

The forms of level crossing protection described so far are the basic forms as they appear to the road user. Indeed, this classification is not fully correct since there is hardly any difference left between the original manually controlled full barrier and the later developed automatic full barrier crossing. Therefore the classification used can also be seen as a step in the evolution of the level crossing protection.

Further development of the manually controlled full barrier crossing has led to remote control with the aid of CCTV and automatic raising of the barriers after the passage of the train. Even automatic lowering of the barriers is employed, whereby the confirmation that the crossing is clear is the only manual aspect left. On the other hand, in some cases fully automatic control of level crossing protection has been changed back into semi-automatic or fully manual control in order to overcome local problems.

12.4 Level crossing protection in Europe – road side

Since the basic forms of level crossing protection have been described in **12.3**, this section will deal with some of the details and application rules that may differ from country to country.

12.4.1 Selection criteria

The form of level crossing protection is in general decided by the Ministry of Transport, which gives general instructions depending on average daily road and train traffic, speeds, road width, number of tracks and local sighting conditions. An overview of the present situation in Europe, divided into the main forms of level crossing protection, is shown in **Fig.12.5**.

ÖBB

Open level crossings without road signals are only accepted on secondary lines with speed restrictions depending on visibility conditions. If visibility is sufficient, then train speeds may not exceed 100km/h or 120km/h in the case of footpath crossings. Otherwise these speeds are 60km/h and 100km/h respectively.

Manually controlled full barrier crossings are no longer installed. Existing ones are replaced by automatic barrier crossings.

NMBS/SNCB

If the level crossing is considered to be part of a road crossing then the road signals are sometimes replaced by standard traffic lights at the crossroads. These are automatically controlled in the same way as level crossing road signals.

SBB

Open level crossings without road signals are only allowed for footpath and low traffic secondary road crossings.

DB AG

Generally, the existing automatic open crossings have to be converted to automatic half barrier crossings. The construction of new level crossings is no longer allowed and existing crossings on lines with speeds between 160 and 200km/h will soon be abolished. Manually controlled full barrier crossings can be of the normally open or normally closed types.

RENFE

Level crossings are only permitted if the product of number of trains and number of cars per day does not exceed 24 000. Manually controlled full barrier crossings are installed if this product is bigger than 2500.

SNCF

The automatic open crossing is no longer installed. The remaining level crossings of this type are being converted to automatic half barrier crossings. Manually controlled full barrier crossings can be of the normally open or normally closed types. Staffing of these crossings may be continuous or intermittent, depending on the importance of the traffic. New crossings of this type are not installed and existing ones are either removed or replaced by automatic half or doubled half barrier crossings.

In order to be able to automate level crossings on lines with speeds higher than 160km/h, it is now a legal requirement that these be equipped with obstacle detectors. However, trials on these have never provided conclusive results. Automatic level crossing protection for higher speeds is therefore not employed.

BR

The use of automatic open crossings is only allowed with a speed restriction dependent upon local sighting conditions, but always within maximum of 88km/h.

For automatic half barrier crossings:

- maximum line speed must not exceed 160km/h;
- 50% of trains must arrive within 50 seconds of the warning being initiated and 95% within 75 seconds;
- the road layout on the approach side must be suitable for vehicles to pass to prevent blocking the crossing;
- not more than two running lines;
- vertical profile of the road should not give the risk of long or low vehicles grounding.

There are no restrictions for manually controlled level crossings, which are therefore used where other crossings are unsuitable.

FS

Open level crossings without road signals are only accepted on secondary lines if visibility conditions are sufficient. Automatic open crossings with road signals are only accepted on single lines with low traffic and maximum speed up to 90km/h.

The other forms of level crossing protection are applicable up to 200km/h. The only exception is the automatic full barrier crossing when monitored by a special signal on the railway side. This signal was designed to simplify exploitation procedures in the case of failures and is used when, due to the distance between the crossing and an existing standard main signal, the latter is not suitable for crossing protection. When the special signal is installed, due to visibility reasons of the signal itself, the maximum line speed cannot exceed 140km/h.

CFL

Automatic open crossings are only used on single track lines with low traffic.

NSB

All new and refurbished level crossings are equipped with half barriers or double half (rarely full) barriers. Level crossings are not permitted on double track lines and on sections with speeds of over 160km/h.

NS

Manually controlled full barrier crossings are only used on crossings with more than three tracks, in which case automatic level crossings are not allowed. It is hoped that they will be replaced by a subway or flyover.

BV

Manually controlled barrier crossings are very seldom used. The type of protection is normally decided from a risk factor, which is the number of trains times the number of road vehicles per day and types of road vehicles. For example, a factor > 1000 or slow vehicles indicates barriers instead of an open crossing. Full barriers or doubled half barriers are used at all level crossings where the allowed train speed exceeds 140km/h or at crossings with heavy road traffic and/or big pedestrian and cyclist traffic. Half barriers are used where full barriers are not required.

12.4.2 Road signals and audible warnings

The type and number of road signals are as follows:

ÖBB

All new protected level crossings are provided with a yellow and red warning signal at both sides of the road. More signals are installed if required by local conditions. Older level crossings are provided with red flashing (60 flashes per minute) lights and audible warnings, which stop when barriers are in the closed position.

NMBS/SNCB

The manually controlled full barrier crossings can be provided with twin red flashing lights. All other protected level crossings are provided with twin red flashing lights which can be completed with an optional third white light that shows as long as the red lights are switched off. Additional lights will be placed at the other side of the road or above the road if required by local conditions. If the level crossing is considered to be part of a road crossing the road signals will be replaced by normal traffic lights.

Full and doubled half barrier crossings are equipped with audible warnings, which are switched on during the closing of the barriers. The audible warnings are optional for the other forms of level crossing protection. All barriers are provided with the white circular road sign with red circle indicating no vehicles.

SBB

All automatic crossings are provided with at least two twin alternating red flashing lights. At locations with poor sighting conditions, additional flashing lights are provided. Barrier crossings are equipped with an additional electronic warning bell which is switched on at the same time as the flashing lights. This bell is switched off as soon as the barriers have been fully lowered. In case of half barriers the bell remains switched on until the barriers start to open. The sound volume of the bell may be adjusted to suit local conditions and be set for day and night volumes.

The level crossing can be interfaced with road traffic lights installation. In this case the flashing lights can be replaced by traffic lights provided that detection and proving of those traffic lights which allow crossing of railway tracks are according to the Federal Law and Regulations for Railways.

DB AG

Old level crossings are fitted with single red flashing (60 flashes per minute) lights at both sides of the road. New level crossings are provided with a yellow and red warning signal instead. Audible warnings are only provided at level crossings that are frequently used by pedestrians.

RENFE

Level crossings are fitted with flashing (60 flashes per minute) twin red road signals and electronic bells. The electronic bells stop when barriers are in the closed position.

SNCF

The level crossing is provided with flashing (70 flashes per minute) red road signals and audible warnings. The audible warning stops when barriers are in the closed position.

BR

Road signals at level crossings, except for miniature warning lights (MWL) crossings, comprise one yellow and twin red flashing (80 flashes per minute) lights located in each of the four corners of the crossing facing the oncoming traffic. Additional units are provided for side roads as necessary. An audible warning sounds during the time that the crossing is operating. The tone of the warning changes if the crossing remains closed for a second train. MWL crossings are fitted with a red and green light on each side of the crossing. This type of installation is used for footpath crossings or minor level crossings where user operated gates or barriers are fitted.

FS

Automatic open and automatic half barrier crossings are provided with two alternating red flashing lights and an audible warning. Lights and audible warning are on until trains have cleared the crossing.

Automatic full barrier crossings and manually controlled full barrier crossings not visible by the crossing keeper are provided with one or more steady red lights, which are on until barriers have been reopened, and an audible warning, which is on until the lowering of the barriers has been completed.

Neither lights nor audible warnings are necessary for manually controlled full barrier crossings if they are visible by the crossing keeper.

CFL

Level crossings are fitted with alternating red flashing (50 flashes per minute) lights and bells or gong. The audible warning stops as soon as all barriers are fully lowered.

NSB

All automatic level crossings are equipped with red (90 flashes per minute) flashing and white (45 flashes per minute) flashing road signals. Automatic half and full barrier crossings are provided with audible warnings before and during lowering of the barriers and audible warnings which stop when barriers are in the closed position.

NS

The automatic open level crossing is provided with alternating twin red flashing (45 flashes per minute) and white flashing (45 flashes per minute) road signals and audible warnings (bell) at the nearside of the road on both sides of the line. All road signals are doubled. One set is shown in the direction of the oncoming traffic, the other set is showing backwards.

In this way road signals are also shown at the offside of the road before the crossing, if desired by local conditions. Yellow flashing (45 flashes per minute) pre-warning road signals are used in the case of nearby side roads or bends in the road.

The half barrier level crossing is distinguished from the open crossing by the absence of the white flashing road signals. Red flashing back side road signals can be used, similar to the open crossing. The half barriers are equipped with barrier lamps, of which the top lamp at the end of the boom shows steady red when switched on. The other two barrier lamps will flash red alternately, in phase with the main road signals. All barrier lamps are shown in both directions. Additional single or alternating twin red (45 flashes per minute) flashing road signals with optional audible warning are placed at the offside of the road before the crossing. Yellow flashing (45 flashes per minute) pre-warning road signals are used in the case of nearby side roads or bends in the road.

Manually controlled full barrier crossings are only equipped with red flashing signals and optional yellow flashing prewarning signals.

BV

All automatic level crossings are provided with alternating twin red (80 flashes per minute) flashing and white (40 flashes per minute) flashing road signals and audible bell warnings at the nearside of the road. If the road is more than 5m wide or if the visibility is considerably better from the left side of the road, a second offside signal will be arranged. The offside signal does not need the white light. Barriers are provided with red barrier lamps, positioned over the middle of the approach lane.

12.4.3 Normal operation sequence

The normal sequence of operation of automatically controlled level crossings is shown in **Fig.12.6**. Where applicable the details for different types are given. All warning times are minimum times for the fastest train and normal length of the crossing. The mentioned times can be increased dependent on the length of the crossing.

12.4.4 Minimum open time

Some railways employ the feature of a minimum road open time. If the approach of a second train would initiate a new warning sequence within a certain time after the automatic level crossing has finished the first sequence, then the crossing will stay closed until the second train has cleared the crossing:

- NMBS/SNCB: 15s: only doubled half barriers.
- SBB: 10s.
- DB AG: 6s.
- BR: 10s.
- FS: 15s: only half barriers.
  30s: only automatic full barriers in this case if the crossing is released, a new closing sequence is permitted only after 30s.
- NS: 10–15s: only occasionally for half barriers.

12.4.5 Second train warning

DB AG employs an active system for second train warning at automatic open crossings only. In the case of a second train an additional indicator is illuminated.

BR employs an active system for second train warning at all automatic crossings. The tone of the audible warning changes if the crossing remains closed for a second train. At automatic open crossings an additional second train indicator is illuminated.

CFL uses fixed warning signs to remind road users that a second train may be hidden by the first having just passed the level crossing.

NS also employs fixed warning signs to warn road users not to start before the red signals have extinguished, since a second train could possibly be approaching.

12.5 Level crossing protection in Europe – railway side

The requirements on the railway side of the level crossing are, in general, determined by the technology used in order to provide a fail-safe system. This applies to the approach detection and the level crossing control logic, as well as the actual level crossing warning system itself. If one or more of these functions cannot be considered to be inherently fail-safe then a form of monitoring is necessary. Usually this works out in the need to have a signal before the level crossing to stop the train if the correct functioning of the level crossing warning system cannot be guaranteed. For this reason several railways employ special level crossing signals. These signals can be omitted if the monitoring function can be performed by a main signal, which is often the case on open lines with a block system and in the station interlocking area.

12.5.1 Technology used by European railways

ÖBB
The level crossing can be controlled by route setting, approach detection or a combination of the two. Approach detection is in general performed by treadles. These devices are duplicated or triplicated for bidirectional working on the open line. Therefore, and with the use of inherently fail-safe technology, no monitoring of signals is normally provided except if the level crossing is in a route or if there is a signal between the point of approach detection and the level crossing. Only on side lines with low traffic and a maximum speed of 80km/h are level crossing supervision signals used to indicate to the driver that the level crossing works correctly.

NMBS/SNCB
The automatic level crossing is controlled on the basis of inherently fail-safe technology. Thus no monitoring of signals is provided. Approach detection is performed by track circuits or treadles, the treadles being doubled for safety reasons. The control logic is integrated in the block system, depending on the driving direction.

SBB
Automatic control of level crossings within station areas is performed by route setting. Level crossings on the open line are controlled by information transmitted from the adjacent station or triggered by train operated switching elements. The proper functioning of the level crossing installation is checked on existing main signals or special flashing level crossing signals and/or the ATP system.

DB AG
Several systems are used for approach detections; for example, track circuits, axle counters, magnetic rail contacts, wheel sensors and inductive loops. Monitoring by means of a level crossing signal is only used on single track lines with speeds of up to 120km/h. This signal is placed at braking distance before the level crossing and is normally extinguished. On the approach of the train the signal only shows a flashing aspect if the level crossing installation is switched on properly. If not, the driver must stop the train before the level crossing. For that reason an ATP INDUSI magnet is installed near the signal, which will invoke an emergency braking.

Other level crossings are either monitored by the signalman or interlocked with main signals.

RENFE
Approach detection is performed by direction sensitive electromagnetic treadles. In spite of the technology used being of the inherently fail-safe type, for monitoring purposes an indicator is used which can show the following aspects:

- steady white vertical line, indicating *level crossing secured, barriers closed*.
- flashing white vertical line, indicating *that there is a disturbance without safety risk whilst barriers are closed*.
- flashing yellow cross, indicating a *disturbance of the level crossing, for example a broken barrier or two or more extinguished road signals*.

SNCF
Approach detection is usually performed by electromechanical or electronic treadles. In addition, on tracks fitted with track circuits the approach detection is maintained by occupation of the track circuits between treadle and level crossing. On lines that are equipped with an automatic block system only track circuits are used. Signal monitoring is not provided since the technology used is of the inherently fail-safe type. However, on certain low traffic lines level crossings are radio controlled from the driver's cab, in which case the closure of the barrier is monitored by a signal ahead of the level crossing.

BR
Approach detection comprises a combination of track circuits and treadles. Directional stick circuits are used to facilitate bidirectional working. Locally monitored automatic open and barrier crossings are monitored by the train driver through a rail signal. This normally exhibits a flashing red indication which changes to flashing white once the red road signals are correctly lit. Should the red lamps in one road signal fail, or a failure of the main power supply occur, the rail signal will not show the proceed aspect. The driver is then authorized to pass over the crossing at caution once he is sure it is safe to do so and report the failure to the signalman.

FS
Approach detection is performed either by route

setting, electromechanical treadles or track circuits. On lines equipped with coded current automatic block, the approach detection is maintained by the block track circuits. On these lines, when needed, an audio frequency jointless track circuit is used to speed up the release of the level crossing.

On automatic half barrier crossings the devices that are not fail-safe, such as treadles and road lights, are duplicated and checked at every passing of a train. Automatic open crossings with road signals are supervised by a special signal which indicates to the driver that the level crossing is working correctly.

All other automatic level crossings are monitored by either a special signal or a standard main signal.

CFL

On the open line approach detection is performed by treadles or inductive loops. Level crossing control by means of route setting is used if the level crossing protection is integrated into a station interlocking. However, in order to avoid overlong warning times due to early route setting, route controlled inductive loops for approach detection may be used. For monitoring of automatic open crossings a white flashing indication is provided. The train driver must stop before the level crossing if this indication is not shown.

NSB

Approach detection is based either on the occupation of a 10kHz and a 50kHz jointless track circuit in the correct sequence or, in station areas, a combination of occupation of track circuits and route setting. The train driver is warned with a distant signal, which should be seen 700m before the level crossing. There is a mark where trains with line speed have 30 seconds left to stop the train if the signal does not clear. The level crossing signal is placed at a maximum of 5m before the crossing. The train driver must try to stop the train if the distant signal does not clear 20 seconds before the train arrives at the level crossing.

NS

All approach detection, with the exception of a few secondary lines where mechanical treadles or electronic wheel and vehicle sensors are used, is based on the occupation of track circuits only. In station areas the approach detection is interlinked with the route setting. On the open line the approach detection is independent of the block system. All automatic level crossings are fitted for bidirectional working for which purpose directional stick circuits or driving direction information from the block system are used. The level crossings are controlled on the basis of inherently fail-safe technology. Any failure will result in the closing of the level crossing. Monitoring in signals is therefore not provided.

BV

Approach detection is based on the occupation of track circuits in station areas interlinked with route setting. On open line all automatic level crossings are fitted for bidirectional working, independent of the driving direction of the block system. All protected level crossings are provided with a signal showing either a red or a white aspect, depending on the status of the level crossing. Full barriers are preferably monitored through main signals. If this is not suitable or possible a special distant signal may be used. This signal shows three orange–yellow lights in V–formation. The lamps are flashing when the barriers are open and show a steady light when they are lowered. The signal is positioned at least 300m before the crossing and should be fully seen from the braking distance. For half barriers this signal is cleared when the booms have reached 75° from the horizontal, for open crossings immediately when the warning starts.

Level crossings on lines with speeds exceeding 140km/h are monitored by the ATP system. In the case of a failure the train will be stopped before the level crossing.

12.5.2 Measures taken to avoid overlong warning times

The approach detection guarantees the nominal warning time according to the fastest train. As a result the warning time will be much longer if the train comes to a stop before the level crossing or if the train is running at much lower speed than line speed.

To avoid very long warning times in the case of stopping trains in station areas, most railways make use of a protecting signal before the level crossing. In this case initiation will start when the route is set. If necessary signal clearance will be delayed in order to guarantee the minimum warning time.

In general, on the open line, longer warning times due to running at lower speed than line speed are accepted.

However, an exception is DB AG, which takes measures on lines on which trains run with very different speeds. Wheel sensors, which are monitored in a main signal, are used in order to measure the speed of the approaching train. The switching on of the level crossing installation and the signal clearance is delayed if the speed is below a certain level.

BV has taken measures on lines with speeds above 140km/h. The normal track circuit approach detection on these lines is fitted for a speed of 140km/h. All trains with a maximum speed above 140km/h initiate the level crossing protection system by means of a train class detector located about 6000m from the crossing. This equipment is not fail-safe but this fact is not dangerous as the crossings are supervised by the ATC system.

Fig. 12.1 Automatic level crossing on NS

Fig. 12.2 Automatic half barrier level crossing with separate barriers for the adjacent cycle path on NS

Fig. 12.3 Automatic half barrier level crossing on ÖBB

Fig. 12.4 Automatic half barrier crossing on FS

Railway	Protected						Unprotected		Total
	Manually/ semi-automatically controlled[1]	Automatically controlled Barriers			Open	%	Total	%	
		Full	Double Half	Half					
ÖBB	1440	107	30	190	180	27	5153	73	7100
NMBS/SNCB	172	–	18	1213	426	80	461	20	2290
SBB	19	632	–	26	33	35	1329	65	2039
DB AG	3100	200	600	4000	2100	50	10 000	50	20 000
RENFE	1564	–	–	149	–	32	3660	68	5373
SNCF	3570	–	622	10 908	80	71	6088	29	21 268
BR	1108	–	–	417	214	19	7492	81	9231
FS	6280	450	–	400	141	97	200	3	7471
CFL	33	1	37	36	8	74	41	26	156
NSB	–	90		375	25	9	5000	91	5490
NS	64	–	–	856	825	80	426	20	2171
BV	50	720	480	860	1790	30	9200	70	13 000

[1] All types of level crossing which are not fully automatically controlled
NB: **Fig. 12.5** shows 1992/93 equipment levels

Fig. 12.5 Level crossings in Europe

Railway	Type	Yellow	Red	Barriers down	Minimum warning time
ÖBB	HB	4	8	6–12	29
	FB/DHB	4	13	8–12	34
NMBS/SNCB	HB	–	15	9–15	37
	DHB	–	15	18–24	49
SBB	FB/HB	–	12	10–15	25
DB AG	FB/DHB/HB	3–5	12	6–10	26
RENFE	HB	–	5	7–10	30
SNCF	HB	–	7	8–10	140km/h 25 160km/h 30
	DHB	–	7	16–20	140km/h 35 160km/h 45
BR	HB	3	4–8	6–8	27
FS	HB	–	Single line 5 Double line 7	8	Single line 30 Double line 32
	FB	–	15	10	90 (if monitored by special signal)
CFL	HB	–	7	12	20
	DHB	–	7	23	30
NSB	HB	–	7	5–7	30
	FB/DHB	–	10	10	30
NS	HB	–	5–8, 5	12–15	25
BV	HB	–	5	10–12	Single line 20 Double line 25
	FB/DHB	–	Single line 10 Double line 15	10–12	Single line 24 Double line 29

Fig. 12.6 Normal operating sequence in seconds

13 Other Safety Systems

J. HAMMARGREN

Contents

13.1 General	357	
13.2 Hot axle box detectors	357	
13.3 Hot wheel detectors	358	
13.4 Flat wheel detectors	358	
13.5 Wheel weighing systems	358	
13.6 Train gauge detectors	358	
13.7 Dropped object detectors	359	

13.8 Avalanche and landslide detectors — 359

13.9 Tail detectors — 359

13.10 Staff warning systems — 359
    13.10.1 Movable automatic warning devices (AWAS) – NS — 359
    13.10.2 Fixed automatic warning devices — 359

13.11 Conclusion — 361

Illustrations

Fig. 13.1 *Hot box detector with electronic treadles* — 357
Fig. 13.2 *Example of a hot axle box detector installation* — 357
Fig. 13.3 *Detector site in Sweden with profile supervision detector, hot wheel detector and hot box detector* — 358
Fig. 13.4 *Flat wheel detector (WILD)* — 358
Fig 13.5 *European table of other safety systems* — 361

13.1 General

The title of this chapter applies to those systems not considered fail-safe but which make a contribution to the general safety conditions of a railway.

The following systems will be described:

- hot axle box detectors
- hot wheel/blocked braking detectors
- flat wheel detectors
- wheel weighing systems
- profile supervision/gauge detectors
- dropped object detectors
- avalanche and landslide detectors
- tail detectors
- staff warning systems

13.2 Hot axle box detectors

To compensate for the loss of visual observation of trains due to centralized control, hot axle box detectors (HBDs) are located at regular intervals on main lines.

Fig. 13.1 Hot box detector with electronic treadles

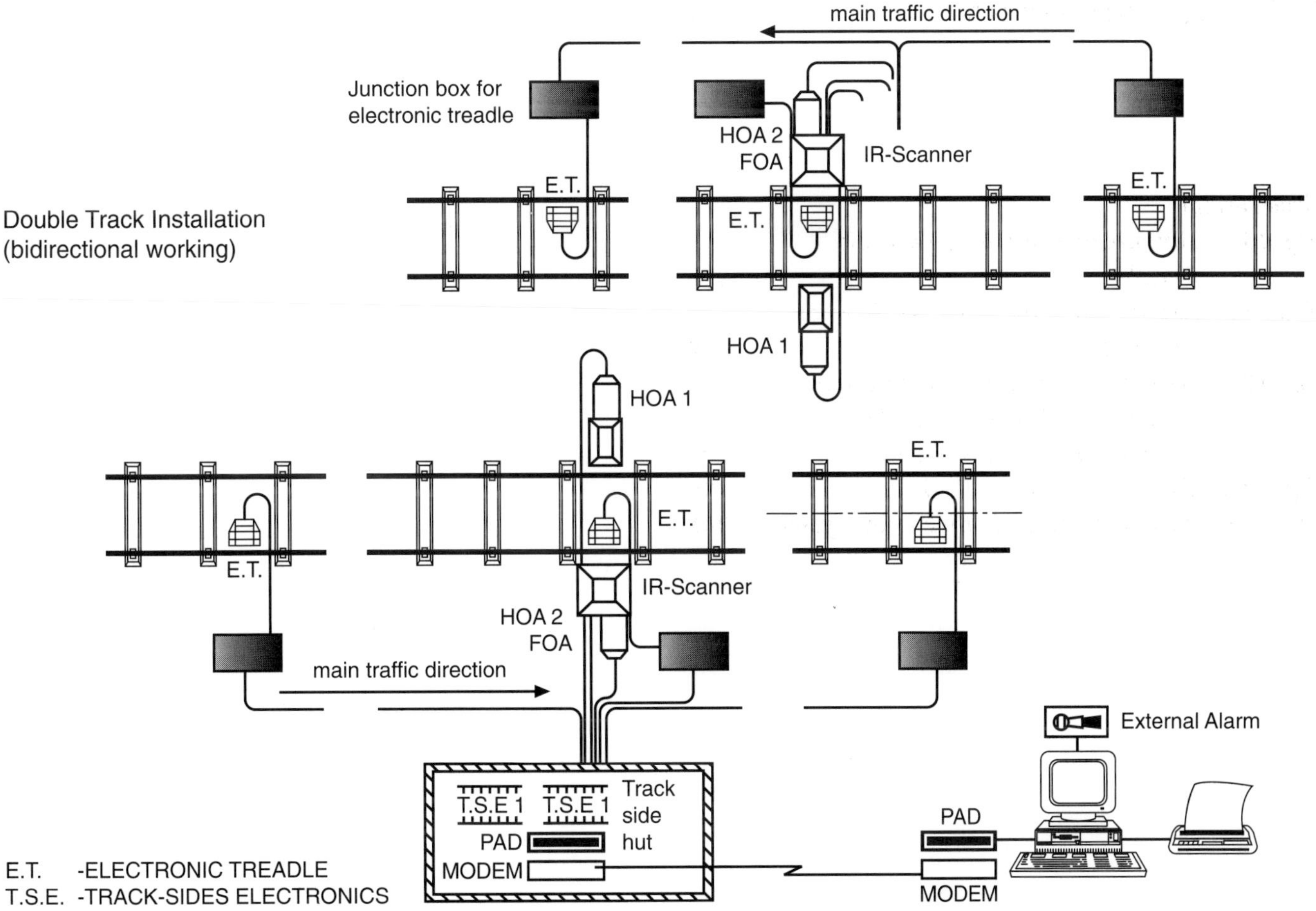

Fig. 13.2 Example of a hot axle box detector installation

E.T. - ELECTRONIC TREADLE
T.S.E. - TRACK-SIDES ELECTRONICS

The detectors are usually located a few kilometres prior to a suitable loop or siding where a defective vehicle, if found, can be detached. They must also be sufficiently far apart and beyond normal stopping places to allow a faulty axle box to heat up and be detected by the equipment.

Typically, installations located at 40km intervals on a main line. However, it is proven that an axle box can be heated up in 25km if a roller is broken in a roller bearing.

The trackside equipment directs a beam of infrared rays at each axle box of a passing train. If a temperature above a prescribed limit is recorded, the equipment activates an alarm at the control centre.

The position of any overheated boxes on the train is available in hard copy form to assist staff in making a visual examination of the train.

Where multiple detectors are located within the area of a single control centre the outputs may be multiplexed via a common terminal in order to simplify the interface with the signalman and to allow for the creation of a common database and recording medium.

13.3 Hot wheel detectors

The hot wheel detector (HWD) system operates along the same principles as the hot axle box detectors and these two types of detectors are often combined together in the same location.

The HWD system prevents loss of efficiency of braking effort and flat wheel occurrence. The HWD scanners are mounted so as to be in sight of the brake shoes.

Fig. 13.3 Detector site in Sweden with profile supervision detector, hot wheel detector and hot box detector

Fig. 13.4 Flat wheel detector (WILD) [Load circuits with strain gauge patterns mounted in the web of the rail and covered by a plastic cover]

13.4 Flat wheel detectors

The flat wheel detectors work by using one of two different principles: either with sensitive track circuits which feel and measure the time of interruption when a flat wheel is passing; or by measuring the stroke in the rail by the flat wheel. Trains therefore have to have attained a certain speed over the measuring spot.

Fault detection is transmitted to a control centre in the same way as described for the HBD, indicating which wheel in the train set has been detected as faulty.

These systems are currently not in general service on many railways but are on trial with several administrations.

13.5 Wheel weighing systems

The wheel weighing system weighs individual wheel loads and detects unbalanced loads, which are mostly a result of untidy loading.

Systems of this type have been installed for testing in some administrations.

13.6 Train gauge detectors

Train gauge detectors can be infrared, optical or mechanical in design. They measure the loading gauge to ensure that the load is within the permitted clearance gauge.

13.7 Dropped object detectors

These detect objects that may have dropped from loaded wagons or road vehicles onto the track at critical locations such as bridges at high speed lines.

The detection is achieved by breaking a double network of wires. The detection causes a stop signal for trains heading towards the location which is reported to the control centre and/or the local signalbox.

A stop signal is only given when both networks are interrupted. The breaking of only one wire, however, is reported to the control centre and/or the local signalbox.

13.8 Avalanche and landslide detectors

Avalanche detection, necessitated when stones or deep snow slide onto the track and cover it, is effected by special fences or trip wires placed along the track.

Landslide detectors are needed when the roadbed itself (and the track) is in danger of sliding away. Detection is made by weak cables strung in zigzag formation within the risk zone from fixed points.

Breakage of the wire causes protecting signals to be placed to danger. These signals are either specially provided or, where other signals or ATP exist, additional controls are included in their operation. The detector may also send an alarm to the control centre or nearest station or signalbox.

Similarly, where the flight path at the perimeter of an airfield passes low over an adjacent railway, trip wires are provided. Should an aircraft break these wires protecting signals are held at red in case debris has landed on the railway or damage has been caused to railway equipment. In addition, direct telephone communication is available between the railway control centre and airfield control tower in the event of an emergency.

Another solution for solving this type of problem is the installation of TV cameras, which give the traffic operator a view of the area, enabling him to stop the traffic if a dangerous situation occurs.

13.9 Tail detectors

Tail detectors are used for simplified block working on lines with low traffic density.

The system consists of a permanent magnet placed on the coupling hook of the last car and working as a transmitter. A receiver is fixed between the rails, which generates an impulse, when the transmitter passes over the receiver. This information is stored until it is set off by the signalman.

13.10 Staff warning systems

Many types of staff warning systems are currently in operational use, some of which are considered below.

Railway personnel are often in danger of being hit by trains when duties are being carried out in certain railway locations. In order to reduce this risk a security man is positioned so that he has a good view of approaching trains.

In the event that the visibility is less than a corresponding minimum travel time – for example 30 seconds – use is made of an automatic warning device.

There are various types of automatic warning devices whose use is dependent on the local situation. The purpose of all such equipment is to give early warning to the security man of approaching trains.

The following types of staff warning systems are currently available:

13.10.1 Movable automatic warning devices (AWAS) – NS

This system is meant to be installed at work locations where approaching trains may not be completely visible.

The operation is based on the release of an acoustical and an optical signal by one of the alarm boxes located at the work site. The alarm box is activated by one of the detectors attached to the rail.

13.10.2 Fixed automatic warning devices

13.10.2.1 WUBO for view obstructing objects – NS

This device is used when the view of an approaching train is obstructed by an object, for example a construction site, engineering work, etc over a large area of the track (a few hundred metres).

The WUBO is a fixed warning signal consisting of two white lights at the same height which are constantly lit in the event of no train approaching.

When a train approaches, the two white lights alternately go on at the same height. The WUBO warning device is used for each track and direction separately.

13.10.2.2 WIBR on bridges and WIT in tunnels – NS

These devices are used when there is an obstructed view of approaching trains at a fixed work site such as the construction or laying of a bridge, or the construction of a tunnel.

The signals indicate on the bridge or in the tunnel that a train is approaching on the track in question from one direction or the other.

It consists of two white lights per track, positioned in a vertical line which are constantly lit when no train is approaching.

In the event of a train approaching, the two white lights are lit alternately.

13.10.2.3 Train operated warning system (TOWS) – BR

In its basic form this system utilizes track circuits on the approach to and through the section of line to control permanently positioned audible warning devices fitted on posts at regular intervals alongside the line.

The equipment can be switched off when no work is under way in the area to reduce nuisance to the public. When switched on a safetone sounds while all the track circuits are clear to inform staff that the system is in working order. When a track circuit becomes occupied, the continuous alarm sound is given to warn of the train's approach.

A variation of this system is used in tunnels whereby the lighting system is used to convey the safe and warning messages thereby avoiding the noise nuisance. Full lighting is used as an equivalent to the safetone and a flashing effect is given for the warning.

13.10.2.4 Inductive loop warning system (ILWS) – BR

This system is an improvement on TOWS and makes use of track circuit and signalling inputs in a similar manner to warn of the approach of trains.

However, instead of permanently installed audible warning equipment, inductive loops are laid beneath the track. These loops are in sections up to 1km in length and carry coded messages driven from a permanently installed control unit.

The lookout man carries a portable warning issuing device (PWID), which, when held within the area of the loop, receives these messages and converts them into warning sounds.

In the absence of trains a non-urgent sound – the safetone – is emitted. This indicates that the equipment is functioning correctly.

On the approach of a train the alarm indication sounds, indicating that staff must move clear of the line. This indication may be relaxed by operation of a button to a less urgent sound known as the reminder warning.

Should a second train approach during reminder warning, the alarm sounds again and cannot be relaxed.

When work is being carried out that involves trains standing within the work site, inhibitions may be initiated on particular lines to qualify the warnings given. Similarly, inhibitions can be initiated automatically where trains stand for long periods such as in a goods loop.

An advanced version of ILWS is to be produced. This will give both spoken and displayed messages, thereby informing staff of the line upon which trains are approaching and their direction of travel.

13.10.2.5 Signal dependent warning system – ÖBB

The control equipment is installed at the location of interlockings. The movable equipment is at the location of staff between stations. It is installed on mainlines with longer or even continuous works on the track.

Basic functions
Example: The maintenance staff working on a track section are protected by a departure signal.

The staff warning system is adapted to the interlocking system in such a way that an activated control system inhibits the clearing of the departure signal.

If a route is lined up and the departure signal has to be cleared a warning signal is sent to the staff.

The exchange of signals between the staff and the interlocking equipment is carried out via existing telephone lines, thus an additional line is unnecessary.

The staff now have to leave the track and afterwards to send a clearance signal to the interlocking. The departure signal is finally cleared for one train route.

One system allows up to four independent maintenance teams between two stations, all working in one block section or split into four sections or mixed.

The protection by a block signal is handled in a similar way.

Initiated by a track occupation or by a treadle a warning signal is sent to the staff. The block signal is cleared only when the clearance signal from the staff is received.

13.10.2.6 Railway crossings for pedestrians – SNCF
Devices to advise of railway traffic by means of illuminated boards and light units are used when there is a need to improve public and staff safety at the crossing point of main railway lines; and more particularly in cases of heavy traffic, insufficient visibility and where several tracks are to be crossed.

The approach of oncoming traffic is announced for a minimum of 15 seconds, by means of a light signal installed on each platform.

Where the devices are intended for the public, the pedestrians crossing pictograms (or pictographs), which are used to resemble public road crossing points, should be used. Pictograms are well known to users and internationally understood.

When the devices are reserved for the staff only, they can be less sophisticated and less expensive.

Where high speed lines are concerned, crossings are arranged in signalbox areas to enable maintenance crews to have access to their work place, for example points zone. Each crossing is equipped with two green lights on each side of the tracks; one green light for track 1 and one for track 2.

13.11 Conclusion

Fig. 13.5 details the employment of other safety systems by European railway administrations included in this book.

Railway	Hot axle box detector	Hot wheel detector	Flat wheel detector	Wheel weighing system	Train gauge detector	Dropped object detector	Slide area (landslide) detector	Tail detector	Staff warning system
ÖBB	X	X	X	X¹	–	–	–	X	X
NMBS/SNCB	X	–	–	–	–	–	–	–	X
SBB	X	X	X¹	X¹	X	–	–	–	X
DB AG	X	X	–	–	X	–	X	X	–
RENFE	X	X	–	–	–	X	–	–	–
SNCF	X	X	–	–	–	X	X	–	X
BR	X	–	–	X	–	–	X	–	X
FS	X	–	–	–	–	–	X	–	X
CFL	–	–	–	–	–	–	–	–	X
NSB	–	–	–	–	–	–	–	X	–
NS	X	X	–	–	–	–	–	–	X
BV	X	X	X¹	X¹	–	–	X	–	X

X Installations in service
X¹ Test installations
– No installations

Fig. 13.5 European table of other safety systems

14 The Future

J. CATRAIN

Contents

14.1 Integration	363	
14.2 Train detection	363	
14.3 Points	364	
14.4 Signals	364	
14.5 Interlockings	364	

14.6 Communications	365	
14.7 Automatic train control	365	
14.8 Level crossings	366	
14.9 Availability	366	
14.10 Standards	366	

Illustrations

Fig.14.1   European projects   **367**

From its early days, the history of railway signalling has been influenced by the need to inform drivers of trains on open sections of track or from one track section to another of their responsibility for adjusting their speed, in accordance with the information signalled to them. Signals, track circuits, point operation and signalboxes were developed as a result of this need. At the same time, it was obvious that drivers had to be secure in their task by some safety net such as an on board signal repeater with audible and visual alarms; for example, Crocodile or AWS.

With new techniques and technological developments, basic signalling has been enlarged with the automation of operational tasks. More and more available data relating to train numbers, timetables and files, is now processed to perform routine tasks such as route setting, commands release, graphic display, conflict prediction, etc. In addition, information to passengers is derived from existing signalling databased passenger information displays and voice recorded announcements.

More recently, the development of high speed rail travel has led to the elimination of trackside signalling, to be replaced by on board cab signalling. Furthermore, the increase in traffic for a given infrastructure has led to the development of speed supervision and flexible block systems to increase line capacity.

14.1 Integration

The future lies therefore in the design of integrated transport systems offering a greater range of customer services in terms of speed, frequency and punctuality, together with improved passenger services on board trains, to include telephones, facsimile, data transmission, video facilities, etc. Moreover, trains will also need to be connected with other forms of transport, for example airports, buses, metros and private cars. Many of the facilities described herein are already developed, and will require extensive provision in the future.

For signalling, such integrated systems mean:

- the development of mobile to fixed installation communication and vice versa in terms of telephone, data and video facilities.
- a greater degree of the functionalism of such systems in terms of a better service to customers by the reduction of interruption to signalling and telecommunications systems.
- a better standardization at the European level, ensuring the compatibility and exchange of information beyond national borders. This aspect should include signalling, operating aid and rolling stock identification.

The future should also see the clarification of the human role in an integrated rail transport system, by the gradual introduction of European operating regulations covering the role played by man in the safety of train movements, in the event of serious incidents within the integrated system, and the resources and procedures available to him to increase the availability of the system.

It is always difficult to predict what the future has in store. However, major advances are being made in the direction of research and development into the production of future railway signalling systems.

High technology, particularly in the field of communications, signal and data processing, together with enhanced protection of transmission from the environment, forms the basis for present and future developments. The design approach applied to these technologies will significantly modify the conventional approach to signalling. The concept of interaction will dominate a more integrated system with a high level of automation.

It is also probable that the cost of existing infrastructures will limit the speedy implementation of such integrated concepts. Accordingly, a sector based approach, taking the future into account, will continue to be necessary for the networks concerned.

One of the major trends at present is the need for intelligent locomotives linked to advanced real time information centres, whilst retaining the option of the continued existence of rolling stock which is not equipped with any modern equipment within this highly interactive environment.

14.2 Train detection

The basic signalling functions will remain fundamental, whatever the form of the future integrated rail transport system. These functions correspond to those described in previous chapters and will be subject to the development predictions below.

Train detection is a vital element of any transportation system. Up to the present the basic equipment for this function remains the track circuit, with some alternatives such as axle counters and train end detection.

Track circuits have to face more and more difficult

problems arising with modern technology. Severe disturbances are caused by modern powerful locomotives, leaf falls and light weight rolling stock.

Thus track circuits are always being evolved in order to maintain the basic functions of non-occupation of track section and broken rail detection. In the future those two functions may be split and achieved by separate types of equipment. It is worthwhile to note that broken rail detection is not required by all railways. However, it may become a requirement with higher speeds. DB AG and BV are considering it for the future, and BR consider that additional safeguards to detect broken rails for lines with speeds over 200km/h would be required by the British Rail regulatory authority.

Automatic vehicle identification coupled with a train integrity system could be a substitute for the track circuit and axle counter in the near future. This factor makes it necessary for Europe to standardize as a priority automatic vehicle identification equipment and for all rolling stock to be suitably equipped.

14.3 Points

The detection and control of points does not appear to need developing significantly other than in terms of technology for improved availability and greater flexibility of trackside operation, such as levers instead of cranks, for use in manual operation. Since the points motor is an essential factor in the availability of the transport system, the prediction of failures is a function that could eventually be integrated into these machines.

Permanent supervision by electronic survey of turnouts is also under development. It will include such functions as average reversing current and time, position of tongue and passage between tongue and stock rail.

14.4 Signals

Although the use of signals will be phased out in the long term, they will remain in operation for several decades to come. The tendency is therefore to find ways of improving their visibility, with the use of halogen lamps and optical fibre as well as by reducing their power consumption. Research into European standardization, based on UIC Recommendation 732 (1992), which sets out the principles to be followed for a future international signalling system, will also be useful.

Such improvements in signals are under consideration by CFL, with the use of compact 640 DB AG-type, so called because it was first installed on DB AG's network.

BR are experimenting with LED and fibre optic technology. Reduction of lamp power and use of halogen lamps together with fibre optic are the main trends of ÖBB and DB AG. The NS policy is to install halogen lamps and optical fibres in signals with difficult access in order to facilitate maintenance.

14.5 Interlockings

Interlockings now use computer technology which provides certain economic advantages and enables this important aspect of signalling to be integrated in the future overall safety information system. Considerable efforts should be made to ensure that the software – a major investment – can be recovered to be run on different hardware, subject to extensive development; if not, the economic benefits of data processing will be greatly reduced. Accordingly, the use of high level languages and European standardization would facilitate the use of software packages on the network or another. The work carried out by the technical committee of the IRSE in this area is to be continued.

All 12 railway administrations have already experimented with electronic interlockings and are now installing them in preference to relay technology, with the cost benefit being the first main consideration. The second consideration is their availability to be connected to the most modern telecommunications channels. BR is currently the leading railway equipped with this type of technology.

At the same time, this technology allows improved electronic interlocking data preparation at a high level of automation, thus reducing time required and human error. Such improvements are under consideration by DB AG, SNCF, BR, FS and NS, and already achieved at ÖBB and BV. With this data preparation, these railways also include that relating to computerized central control. Such facilities should constitute a safety procedure as well as being complete with automatic test commissioning, computer aided engineering and well documented files. Furthermore, the modular design of the architecture should be developed to allow for technical development, without jeopardizing overall investment.

Automatic systems are already performing such

routine tasks as automatic route setting and release, with human intervention only in the event of incident. It is therefore essential that the man–machine interface in signalboxes be transparent in order to retain the traditional form of railway operation.

With the availability of these facilities the areas controlled by signalboxes are being extended and, accordingly, the data collected by the detectors concerning points, signals and track circuits, should be capable of being connected to signalboxes by multichannel safety or open network transmission systems, with a backup loop for increased availability. Such trends are also extended to the rail traffic management centres:

- ÖBB – under consideration
- DB AG – CIR
- SNCF – PAR
- BR – IECC
- BV – ELVIS project
- FS – DCO

14.6 Communications

In the future, block systems whether automatic or not, should be developed alongside telecommunications with trains and become increasingly integrated within the areas controlled by signalboxes.

Single track lines, the continued operation of which is often at the limits of profitability, should be equipped with simplified signalling systems based on radio communications between the line supervision centre and the trains the centre is responsible for (except DB AG). Trackside equipment should be limited to that which is absolutely necessary, such as automatic return points, markers, elimination of trackside signals, with authorization being given on boards, and in the absence of station staff, by RETB, radio block and ECLAIR type systems.

The most profound development over the next few years will concern the role played by telecommunications in the provision of channels which will allow the coherence of the integrated information system required for advanced automation of railway networks. The various human links in the chain will be replaced, for enhanced supervision, by the provision of a broadly based information network.

Furthermore, telecommunications will enable the transmission of information to be extended beyond that which is essential for safety and railway operations in order to improve passengers services, such as telephones, facsimile, data transmission, reservation of taxis or car rental. These will be provided by the use of common channel communications.

Transmission via optical fibre line systems will be further developed with increased new transmission capacities as well as enhanced protection against environmental disturbance.

SBB, DB AG, RENFE, SNCF, NS and BV are planning computer integrated railroading in which a continuous information flow will involve all functions and levels in an overall system. Here, too, standardization of telecom protocols will be of great interest for an improved European railway network integration. Indeed, ÖBB has already standardized the X 25 protocol on its network which connects interlockings, operation control centres (BOS), computer assisted train control (RZu) and other peripheral systems.

14.7 Automatic train control

Safety can only be assured if trains comply with the information provided by the signalling system. This task, which is currently performed by the human driver, will be transferred directly to automatic control systems, the reliability of which in performing repetitive tasks is much greater. This development will be accompanied by the regulation of trains in real time, to achieve greater savings for the transport system. The installation of ATP and ATO systems will be in the network investment priorities for the years to come. These investments, together with those that are essential for telecommunications with trains, will gradually free lines of equipment that are difficult to maintain, such as signals, treadles and power supply cables.

In future, Austria will extend the continuous train control LZB on its main line but may consider the European standardization if available in time. Similarly, SBB and CFL, bearing in mind their geographical concerns, should find that ETCS is the solution. BR is currently investigating discontinuous ATP; so too are FS and NMBS/SNCB. NS have continuous ATP and will develop a discontinuous system with extended functionality, awaiting ETCS. Discontinuous and continuous ATP are fully introduced now on SNCF, DB AG and RENFE.

Because of their obvious advantages for the environment and their economic impact in terms of energy saving, railways will increase their share of the transport market in the future, by offering high speed services operating at speeds of up to 200km/h and very high speed services of up to 350km/h. Success in achieving these objectives implies increased reliability of equipment (particularly track equipment), the elimination of trackside signalling for these trains, or if not totally then with a large reduction whilst increasing line capacity using shortened block without line side signals (DB AG), together with real time information for passengers concerning their journey and punctuality – in other words, quality of service. Quality service must be provided despite the complexity of the problems associated with a limited infrastructure and the density of traffic required by the need for profitability.

It is therefore certain that the importance of the role played by traffic management will increase in the future. The quality of traffic management will depend on the network of information in an extended field of action which is compatible with the capabilities of the systems, in terms of the backup procedures to be used in the event of an incident.

14.8 Level crossings

Because of the increase of both road and rail traffic, it is clear that level crossings are becoming obsolete. All rail networks have begun by fitting level crossings with half barriers and subsequently implementing work programmes to remove them altogether. This is the position for ÖBB, which will reduce the total number by half during the next 20 years.

As for signals, these will only be removed slowly throughout this long period. It is probable that microprocessor based systems will replace the present relay based systems, thereby improving functions such as the warning period and the detection of obstacles.

Such computerization is under evaluation by ÖBB and BR. It already exists on DB AG and CFL. In addition, the integrated safety system will embody other, at present, stand alone devices such as the automatic protection of staff working on the track and hot box detectors.

14.9 Availability

In the future, signal engineers will have to pay more attention to the overall system engineering in order to maintain the safe running of trains whatever the degraded situation. In such a concept, signal engineers will have to supply the operations manager with enough reliable information so that the overall safety of trains is not left as the total responsibility of the operations manager.

In addition, they will integrate within their overall system the most advanced diagnostic techniques in order to minimize the degraded situation at the lower level, such as telediagnostic, alternative strategies and fault tolerant systems. The application of artificial intelligence and expert system will be of great benefit in solving these problems.

Another very important task will be the software validation with very formal procedures as well as the necessity of having very precise and detailed system specifications.

14.10 Standards

European railways should make specific commitments to the standardization of the conditions laid down for safety devices and systems. The work already carried out into the standardization of a common architecture for the European train control system (ETCS) is encouraging but must be combined with work in other fields. The use of standards in communication systems, such as LANs (local area networks) and wide band, are a priority in facilitating the establishment of bridges between subsystems in order to benefit the serial production at low cost.

These approaches imply changes to the organization of companies as well as in the railways, and the setting up of a high level quality assurance system for all concerned. This will also lead to the rationalization of available signalling systems, due to the very high level of investment to be made for research and development into these complex integrated systems.

In December 1989, the European Commission set up a working group to draft a master plan for a European high speed train network. The master plan, drafted for 2010, was released in December 1990 and the Council set out its basic approach in its resolution of 17 December 1990, to be formalized by a directive from the Commission.

The technical conditions designed to ensure the compatibility of the infrastructure, equipment and rolling stock of a European railway network are bound to affect signal specialists in terms of control and supervision systems. Over the next 10 years, an integrated control and supervision system with rising compatibility should be designed. This requires the standardization of national regulations and a single certification procedure for the new system.

The international technical committee of the IRSE

can play a part in this process by bringing railways and manufacturers together. The first report from this committee, *'Safety System Validation with regard to Cross Acceptance of Signalling Systems by the Railways'*, clearly demonstrates the will of the IRSE to overcome the problems of the profession. A second report on *'Operational Availability'* has highlighted the gaps between different railway cultures and the need to harmonize the database definition for a better understanding of each other's problems.

Fig. 14.1 European projects

Railway	Name of Project	Definition	Benefit	Target
ÖBB	*Die neue Bahn*	Improvements to high speed and high performance lines	Reduced travel time and increased capacity on main lines	1995–2000
NMBS/ SNCB	Concentration Plan	Computerized control centre	Fewer signalboxes	2000
SBB	Rail 2000	Fixed timetable intervals for all main lines	Improved passenger service	2000
		New signals	Improved level of information to driver	
			Increased traffic capacity	
DB AG	DIBMOF	Railway services integrated mobile radio system	Economic radio system	1996
DB AG/ SNCF	DEUFRAKO–M	System for controlling, commanding and management of trains	Interoperability of trains on European networks	1996–2002
DB AG	CIR-ELKE	Increased efficiency on the core network	Increased traffic capacity by continuous train running control (LZB) combined with high capacity block system	Pilot project 1995 Complete core network 2000
DB AG	CIR–net	Open network for both telecommunications and signalling	Usability of commercial networks	Pilot project 1995
DB AG	CIRROD	Computer controlled integrated operating and dispatching system	Automated operating and disposition over wider areas	2000
DB AG	AzM	Centralized multiple section axle counting	Higher availability of train detection systems even in cases of failure or destruction	Pilot project 1996
SNCF	ASTREE	Real time train control	Increased line capacity and improved traffic management	2010

Railway	Name of Project	Definition	Benefit	Target
SNCF	KVB	Automatic train protection	Improved safety on all the network	1991–1996
SNCF	ANTARES	Automatic train protection	Improved safety on all the network	Under evaluation
BR	ATP	Automatic train protection	Improved safety on all the network	Under development and evaluation 1996
FS	SISTRACO	Integrated system for high speed line signalling–transmission and control	Higher performances in managing and controlling high speed line traffic	1998–2000
FS	Track to train radio link	Integrated project for service, emergency and passenger communications	Improvements in service punctuality and quality by an economic radio based system	1994–1995
FS	SISCT	Integrated system for supervision and management of traffic	Reduction in costs, improvements in service, quality and better use of infrastructure	1994–1995
FS	Discontinuous ATP	Integration of systems for the spacing of trains by on board discontinuous cab signalling and speed control	Global ATP	1993–1999
NS	Rail 21	Doubling the capacity and higher speeds	Improving quality	2005–2015
NS	VPT Post 21	Traffic management	Automated operating and disposition for wide areas	1993–1998
NS	BB21	Signalling and train control 21st century	Improved traffic capacity	2010
BV	ELVIS	Distributed geographical computerized system	Increased availability and traffic capacity	1991–1996
BV	Radio block	ATC based safety system for low traffic lines	Low investment and maintenance cost	1994–1995
ECC/ UIC	ETCS	European train control system	European Specification for ATC system	1994–1995

Sources

Title	Authors	Reference

Chapter 1
Signalling systems of Swiss Federal Railways	C. Zufferey	IRSE Proc. 91/92
Italian State Railways Signalling	G. Cerullo	IRSE Proc. 90/91
Basic Principles of Signalling Practice on BR	B. Heard	IRSE Proc. 90/91
Signalling Technology of the German Federal Railways	A. Bidinger	IRSE Proc. 90/91
Safety and Progress	J. P. Guilloux	IRSE Proc. 90/91

Chapter 2
The SNCF Approach to Track Circuits	F. Van Deth	IRSE Proc. 92/93

Chapter 4
S60 Planungsvorschrift Allgemeines Teil 1	ÖBB	
V2 Signalvorschrift ÖBB		
Eisenbahnsicherungstechnik in der Schweiz	K. Oehler	Die Entwicklung der Elektrischen Einrichtungen
Stellwerke der Schweizer Bahnen	E. Palm	EBV
Die Lichtsignale bei der Deutschen Bundesbahn	Lütgert	Signal und Draht 4–61
Die Scholtung des Spurplanstellwerks SpDrL60	Wehner	Signal und Draht 72–73
La signalisation ferroviair	R. Rétiveau	Presse des Ponts et Chaussées

Chapter 5
The Elektra System	H. Steinbrecher	IRSE Aspect 91
Das Spurplanstellwerk SPDrS 600	H. Walther/H. Appel	Signal und Draht 6–79
MC L 84 das neue Stellwerk	K. Piontek	Deine Bahn 8–88
Microcomputer Interlocking in Chiasso	H. Hayoz	Schweizer Eisenbahn Revue
Das elektronische Stellwerk ESTW L90	–	SEL, Stuttgart
Modern Technology in Fail-Safe Systems	K.-H. Wobig	IRSE Proc. 86–87
Die Signaltechnik der ÖBB	H. Steindl	Signal und Draht 91
Videopult – A Video Display Control and Management System for Interlockings	H. Steinbrecher	IRSE Proc. 86–87
Apparrati Controli Elettrici a Pulsanti di Itinerario	G. Cerullo\B. Costa	CIFI

Chapter 6
Compact Centralized Traffic Control SIG L90	A. Knight	IRSE Aspect 91

Chapter 7
Radiocommunications in Railways	Sausins	IRSE Aspect 91

Chapter 9
Speed Control System on SNCF	J. P. Guilloux	IRSE Aspect 91
Transmission Based Train Control System	H. Uebel	IRSE Aspect 91
Selcab Automatic Train Protection	R. E. B. Barnard	IRSE Aspect 91
Flexible Automatic Train Control	Rose/Fischer	IRSE Proc. 89–90
The NS Approach to the Development of a Second Generation ATC System	P. Middelraad	IRSE Proc. 89–90
BR Automatic Train Protection Trials	Binard/Van de Voorde/ Barnard/Viebel	IRSE Proc. 91–92

Chapter 10
Experience with the Introduction and Operation of High and Very High Speed Lines	UIC	Arrezo 87

Chapter 11
The IECC and the Next Ten Years	Raynor/Bell/Edger	IRSE Aspect 91
The British Rail Automatic Route Setting System	Hurley	IRSE Aspect 91
Integrated Electronic Central Control at NMBS on the Belgian State Railway	Thielemans	IRSE Aspect 91
The Zurich Remote Control Systems on Swiss Federal Railways	Kaupp/Wendt	Signal und Draht 92
Lucerne Signalling Centre	Integra	91
Computer Aided Train Supervisory System in Nuremberg	Kuhbier	Internationales Verkehrswesen 90
Das Integrierte System der Sicherungs und Telekommunikation für die Schnellfahrstrecke Madrid Sevilla	Beato/Mura	Eisenbahn technische Rundschau 12–90

Chapter 14

The UIC Project for Developing the Specification for a European Train Control System	P. Winter	IRSE Aspect 91
ETCS: The European Train Control System	Thomas/Coenrad	IRSE Proc. 92–93
Cross Acceptance of Vital Signalling System	IRSE Technical Committee	Rapport No. 1
Rail 2000 Switzerland: Exciting Project for the 21st Century	O. Stalder	IRSE Proc. 91–92

Chapter-by-chapter Index

1 Railway signalling principles
glossary 3
basis of signalling systems 5
  key elements 5
  application of system 5
  two and three aspect system 6
  four aspect system 6
  moving block 6
parameters of European signalling systems (table) 7
speed signalling and route signalling 8
  speed signalling 8
  route signalling 9
  permanent speed restrictions 9
  diverging junctions 9
  comparative merits 10
signal profiles 10
  running signals 10
  stop signals 10
  warning signals 12
  repeating signals 13
  contraflow signals 14
number plates and signs 15
countdown boards 15
shunting signals 15
  shunting aspects on running lines 15
  preceding shunts 16
  shunting to and from sidings 16
  shunting signal types (table) 17
  trains starting from sidings 18
other signals 18
  protection signals 18
  substitution signals 18
  platform duties signals 19
  level crossing rail signals 19
  signalling for freight trains 19
  fixed signs 20
speed restriction signs 20
permanent speed restrictions 20
temporary speed restrictions 20
signal aspects and aspect sequences 21
UIC recommendations 21
aspects in use in Europe 21
  single green (and table) 22/23
  double green (and table) 22/24
  single yellow (and table) 22/25
  double yellow (and table) 26
  green and yellow together (and table) 27/28
  single red (and table) 29
  single white (and table) 30
  other combinations (and table) 31
uses of flashing aspects (and table) 33
policy on lamp failure 35
aspect sequences 35
signalling for following trains 35
signalling for diverging junctions 37
signalling in stations 37
signal aspect and aspect sequence diagrams 37
  ÖBB 38
  NMBS/SNCB 43
  SBB 49
  DB AG 56
  RENFE 62
  SNCF 65
  BR 73
  FS 80
  CFL 87
  NSB 91
  NS 94
  BV 101
bidirectional working 105
principles of interlocking and controls 105
  type of system 105
  method of operation 106
  route calling 107
  route setting 107
  route locking 107
  approach locking 107
method of route release (table) 108
aspect controls - stop signals 108
line of route controls 108
signal ahead controls 109
foul track and flank controls 109
overlap controls 109
preceding shunts 110
approach control of aspect 110
aspect controls - shunting signals 111
shunting signal controls 111
aspect controls - warning signals 112
aspect controls - other signals 113
  repeating signals 113
  substitution signals 113
  platform duties signals 113
aspect replacement 113
route release 113
release of approach locking 114
train operated route release 114
sectional release route locking 114
release of overlap locking 115
failure of route to release 115
point operation 115
level crossing controls 116
controlled level crossings 116
automatic level crossings 116
limitations on the use of level crossings 117
bidirectional and single line controls 117
remote control of interlockings 117
special controls 118
  tunnel 118
  freight train 118
  staff protection 120
  trackside warning 120
  patrolman's lockout 120
  staff working on trains 120

2 Train detection systems
track circuits - survey of European situation 122
types of track circuits selectively in use across Europe (with table) 124
DC track circuits 127
AC track circuits 128
reed track circuits 128
characteristics of AC TC employed in Europe (table) 130
HVI - track circuits 132
short jointless track circuits 132
audio frequency jointless track circuits 133
axle counters 136
treadles 138
characteristics of treadles employed in Europe (table) 139
vehicle sensors 141
miscellaneous
  magnetic relay 142
  tail detectors 142

3 Switch operating and proving systems
switch equipment
  definition 144
  composition 144
  description 144
  switching 145
  angle of crossing 145
points
  history 145
  development 146
  articulated and flexible points 146
  profiles 146
  sleepers 146

fittings 146
control: local and remote 146
trailability 146
points situation on some of the European networks 147
clamping or locking 148
permissive control 148
control, locking and detection 148
mechanical points
   operation 148
   double key operated points 149
   single key operated points 149
electrically powered points
   operation 149
   principle of operation 150
   special note 150
   typical solution 150
electrohydraulic point machine 150
   operation 151
   with rail clamp locks 151
   with hydraulic power unit 151
locking devices 151
   clamp locking 152
   claw locking 152
   clamp locking, VCC 152
point detection 153
derailer 154
command and control circuits 154
summary of the safety conditions 154

4 Signals
general - multi aspect signals 161
positioning of signals (table) 161
sizes of signal lights 163
sizes of signal lamps (table) 165
types of lamp 167
double filaments (table) 167
types of signal lamp (table) 169
power supply 171
signal power supply (table) 173
flashing frequency standards (table) 175
control and proving systems 175
main signal lamp failure 176
visibility distance 178
distance between main signal and distant signal 178
position of signal posts 180
distance between signals and control circuits 180

5 Interlocking cabin
early interlockings 183
modern interlockings 183
early relay interlockings 184
connections to points and signals 185

checking and testing 185
summary of interlockings (table) 186
geographical circuitry interlockings 187
equipment location 189
computer interlockings 189
routes and route conditions 196
   train routes 196
   shunt routes 196
   route conditions 196
interlocking functions 197
operation interface 205
power supply 213
interlocking and associated power supply (table) 216
standards and legal obligations 219

6 Block systems
operating conditions
   single track 222
   double track 222
block system choice criteria 222
survey of block systems: definitions 222
characteristics
   single lines 223
   double or multiple lines 223
   wrong line working 223
   bidirectional working 223
block system application (chart) 224
description of block systems 226
   token working block 226
   electric token working 226
   radio electronic token block 226
   telephone block 227
ÖBB: Zugleitbetrieb 228
SNCF: ECLAIR 228
NS: Centraal Telecomblokstelsel 228
DB AG: Zugleitbetrieb 228
NS: tokenless working block 229
manual block - operation sequence 229
relay block 231
automatic block systems 231
   non-centralized automatic block 231
   centralized automatic block 232
   coded current automatic block 232
disturbance procedures 233
typical applications 233

7 Radio systems in signalling
definitions of radio systems 236
Very High Frequency
   (VHF 70-88 MHz) 236
   (VHF 155-220 MHz) 236

Ultra High Frequency (UHF 240-470 MHz) 236
application of radio systems 236
current usage of radio systems 236
voice communications and non safety data transmission 238
driver only operation 238
shunting operation 238
central control operation 238
remote control by radio 238
track to train 238
signalling 239
warning systems 239
network requirements 239
the future 239

8 Internal and external safety conditions
general - safety design 241
system design 242
   safety and dependability 242
   system lifecycle 242
safety 243
   safety organization 243
   safety integrity 243
   risk analysis 246
   technical safety assurance 248
   software case 248
traditional relay circuitry 248
design of circuits 250
tests 251
   track circuits 251
   signals 251
   points and crossings 251
   other measures 251
reliability 252
duplication of elements 252
   transmission lines 252
   duplication of control 252
   duplication of power supply 252
   duplication of hardware 253
maintenance 253
homologation 253
quality control 254
other preventive measures 254
standardization 254
relays 255
signalling cables 258
cable laying systems 258
   suspended cable 259
   underground cable 259
   service tunnels 259
   layout of conduits 259
   trenches 260
   safety measures 261
staff licensing 262

analysis of the different
  administrations 262
cable laying 262
cables used 263
installation of local control panels 263
installation of equipment and systems 264

9 Automatic train protection and control
need for protection and control 266
intermittent systems 268
signal repetition on board 268
Crocodile 268
Automatic Warning System (BR) 269
Signum (SBB) 269
INDUSI (ÖBB, DB AG) 269
ASFA (RENFE) 270
speed supervision system 270
transmission beacon (NMBS/SNCB) 270
Contrôle de vitesse à balises KVB (SNCF) 271
Ebicab 10 000 (NSB, BV) 271
continuous systems 272
centralized systems 272
LinienZugBeeinflussung LZB (ÖBB, DB AG, RENFE) 272
decentralized systems 273
Transmission Voie Machine (track to train transmission) TVM (SNCF) 273
Automatische Trein Beïnvloeding ATB (NS) 273
coded track circuit automatic block BACC (FS) 274
European Train Control System ETCS 274
Automatic Train Protection Specification ATPS (BR) 275
present systems in Europe 275
summary of ATP control systems in current use (table) 276

10 High speed line signalling system
basic principles - traffic features 283
high speed signalling 160-250 km/h 285
Very High Speed lines - VHS - speeds 250 km/h 286
functional organization (VHS) 287
general operating modes (VHS) 290
safety and availability aspects 294
maintenance 294
general signalling system description (VHS) 295

main technology used on high speed lines 298

11 Rail traffic management
typical rail traffic management structure 312
  general management 312
  super regional centres 312
  regional centres 312
  stations 312
traffic supervision 313
  operational concept 313
  operational planning 316
  customer 316
  timetable 316
  network 317
  rolling stock 317
  tractive units 317
  personnel 317
  energy 317
disposition 317
  customer 317
  timetable 317
  network 317
control 318
  customer 318
  network 318
  personnel 318
  energy 318
  safety 318
timetable 319
train description 319
  summary of technical solutions 320
  functional description of signal information and stepping 321
  operator control - TDS 322
  fringe boxes 323
  link of train describer systems 324
  fault logging and alarms 324
  automatic editing of train numbers 324
  system configuration 324
  safety and other aspects 325
train recording 326
operation 326
  display - visualization of the process 326
  control functions - manual route setting 327
  electronic man-machine interface - 'Videopult' 330
  automatic route setting 330
  shunting operation 332
passenger information 332
  visual information 333
  voice communication 333

information and distress call column 334
closed circuit television - CCTV 334
centralized traffic control 334
traffic supervision office 334
traction voltage supply 336
locomotive management 336
remote control centres 336
other management and information systems 342
  freight traffic information system 342
  vehicle identification system 342

12 Level crossing protection
characteristics of level crossing protection 344
main forms of level crossing protection 344
  manually controlled full barrier crossing 344
  automatic open crossings with road signals 345
  automatic half barrier crossing 345
  automatic doubled half barrier and full barrier crossing 345
  open level crossing without road signals 346
forms of level crossing protection control 346
level crossing protection in Europe - road side 346
  selection criteria 346
  road signals and audible warnings 347
  normal operation sequence 349
  minimum open time 349
  second train warning 349
level crossing protection in Europe - railway side 349
  technology used by European railways 350
measures taken to avoid overlong warning times 351
level crossings in service in Europe (table) 354
European normal operating sequences in seconds (table) 355

13 Other safety systems
hot axle box detectors 357
hot wheel detectors 358
flat wheel detectors 358
wheel weighing systems 358
train gauge detectors 358
dropped object detectors 359

avalanche and landslide detectors 359
tail detectors 359
staff warning systems - European
    application 359
other European safety systems (table)
    361

14 The future
integration 363
train detection 363
points 364
signals 364
interlockings 364
communications 365
automatic train control 365
level crossings 366
system availability 366
standards 366
European projects (table) 367